Topics in Heterocyclic Chemistry

Volume 62

The series Topics in Heterocyclic Chemistry presents critical reviews on present and future trends in the research of heterocyclic compounds. Overall the scope is to cover topics dealing with all areas within heterocyclic chemistry, both experimental and theoretical, of interest to the general heterocyclic chemistry community. The series consists of topic related volumes edited by renowned editors with contributions of experts in the field. All chapters from Topics in Heterocyclic Chemistry are published OnlineFirst with an individual DOI. In references, Topics in Heterocyclic Chemistry is abbreviated as Top Heterocycl Chem and cited as a journal.

Wen-Jing Xiao
Editor

Photocatalysis for Heterocycle Formation and Functionalisation

Springer

Editor
Wen-Jing Xiao
Central China Normal University
Wuhan, China

ISSN 1861-9282 ISSN 1861-9290 (electronic)
Topics in Heterocyclic Chemistry
ISBN 978-3-031-95004-9 ISBN 978-3-031-95005-6 (eBook)
https://doi.org/10.1007/978-3-031-95005-6

This Springer imprint is published by the registered company Springer Nature Switzerland AG
The registered company address is: Gewerbestrasse 11, 6330 Cham, Switzerland

Preface

Heterocyclic compounds constitute the backbone of modern organic chemistry, serving as indispensable structural motifs in pharmaceuticals, agrochemicals, materials science, and bioactive natural products. The continuous demand for sustainable and efficient synthetic methodologies has driven the rapid evolution of photocatalysis as a transformative tool in organic synthesis. By harnessing visible light as a renewable energy source, photoredox and energy transfer catalyses have unlocked unprecedented reactivity under mild conditions, enabling the construction and functionalization of heterocycles with remarkable precision and diversity. This volume, Photocatalysis for Heterocycle Formation and Functionalisation, seeks to capture the cutting-edge advancements in this dynamic field, offering a comprehensive resource for researchers across academia and industry.

The 11 chapters of this book collectively highlight the versatility of photochemical strategies in addressing synthetic challenges. Chapter 1 (Quan-Qing Zhao, Yu Zhao, and others) provides a systematic overview of visible light-induced radical cascade reactions for synthesizing five-membered N-heterocycles, emphasizing their relevance in drug discovery. In Chap. 2, Zhiyong Jiang explores asymmetric photoredox catalysis for enantioselective functionalization of pyridine derivatives, merging hydrogen-bonding interactions with radical reactivity to forge chiral azaarenes. Sungwoo Hong's contribution (Chap. 3) delves into the radical-mediated C–H functionalization of pyridines using N-functionalized pyridinium salts, showcasing their dual roles as radical precursors and regioselective surrogates.

Chapters 4–6 focus on photochemical innovations for specific heterocyclic scaffolds. Alakananda Hajra (Chap. 4) details photoinduced syntheses of imidazopyridines and imidazothiazoles, underscoring their medicinal significance. Liangqiu Lu (Chap. 5) discusses the integration of photoinduced Wolff rearrangement with asymmetric catalysis to access chiral heterocycles, while Jun Xuan (Chap. 6) highlights the utility of 2H-azirines as versatile intermediates under photochemical conditions.

The subsequent chapters address emerging mechanistic paradigms and applications. Yuan-Zheng Cheng, Chao Zheng, and Shu-Li You (Chap. 7) analyze energy transfer-driven dearomatization of heteroarenes, a powerful strategy for constructing spiro-, fused, and bridged architectures. Bing Yu (Chap. 8) reviews photocatalytic arylation of heterocycles, contrasting its advantages over traditional cross-coupling methods. Xuefeng Jiang (Chap. 9) navigates the synthesis and functionalization of fused thiophenes, emphasizing their roles in materials science.

The final chapters bridge photochemistry with interdisciplinary frontiers. In Chap. 10, Wen-Jing Xiao and Jia-Rong Chen outline visible light-driven strategies for constructing axially chiral heterocycles, merging stereocontrol with radical reactivity. Fangrui Zhong (Chap. 11) concludes the volume with a visionary perspective on photobiocatalysis, synergizing enzymatic specificity with photoredox catalysis to expand synthetic possibilities.

Each chapter not only summarizes seminal developments from 2018 to 2023 but also critically evaluates mechanistic insights, substrate scope, and unresolved challenges. The authors—leading experts in their respective domains—have meticulously curated content to balance fundamental principles with practical applications, ensuring relevance for both newcomers and seasoned researchers.

This project owes its success to the invaluable contributions of the chapter authors, whose expertise and dedication have shaped a cohesive narrative of innovation. We extend profound gratitude to the reviewers for their rigorous evaluations and constructive feedback, which significantly enhanced the scientific rigor and clarity of this work.

As photocatalysis continues to redefine the boundaries of synthetic chemistry, this volume aims to inspire further exploration into sustainable methodologies, asymmetric transformations, and interdisciplinary synergies. We hope that it serves as a catalyst for discovery, fostering advancements that address global challenges in healthcare, energy, and environmental sustainability.

Contents

Top Heterocycl Chem (2025) 62: 1–40
https://doi.org/10.1007/7081_2025_87

Published online: 9 May 2025

Recent Advances in Photoinduced Radical Cascade Cyclizations Enabling Five-Membered *N*-Heterocycle Formation

Min Gu, Yan-Zhen Li, Yu-Jie Cao, Xiao-Bin Wang, Yi Jiang, Chen Wang, Quan-Qing Zhao, and Yu Zhao

Contents

Abstract Five-membered *N*-heterocycles are ubiquitous in natural products, pharmaceuticals, and other biologically active compounds. Over the past decades, visible light-induced photoredox catalysis has emerged as a robust and sustainable mode of catalysis in organic synthesis. Consequently, significant attention has been directed towards this field, and numerous mild and efficient methods for synthesizing these structurally diverse five-membered *N*-heterocycles have been disclosed. In this

---

Min Gu, Yan-Zhen Li, and Yu-Jie Cao have made equal contributions and are co-first authors.

M. Gu, Y.-Z. Li, Y.-J. Cao, X.-B. Wang, Y. Jiang, C. Wang, and Q.-Q. Zhao (✉)
Jiangsu Key Laboratory of Marine Pharmaceutical Compound Screening, College of Pharmacy, Jiangsu Ocean University, Lianyungang, Jiangsu, People's Republic of China
e-mail: zhaoquanqing@jou.edu.cn

Y. Zhao (✉)
Shaanxi Key Laboratory of Chemical Reaction Engineering, College of Chemistry and Chemical Engineering, Yan'an University, Yan'an, Shaanxi, People's Republic of China

chapter, we primarily summarize the recent advances in visible light-induced radical cascade reactions for the synthesis of five-membered *N*-heterocycles from 2018 to August 2023, highlighting the formation of various classes of *N*-heterocycles and the involvement of radical intermediates.

Keywords Five-membered *N*-heterocycles · Photoredox catalysis · Radical cascade cyclizations

1 Introduction

Five-membered *N*-heterocycles have been regarded as a significant class of core scaffolds, prevalent in natural products, drugs, and biologically active compounds (Fig. 1) [1]. For instance, numerous pyrrole-derived compounds demonstrate a wide array of biological properties, such as antibacterial, anticancer, and anti-inflammatory effects [1a, b]. Indole derivatives are widely recognized as effective antitubercular agents [1c, d], while pyrroloindoline-based scaffolds are commonly found in alkaloids [1e]. Furthermore, a variety of pyrazole, oxazole, and thiazole-based derivatives exhibit antimalarial activity [1f] or serve as anticancer agents [1g] and function as anthelmintic drugs [1h], respectively. Due to their significant biological profiles and complex structures, continuous research efforts have been devoted to the formation of these scaffolds over the past few decades [2]. However, in line with the principle of green chemistry and sustainable development, there is a continued demand for highly efficient, economic, and mild methods to assemble these compounds.

Recently, the field of visible light-induced photoredox catalysis has evolved into a robust platform for generating highly active radical species. These intermediates are capable of participating sequential radical cascade cyclization reactions to construct target scaffolds and molecules, whether a photocatalyst is used or not [3]. Along these lines, a series of innovative approaches for the synthesis of five-membered *N*-heterocycles, encompassing pyrroles, indoles, and their derivatives, have been disclosed [4]. In this chapter, we focus on summarizing the key examples of the synthesis of five-membered *N*-heterocycles induced by visible light, as reported over the past 5 years. To enhance comprehension for readers, this book chapter is arranged in accordance with the classification of five-membered *N*-heterocycles. Within each category, further subdivisions are made based on the distinct kinds of key radical intermediates that are involved. It is essential to highlight that the manuscript does not cover the synthesis of five-membered *N*-heterocycles via visible light-induced radical covalent catalysis [5].

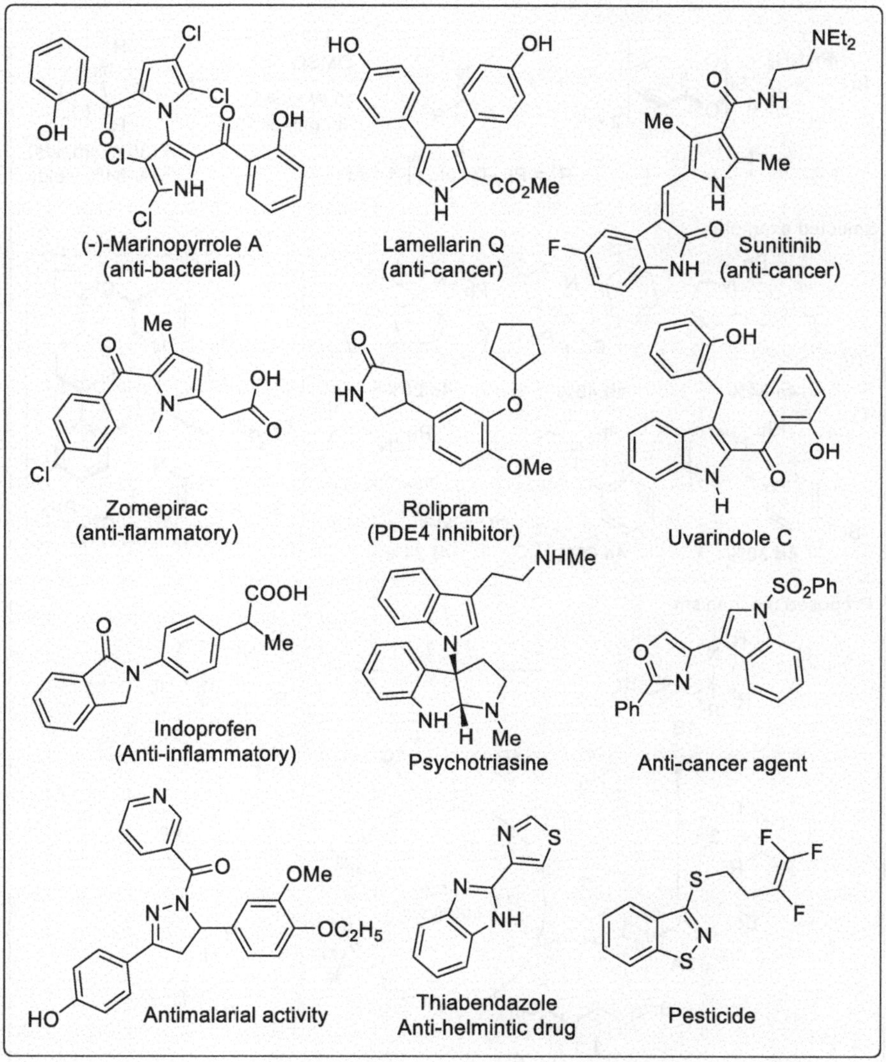

Fig. 1 Selected examples of five-membered *N*-heterocyclic compounds

2 Pyrroles and Their Derivatives

2.1 C-Radical-Mediated Reaction Pathways

Ammonium alkylbis(catecholato)silicates, serving as effective alkyl radical precursors, have been utilized in the radical alkylation of imines and the cyclopropanation of olefins under photoredox catalytic conditions. In 2019, Kelly, Molander, and colleagues reported a photoredox-mediated radical/polar crossover reaction of

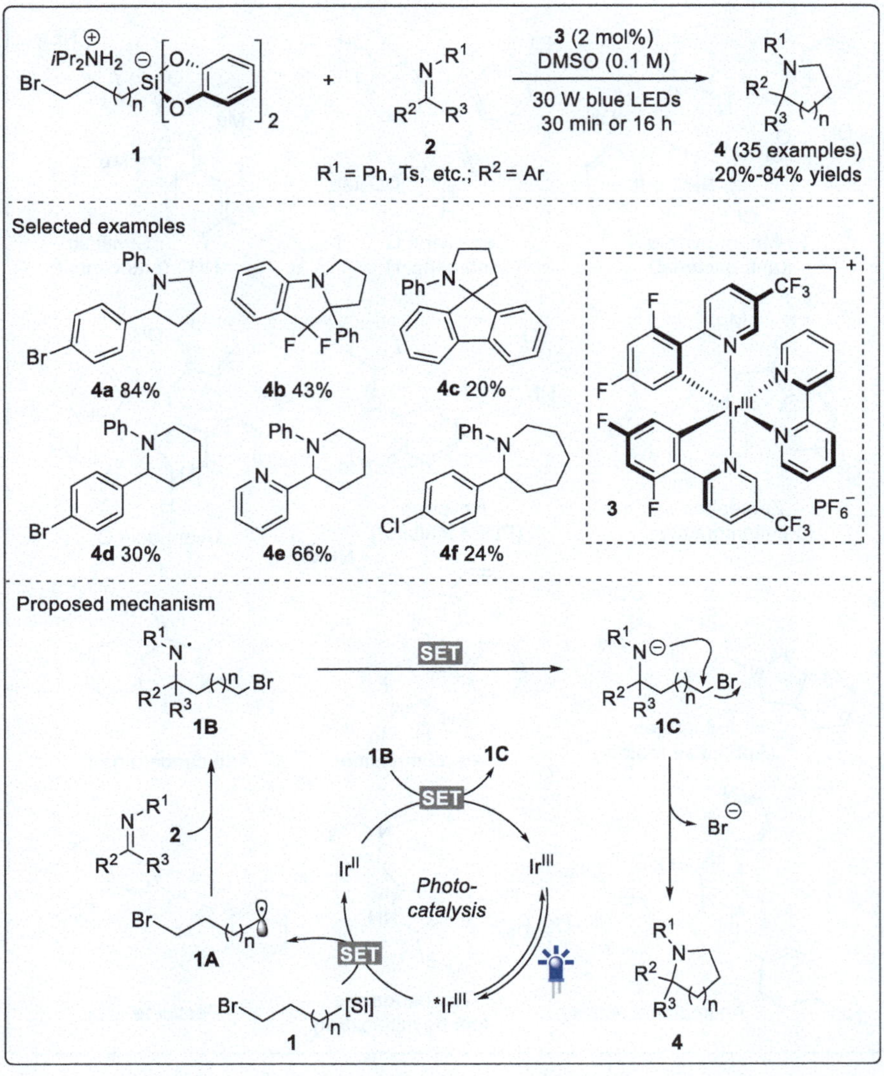

Scheme 1 *C*-radical-mediated cyclization of 3-bromopropylsilicate with aldimines

ammonium alkylbis(catecholato)silicates **1** with *N*-phenyl aldimines **2**, enabling the rapid access to saturated *N*-heterocycles **4** from acyclic precursors under mild conditions (Scheme 1) [6]. In this protocol, numerous silicates **1** and aldimine **2** derived from both electron-rich and electron-poor benzaldehydes were well accommodated to provide the corresponding pyrrolidines, piperidines, and azepanes in generally moderate to good yields (up to 84% yield). In addition to aldimines, ketimines were also compatible with this reaction system, thereby allowing the formation of pyrrolidines with α-quaternary centers (**4b**–**4c**). Remarkably, the

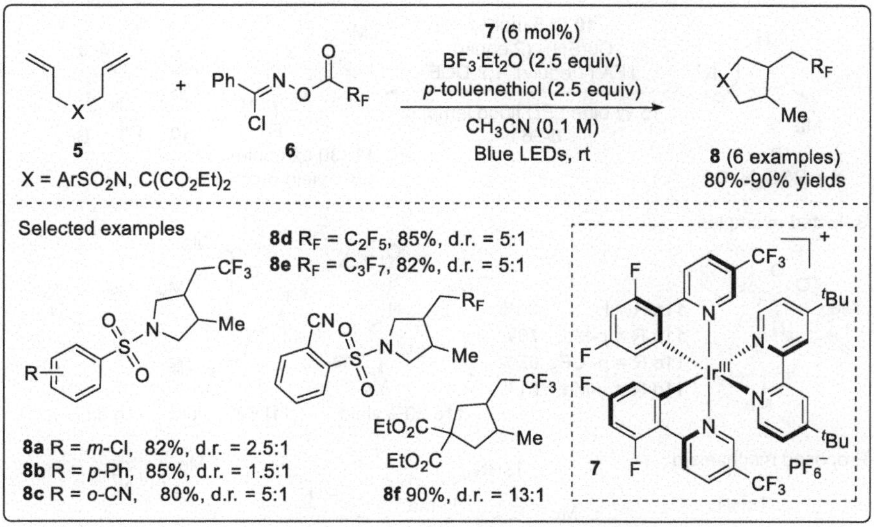

Scheme 2 *C*-radical-mediated cyclization of diallylic substrates with NHBC esters

majority of these reactions were completed within 30 minutes, demonstrating the high reactivity and efficiency of this protocol.

A plausible mechanistic pathway for the reaction was outlined as follows. Initially, visible light-driven photoexcitation of Ir^{III} results in the formation of its excited state $^{*}Ir^{III}$. Then, reductive quenching of $^{*}Ir^{III}$ ($E_{1/2}$ [$^{*}Ir^{III}/Ir^{II}$]: 1.35 V vs. SCE) by alkylsilicate **1** ($E_{1/2} = +0.4$–0.7 vs. SCE for most alkylsilicates) occurs to generate reductive state of Ir^{II} and alkyl radical **1A**, which further undergoes a radical addition to imine **2** to give *N*-centered radical **1B**. Afterwards, single electron reduction of **1B** by Ir^{II} takes place, followed by an anionic cyclization to afford *N*-heterocycle **4**.

Redox-active esters have been frequently employed as alkyl radical precursors to achieve C–C bond formations under photoredox catalytic conditions. In 2019, Wang, Liang, and Wu et al. described a photoinduced fluoroalkylation of diallylic substrates **5**, using *N*-hydroxybenzimidoyl chloride (NHBC) esters **6** as alkyl radical precursors (Scheme 2) [7]. In this reaction, an intramolecular radical-mediated cyclization process of diallyl sulfonamides and diesters was involved to give the corresponding fluorinated pyrrolidines **8a**–**8f** in generally good yields (80–90% yields) and diastereomeric ratios (up to 5:1 d.r.). DFT calculations showed that NHBC esters proceed by the fluorocarbon radical pathway.

Apart from intermolecular radical cyclization reactions mediated by carbon radicals, Yoon and colleagues reported an intramolecular copper-mediated radical-polar crossover of sterically bulky alkenes **9** under photoredox catalytic conditions (Scheme 3) [8]. This approach enables the direct construction of pyrrolidines **11**. This reaction mainly proceeded via an oxidative heterofunctionalization pathway, facilitating the conversion of the electron-rich alkenes and the heteroatomic

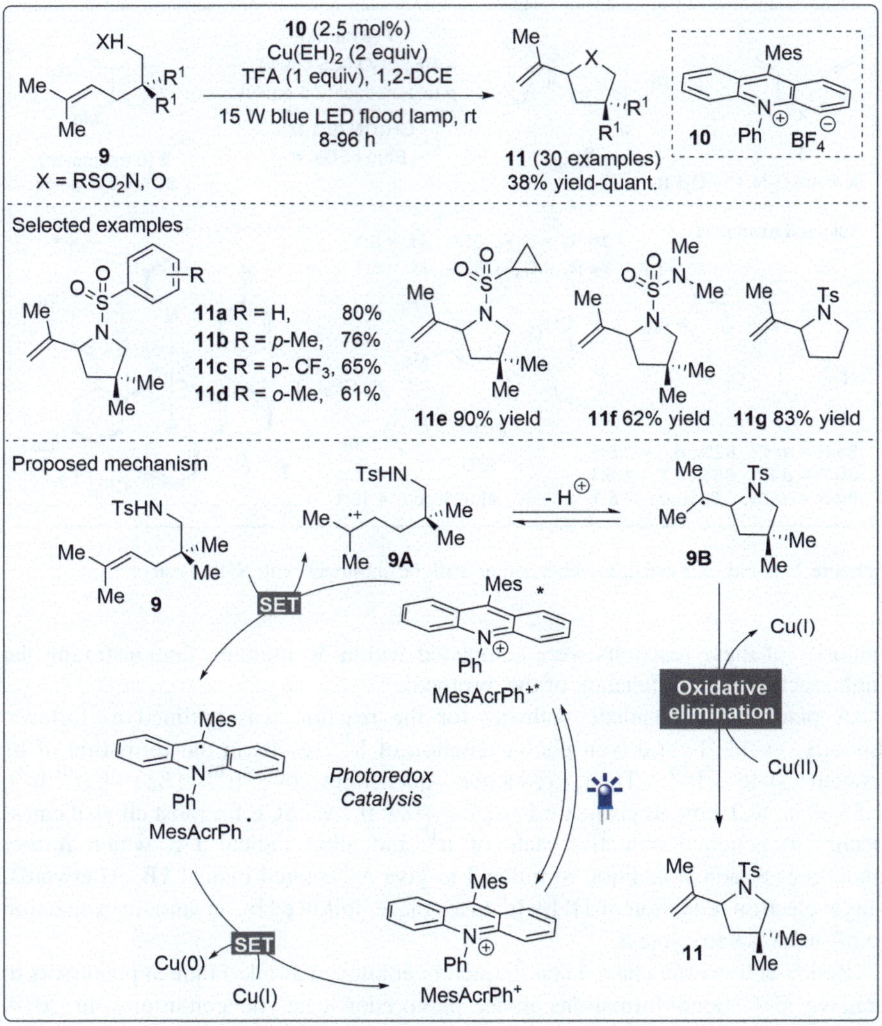

Scheme 3 Cu-mediated radical-polar crossover of sterically bulky alkenes

nucleophiles in **9** into stable heterocycles. It was observed that this protocol tolerated a wide range of bulky tri- and tetra-substituted alkenes bearing varied nucleophilic partners (**11a**–**11f**). Modification to the length and identity of the tethering group is also well accommodated to give the corresponding pyrrolidines **11g**, and six-membered rings could also be accessible, albeit with moderate yield.

A plausible mechanism was depicted in Scheme 3. Initially, visible light-driven photoexcitation of MesAcrPh$^+$ affords a highly oxidizing excited-state MesAcrPh^{+*} ($^*E_{1/2}^{\text{red}}$ = +2.12 V vs. SCE), which then undergoes a rapid reductive quenching by alkene **9** to generate radical cation **9A** and MesAcrPh. Then, radical cation **9A** could

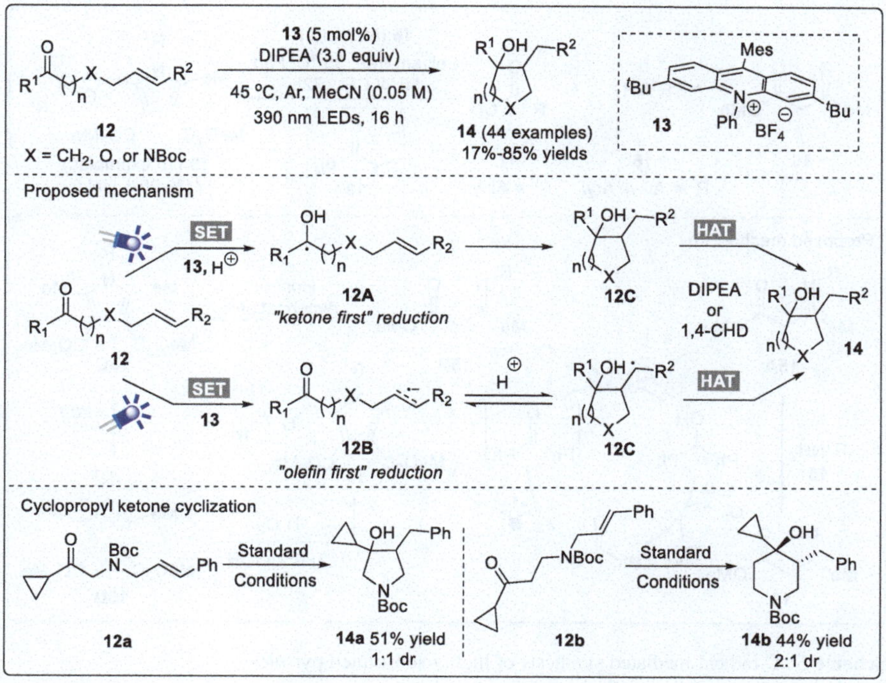

Scheme 4 Ketone-olefin coupling of aliphatic and aromatic carbonyls

be readily trapped by a pendant heteroatomic nucleophile to obtain carbon radical **9B**. Subsequently, radical-mediated oxidative addition of **9B** by Cu^{II} oxidizing agent is expected to generate a Cu^{III} species. Then, this intermediate would then undergo an elimination to yield product **11**. Final oxidation of MesAcrPh by Cu^{I} species would close the photoredox catalytic cycle to regenerate ground-state $MesAcrPh^{+}$.

Distinguished from the aforementioned amide-olefin coupling reactions, Nicewicz and co-workers very recently reported a metal-free ketone-olefin coupling reaction of reactant **12** to synthesize similar pyrrolidines **14** (Scheme 4) [9]. It was found that acridinium organic dyes **13** served as photoredox catalyst, and further mechanistic studies supported an "olefin-first" ketone-olefin coupling mechanism. For example, cyclopropane rings are known to open upon generation of a ketyl radical anion, whereas the corresponding cyclopropane rings on ketones **12a** and **12b** remained intact in both 5-*exo*-trig and 6-*exo*-trig cyclization to furnish cyclic compounds **14a** and **14b**, revealing an "olefin first"-type mechanism.

Recently, the synthesis of pyrroles through photochemical methods, utilizing various chemical feedstocks, has been significantly enriched [10]. In 2021, Salles and co-workers described a multicomponent route to obtain highly substituted pyrroles **19** under metal-free conditions (Scheme 5) [11]. It was found that readily available benzophenone **18** was employed as an efficient organic photocatalyst, which mediated the conversion of methyl acetoacetate **15** into *C*-radical intermediate via a hydrogen atom transfer process. Due to the multicomponent nature, amines **16**

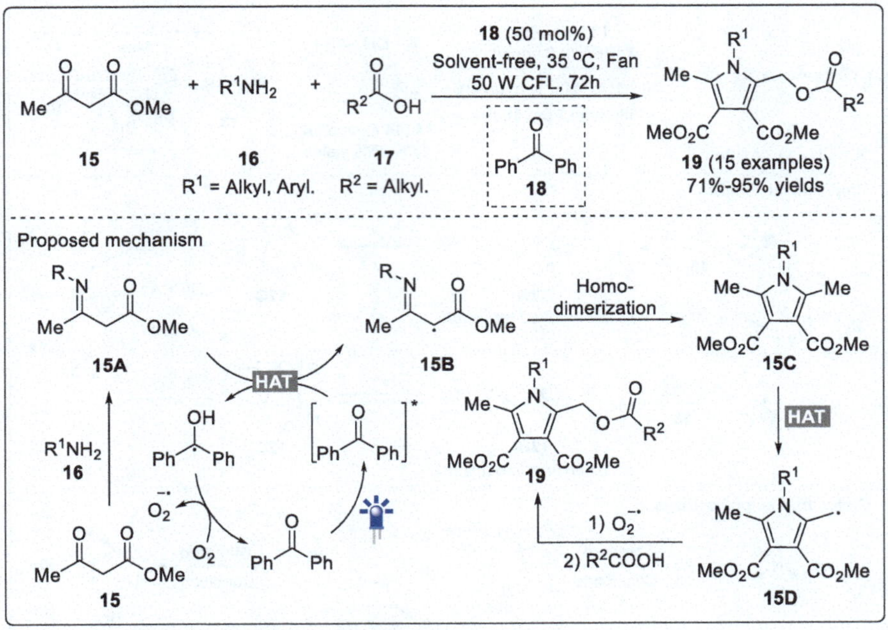

Scheme 5 *C*-radical-mediated synthesis of highly substituted pyrroles

and carboxylic acids **17** along with methyl acetoacetate **15** were found to be competent substrates. Mechanistic investigations have indicated that imine **15A** serves as an in situ-generated intermediate in the reaction pathway.

As illustrated in Scheme 5, a plausible mechanistic pathway was proposed. Photoexcitation of benzophenone takes place to give its triplet state upon irradiation with visible light. Then, a photoinduced hydrogen atom transfer (HAT) from imine **15A**, generated in situ via a condensation of methyl acetoacetate **15** and amines **16**, to benzophenone occurs to afford ketyl radical and carbon radical **15B**. The reverse HAT process between ketyl radical and oxygen in the air generates superoxide radical anion and regenerates benzophenone. Homodimerization of radical **15B** accompanied with an elimination of amine provides pyrrole **15C**. Another HAT process between triplet-state benzophenone and pyrrole **15C** obtains radical **15D**, which then undergoes a single-electron oxidation coupled with a nucleophilic attack by carboxylic acids **17** to release target pyrroles **19**.

The photochemical activation of strong carbon–halide bonds remains a challenging tasks, premarily due to the necessity of employing harsh reducing conditions. In this regard, Crespi and Giedyk et al. disclosed a controllable, photocatalytic C–H arylation method that channels the reaction of chlorinated benzamides **20** to produce isoindolinones **22** (Scheme 6) [12]. Interestingly, the optimal conditions were exceptionally mild, using aqueous micellar solutions as the reaction medium, and methylene blue served as a cost-effective photocatalyst. In this reaction, a series of *o*-chlorinated benzamides were well accommodated to obtain isoindolinones **22**.

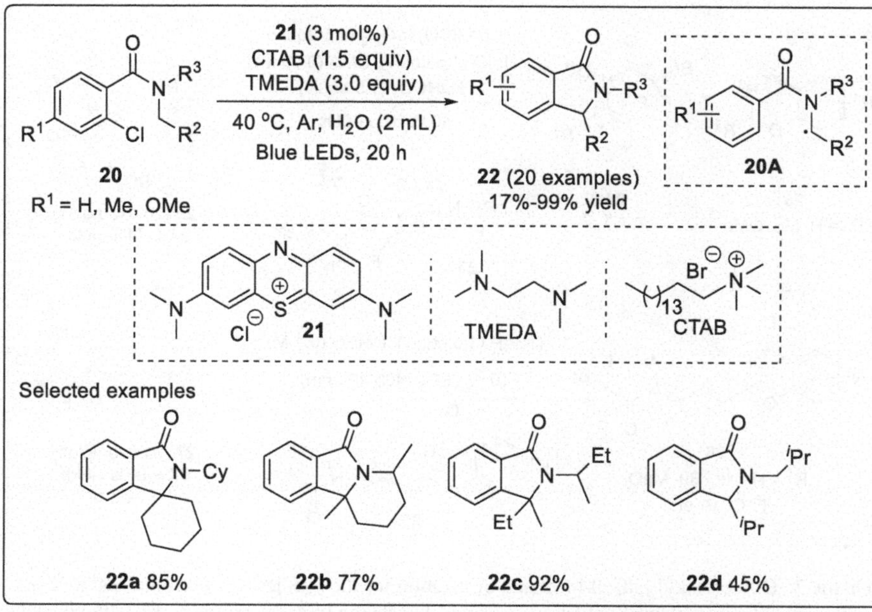

Scheme 6 *C*-radical-mediated C–H arylation for the synthesis of isoindolinones

Notably, it was observed that tertiary C–H bonds of chlorinated benzamides **20** exhibited a marked preference for the C–H arylation, furnishing the final products in excellent yields (**22a–22c**, 77–92% yields). Mechanistic studies revealed that a single-electron reduction of C–Cl bond occurrs, followed by an intramolecular 1,5-HAT process to generate *C*-radical **20A**. Moreover, cyclic voltammetry measurements have shown that the reduction of halogenated benzamides is significantly efficient in a micellar system compared to a standard solution in MeCN.

In addition to the aforementioned photocatalyst-mediated reactions, photocatalyst-free radical cyclization reactions driven by visible light have also risen as an alternative pathway in this field. In 2022, Yu and co-workers achieved a three-component radical tandem bicyclization of *N*-allylbromodifluoroacetamides **24** and terminal alkynes **23** with the insertion of sulfur dioxide (Scheme 7a) [13]. In this protocol, various electron-donating and electron-withdrawing groups on aromatic ring of *N*-allylbromodifluoroacetamides **24** and 2-alkynylarylethers **23** proceeded smoothly to give pyrrolidinone derivatives **25**. Unfortunately, *N*-allylbromodifluoroacetamides **24** with cyclopentyl or pyridyl group on *N*-atom failed to deliver the desired product. According to the radical-trapping experiment and UV–vis experiment, it was proposed that the charge transfer complex **23A** acted as a difluoroalkyl radical precursor.

Subsequently, in 2023, the Sun and Jin group reported a similarly difluoroalkyl radical-mediated radical cyclization cascade approach for the assembly of pyrrolidinone derivatives, offering a similar strategy (Scheme 7b) [14]. This reaction predominantly occurs at wavelengths of 400–405 nm, with no observed reactivity

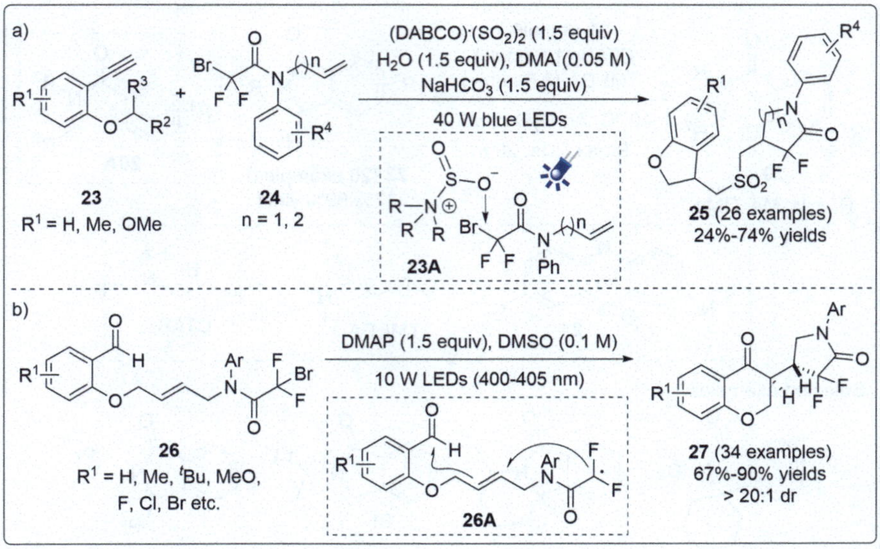

Scheme 7 Difluoroalkyl radical-mediated cyclization for the synthesis of pyrrolinones: a) Three-component difluoroamidosulfonylation mediated by charge-transfer complex; b) Intramolecular radical cyclization cascades initiated by direct homolysis of Csp^3-Br bond.

beyond 450 nm, possibly due to insufficient energy to facilitate the homolysis of the Csp^3–Br bond (BDE = 68 kcal mol^{-1}) homolysis. Moreover, this case represents the first example of visible light-mediated Csp^3–Br bond homolysis-initiated radical cyclization cascade, affording products **27** with commendable regioselectivity and diastereoselectivity.

2.2 N-*Radical-Mediated Reaction Pathways*

N-centered radicals represent a versatile class of intermediates for constructing nitrogen heterocycles under visible-light photocatalytic conditions. Iminyl radicals, an important subclass of nitrogen radicals, can conventionally be generated through N–O bond cleavage, followed by hydroimination process with olefin to yield pyrroline derivatives. In this context, in 2018, Feng and Loh et al. reported a visible light-promoted hydroimination of **28**, using *fac*-Ir(ppy)$_3$ **29** as a photocatalyst (Scheme 8a) [15]. Under optimal conditions, numerous 3,4-pyrroline derivatives **30** could be obtained with good efficiency and regioselectivity (up to 98% yield). It is noteworthy that the use of DMPU as hydrogen donor is significant for enhancing reaction efficiency, and the *O*-acyl oxime moiety on substrate **28** is recognized as an efficient iminyl radical precursor. Yuan and Yu et al. achieved a photoredox-catalyzed iminoalkenylation of alkenes, starting from the similar γ-alkenyl *O*-acyl oximes **28**, for the synthesis of both (*E*)- and (*Z*)-cinnamylpyrrolines [16]. This method utilizes vinyl boronic acids as alkenyl sources [16].

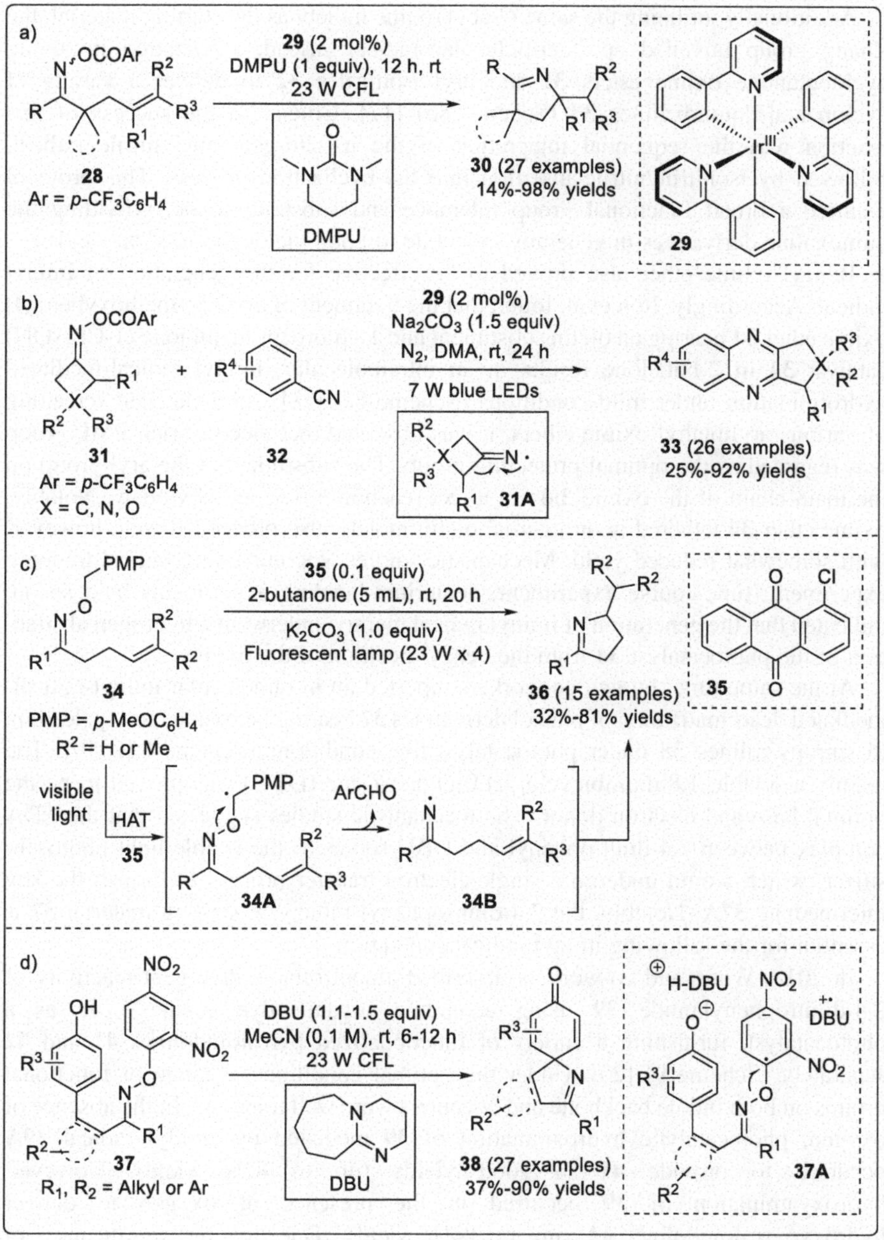

Scheme 8 Iminyl radical-mediated cyclization for the synthesis of pyrrolines: a) Hydroimination of unactivated alkenes for the synthesis of pyrrolines; b) Radical cascade cyclization of oxime esters and aryl isonitriles; c) Synthesis of pyrrolines via hydroimination cyclization of benzyl oxime ether; d) Radical cyclization of ketoximes for the synthesis of spiropyrrolines.

Additionally, utilizing the same *O*-acyl oxime moiety as the starting material, the Zhang group unveiled an intermolecular radical cascade cyclization involving cyclobutanone oxime esters **31** and aryl isonitriles **32** to deliver a variety of cyclopenta[*b*]quinoxalines **33** (Scheme 8b) [17]. Crucial to the success of this reaction was the sequential integration of the in situ-generated nitrile radical, followed by isonitrile/nitrile insertion into the cyclization process. This protocol featured a broad functional group tolerance and substrate scope, providing the quinoxaline derivatives in generally moderate to good yields (up to 92% yield).

Benzyl oxime ether also proved to be effective for the generation of iminyl radical. Accordingly, Itoh et al. found that the treatment of an *O*-(4-methoxybenzyl) oxime ether **34** bearing an olefin substituent and 1-chloroanthraquinone (1-Cl-AQN) catalyst **34** in 2-butanone results in an intramolecular, iminyl radical-mediated hydroimination under mild conditions (Scheme 8c) [18]. After detailed screening of various arylmethyl oxime ethers, it was observed that electron-rich PMP group was regarded as the optimal protecting group. The substituent at the aryl group on the main chain of the oxime did not affect reaction efficiency obviously. Notably, oxime ether **34** tethered in an internal olefin moiety also proceeded well, however, with somewhat reduced yield. Mechanistic studies, encompassing radical-trapping experiment, time-course experiment, deuterium-labeling experiments, and so on, indicated that the generation of iminyl radical mainly underwent a hydrogen abstraction of the photocatalyst **34** from the benzyl position of the oxime.

At the same time, Fu and co-workers reported an intramolecular iminyl-radical-mediated dearomatization of phenol derivatives **37** bearing ketoxime moiety, leading to spiropyrrolines **38** under photocatalyst-free conditions (Scheme 8d) [19]. The readily available 1,8-diazabicyclo[5.4.0]undec-7-ene (DBU) was proved to be the optimal base and electron donor. The mechanistic studies suggested that the EDA complex between 2,4-dinitrophenyl and DBU serves as the visible-light photosensitizer, which would undergo a single-electron transfer process to furnish the key intermediate **37A**. Notably, the 2,4-dinitrophenyl moiety in starting material **37** is essential for the following iminyl radical generation.

In 2018, Wang and co-workers described a controllable divergent reactivity of 2,4-dinitrophenylamide **39** using economic organic dye eosin Y **40** as a photocatalyst, furnishing a variety of functionalized pyrroloindolines **41** and **42** separately (Scheme 9) [20]. Under the optimal conditions, a range of functional groups on both indole backbone and *N*-centers were well tolerated. In the absence of oxygen, photocatalytic hydroamination of **39** mediated by amidyl radical **39A** occurred to provide **41** in good yields (up to 82% yield). However, hydroxyamination of **39** occurred in the presence of oxygen to deliver hydroxypyrroloindoline **42** (up to 84% yield). Due to the significance of pyrroloindoline skeletons in pharmaceutical studies, they achieved a rapid access to (±)-flustraminol B **43** via a photocatalytic hydroxyamination of substrate **39a** followed by a reduction of amide **42a**. Afterwards, starting from the same amide moiety, a photocatalyst-free amidyl radical cyclization of allenes was developed by Schomaker and co-workers in 2021, providing a variety of pyrrolinones [21].

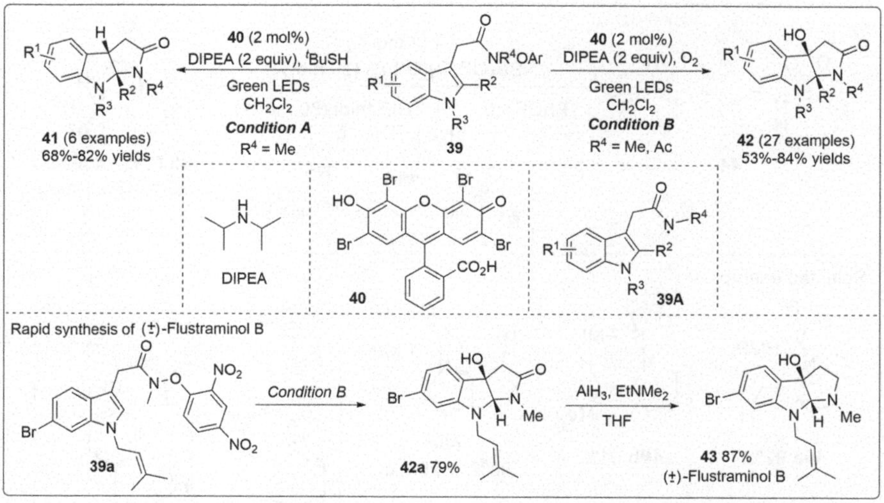

Scheme 9 *N*-radical-mediated synthesis of pyrroline derivatives via N–O bond cleavage

Besides, *N*-centered sulfonamidyl radicals also proved to be effective for the synthesis of pyrrolines. Due to its electrophilicity, sulfonamidyl radicals are expected to react readily with nucleophilic olefins. Accordingly, the Knowles group developed a sulfonamidyl radical-mediated intramolecular cyclization of olefins **44** via proton-coupled electron transfer (PCET) strategy (Scheme 10a) [22]. In this reaction, a number of olefin substitution patterns were accommodated, including terminal, 1,2-disubstituted, and trisubstituted olefins (**46a–46f**, up to 92% yield). Through a PCET process, sulfonamidyl radicals were generated directly from the cleavage of N–H bond and the thiol was employed as HAT catalyst. Using the chiral phosphate base, they further achieved the asymmetric version of this reaction in 2020 [23].

Moreover, those sulfonamidyl radicals could also be generated via nitrogen–halide bond cleavage under visible light-driven photoredox conditions. In 2021, Qu and Zhang et al. reported a hydroamination and carboamination of alkenes **47** via a manganese-catalyzed N–F bond activation strategy (Scheme 10b) [24]. In this reaction, simple manganese complex $Mn_2(CO)_{10}$ functioned as the precatalyst, and an inexpensive silane $(MeO)_3SiH$ was used as both the hydrogen-atom donor and the F-atom acceptor. The optimal reaction conditions were found to be applicable to a broad range of substrates **47**, especially a broad range of aliphatic sulfonamides.

Moreover, Stephenson and co-workers in 2020 disclosed a sulfonamidyl radical-mediated carboamination/dearomatization cascade of **51** for the rapid assembly of cyclohexadiene-fused sultams **52** containing pyrrolinone scaffolds (Scheme 11a) [25a]. This reaction was carried out at room temperature under visible-light irradiation, proceeding smoothly through the combination of an iridium photocatalyst **45** and a dialkyl phosphate base. It was observed that simple γ,δ-unsaturated *N*-arylsulfonyl enamides were converted into highly substituted, three-dimensional

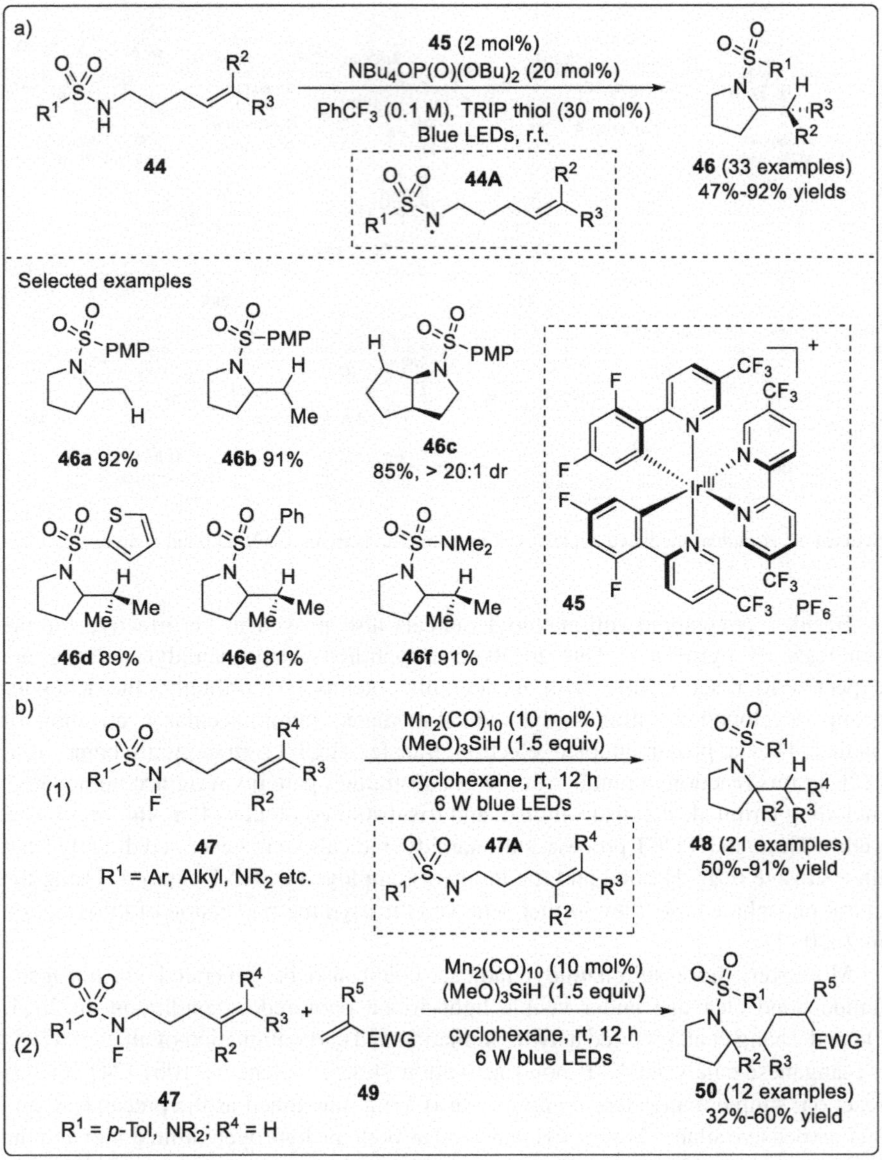

Scheme 10 Sulfonamidyl radical-mediated cyclization for the synthesis of pyrrolidines: a) Intramolecular anti-Markovnikov hydroamination of unactivated alkenes through PCET strategy; b) Intramolecular hydroamination and carboamination of alkenes via N-F bond activation.

compounds in satisfactory yields and good diastereoselectivity. Notably, based on similar sulfonamidyl-radical intermediates, Qin and colleagues have devised a series of efficient and systematic approaches for synthesizing a diverse array of *N*-heterocycle-fused alkaloids [25b].

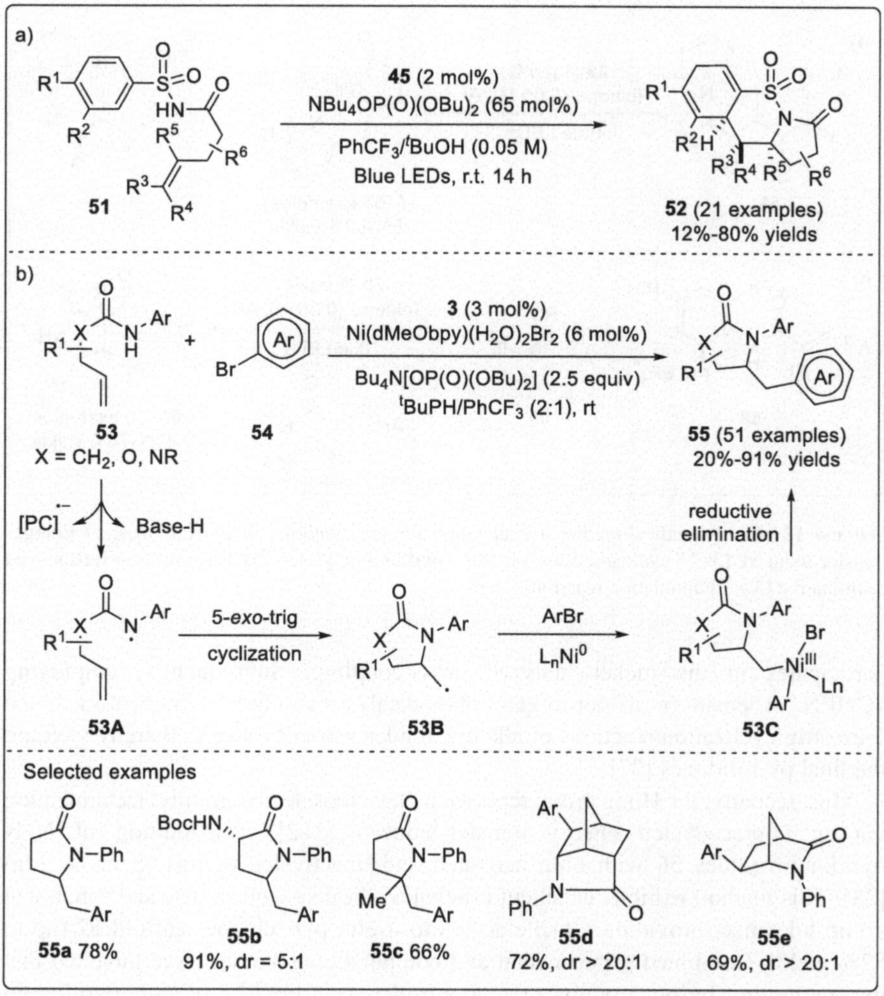

Scheme 11 Amidyl radical-mediated cyclization for the synthesis of pyrrolinones: a) Intramolecular carboamination/dearomatization cascade reaction of alkenes; b) Intermolecular cascade amidoarylation of olefins.

Besides sulfonamidyl radicals, amidyl radicals, which are produced through the cleavage of N–H bonds, have also been extensively used in the synthesis of pyrrole derivatives. In this realm, in 2019, the Molander group described a diastereoselective amidoarylation of unactivated olefins **53** with arylbromide **54** merging photoredox PCET with Ni-catalyzed cross-coupling strategy (Scheme 11b) [26]. This case was not restricted to amides; carbamates and ureas were also compatible. As for the mechanism, through a PCET process, amidyl radicals **53A** could be obtained directly from non-prefunctionalized N–H bonds. Then, radical intermediates **53A** are trapped by pendant olefins with the release of alkyl radicals **53B**, which further

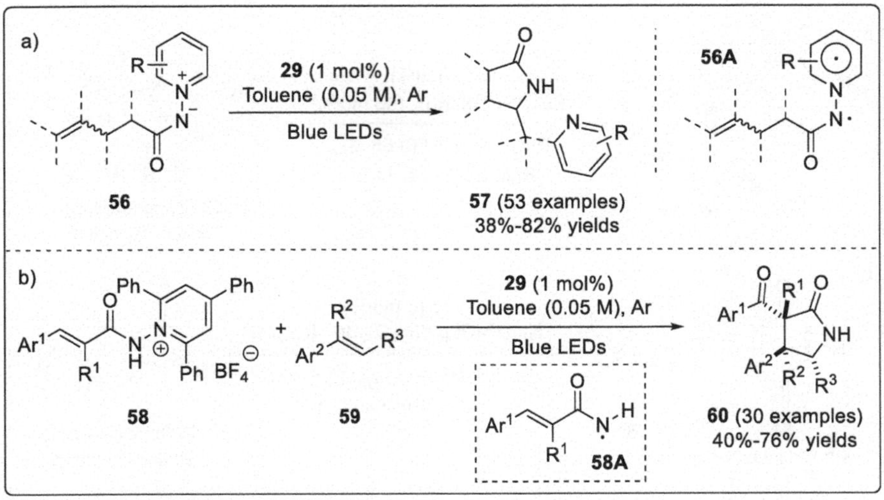

Scheme 12 Amidyl radical-mediated cyclization for the synthesis of pyrrolinones: a) Energy-transfer-induced [3+2] cycloadditions of N–N pyridinium ylides; b) Electron-transfer-induced deaminative [3 + 2] annulation reaction.

participates in the nickel-catalyzed cross-coupling. Subsequently, employing 4CzIPN or eosin Y as an organo-photocatalyst, similarly *N*-radical-mediated 5-*exo*-trig cyclization reactions of alkenyl amides were developed, thereby yielding the final pyrrolidones [27].

Most recently, the Hong group reported a diastereoselective pyridyl lactamization via an unprecedented energy transfer-induced [3+2] cycloaddition of N–N pyridinium ylides **56** with both activated and unactivated olefins (Scheme 12a) [28]. This method exhibits excellent efficiency, diastereoselectivity, and functional group tolerance, providing a facile access to useful pyrrolinone scaffold **57** (up to 82% yield). Combined experimental and computational studies have unveiled that the energy transfer process gives rise to a triplet-state diradical of N–N pyridinium ylides **56A**, which facilitates the stepwise cycloaddition.

In contrast to the aforementioned energy transfer pathway, in 2022, Yang and Xia et al. achieved an intermolecular deaminative [3+2] annulation reaction between compound **58** and olefins **59** via a photoinduced single-electron transfer route (Scheme 12b) [29]. This reaction exhibits good functional group tolerance, affording pyrrolinones **60** with excellent diastereoselectivity (40–76% yield). Preliminary studies, including the TEMPO-trapping experiment and control experiments, suggested that the employed *N*-aminopyridinium salts provide the key amidyl radical **58A** through a photoinduced single-electron reduction of N–N bond cleavage.

Recently, the scope of *N*-radical-mediated cyclization reactions for the synthesis of pyrrole scaffolds has been significantly expanded [30]. In 2019, Protti and Bandini et al. realized an unprecedented *N*-radical-mediated cyclization of

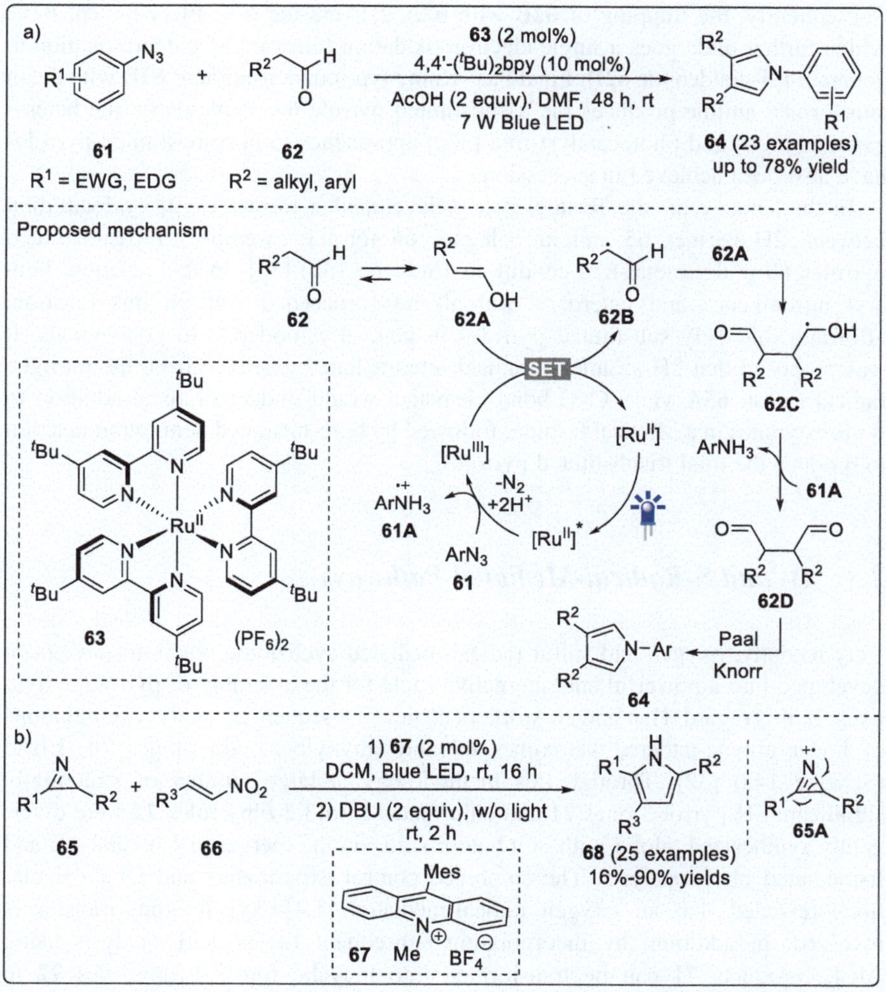

Scheme 13 *N*-radical-mediated cyclization for the synthesis of pyrroles: a) Synthesis of 1,3,4-trisubstituted pyrroles from aryl azides; b) Metal-free synthesis of 1,2,4-trisubstituted pyrroles from 2H-azirines.

condensing enolizable aldehydes **62** and arylazides **61**, resulting in the formation of trisubstituted pyrroles **64** (Scheme 13a) [30a]. This method, by means of catalytic amounts of Ru^{II} photocatalyst **63** and 7 W blue LED irradiation, allows for the replacement of classic stoichiometric oxidants, such as TBHP and Mn(III) salts.

A plausible mechanism was postulated as follows. The photoexcited $Ru^{II}*$ catalyst undergoes an oxidative quenching by the arylazide **61**, generating an anilino radical, which is subsequently protonated to form radical cation **61A**. Then, Ru^{III} species acts as an SET oxidant to oxidize enolized aldehyde **62**, affording the α-carbonyl-radical **62B** with the release of the ground state of Ru^{II} photocatalyst.

Subsequently, the trapping of **62B** with **62A** delivers the α-hydroxy-radical **62C**, which further undergoes a single-electron oxidation followed by a deprotonation to generate 1,4-dialdehyde **62D**. Final Paal-Knorr-type condensation of **62D** with the in situ-formed aniline produces the trisubstituted pyrrole **64**. Particularly, the heterogeneous [30b] and photocatalyst-free [30c] approaches to polysubstituted pyrroles have also been achieved in succession.

In the same year, the Rastogi group developed a formal [3+2] cycloaddition between 2H-azirines **65** and nitroalkenes **66** for the assembly of trisubstituted pyrroles **68** under metal-free conditions (Scheme 13b) [31]. In this reaction, both aryl nitroalkenes and heteroaryl nitroalkenes proceeded well in this reaction, affording diversely substituted pyrroles in generally moderate to good yields. It was proposed that 2H-azaallenyl radical intermediate, generated from the nitrogen radical cation **65A** via a C–C bond cleavage, would undergo radical addition to β-nitrostyrenes in a Michael fashion, followed by base-mediated denitration reaction to produce the final trisubstituted pyrroles.

2.3 O- and S-Radical-Mediated Pathways

Very recently, oxygen and sulfur radical-mediated cyclization reactions have been developed into a powerful and alternative route for the assembly of pyrrole derivatives. In this regard, Han and co-workers achieved a sequential catalytic annulations of linear alkyne-tethered ketoximes **69** and ethynylbenziodoxolones **70** (EBX) (Scheme 14a) [32]. Through this methodology, a large number of structurally significant 3H-pyrrol-3-ones **71** and chlorinated furo[3,2-*b*]pyrroles **72** were divergently synthesized, along with an *O*-atom shift via the merger of Cu catalysis and Ir-mediated photocatalysis. The combined control experiments and DFT calculations revealed that an oxygen radical-mediated [1,4]-oxygen atom transfer is involved. In addition, by incorporating subsequent Lewis acid catalysis using $AlCl_3$, products **71** can be transformed into bicyclic furo[3,2-*b*]pyrroles **72** in generally moderate yields at 50°C (Scheme 14b).

For the mechanism, it was proposed that the initial oxidization of ketoxime **69** by Cu^{II} through single-electron transfer gives Cu^{I} and iminoxyl radical **69A**. This radical then undergoes a 5-*exo*-dig cyclization and is trapped by **70**, followed by a rapid β-fragmentation to afford compounds **69B** via the elimination of the benziodoxolonyl radical (BI). Subsequently, a photoinduced energy transfer of **69B** takes place to provide pyrrol-3-one **71** via an *N,O*-diradical rearrangement. As shown in Scheme 14b, the annulation of the carbonyl and alkynyl by O–C bond formation produces the zwitterion **71B** in the presence of $AlCl_3$. After abstraction of a chlorine from DCE, **71B** is converted into intermediate **71C**, which then undergoes protonation from the surrounding to finally yield chlorinated furo[3,2-*b*]pyrrole **72**. Notably, this strategy introduces a novel mode of oxygen atom shift, which also broadens the new boundary of heteroatom radical rearrangement.

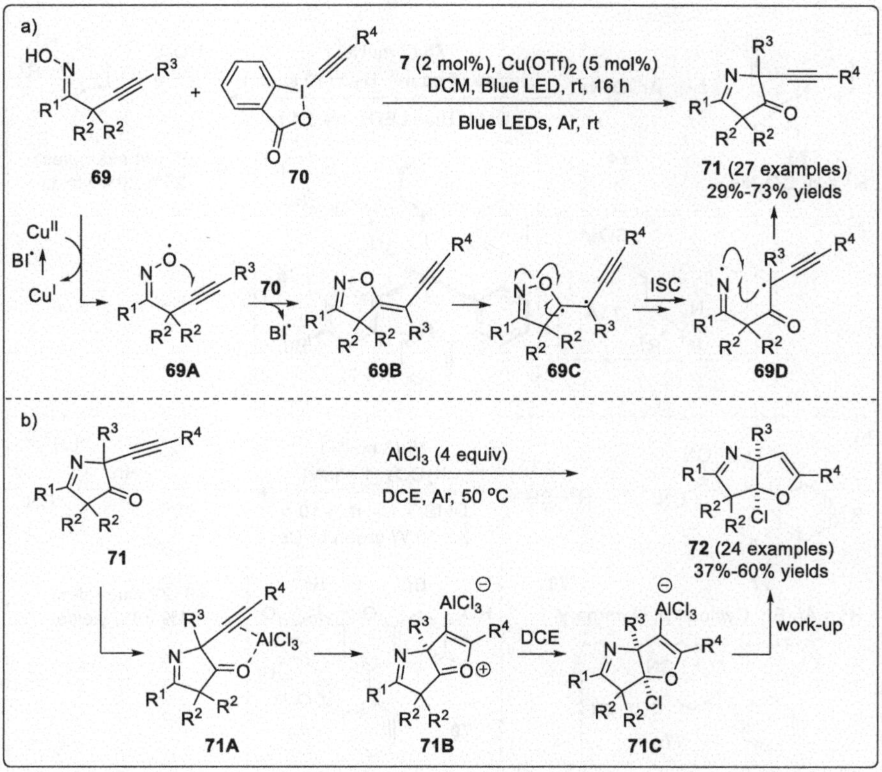

Scheme 14 *O*-radical-mediated [1,4]-oxygen atom transfer for the synthesis of pyrrole derivatives: a) Radical [1,4]oxygen-atom transfer mediated synthesis of pyrrol-3-ones; b) Further synthetic transformations of pyrrol-3-ones.

In the same year, Cai and Jin et al. reported an efficient radical addition translocation-cyclization of *N*-homopropargylic amines **73** with sodium sulfonates **74,** offering sulfonyl-containing pyrrolidine **76** (Scheme 15a) [33]. Under the optimal conditions, sodium sulfonates **74** acted as the optimal sulfonyl radical precursors, while *N*-homopropargylic amines **73** functioned as good radical acceptors to trap the sulfonyl radical, thereby affording intermediate **73A.** This reaction exhibited a broad substrate scope, thus providing a wide range of monocyclic, bicyclic, and polycyclic azacycles.

At the same time, the Patel group accomplished a thiyl radical-triggered approach to thio-functionalized pyrroles **80**, using β-ketodinitriles **77** and thiophenols **78** as reaction partners under green-light irradiation (Scheme 15b) [34]. In this protocol, multiple bonds, including C–N, C–S, and C=C bonds, are constructed simultaneously. This reaction also features a wide range of β-ketodinitriles and thiophenols, affording the corresponding products in generally good yields (up to 78% yield). However, reactant ethyl 2-cyano-4-oxo-4-phenylbutanoate **77a** bearing an ester group failed to give the target product under the standard conditions.

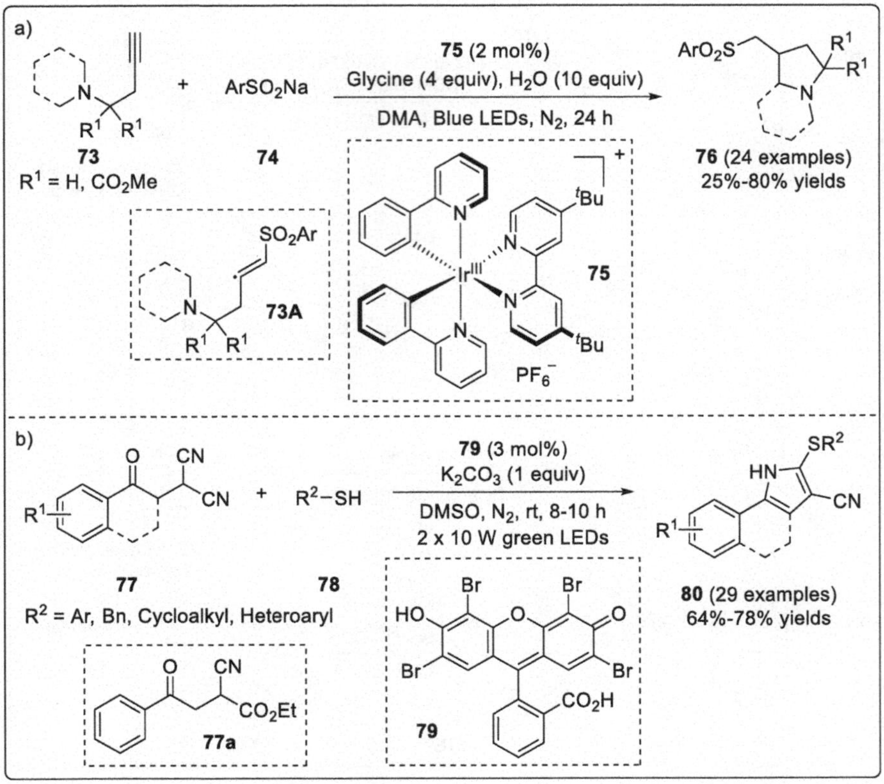

Scheme 15 *S*-radical-mediated cyclization for the synthesis of thio-functionalized pyrroles: a) Sulfonyl radical addition-translocationcyclization to construct pyrrole derivatives; b) Thiyl-radical-mediated synthesis of thio-functionalized pyrroles.

3 Indoles and Their Derivatives

3.1 C-Radical-Mediated Reaction Pathways

Due to their privileged structure [1c, d] and significant role in synthetic chemistry [35], a series of preparative methods for the synthesis of indoles and related heterocycles have been explored. In 2018, Xie and Zhu group reported an intermolecular *C*-radical cyclization cascade approach to indolo[2,1-*a*]isoquinoline derivatives from *o*-tetrahydroisoquinoline-substituted styrenes **81** with *gem*-dibromides **82** (Scheme 16a) [36]. After a detailed screening of bases, it was observed that K_2HPO_4 proved to be the suitable base. Both increasing and decreasing the intensity of the base only led to the formation of desired product in moderate yields. The presence of both electron-withdrawing group and electron-donating groups on the aromatic ring had little impact on the reaction efficiency, showcasing good functional group tolerance. For the mechanism, it was proposed that *gem*-

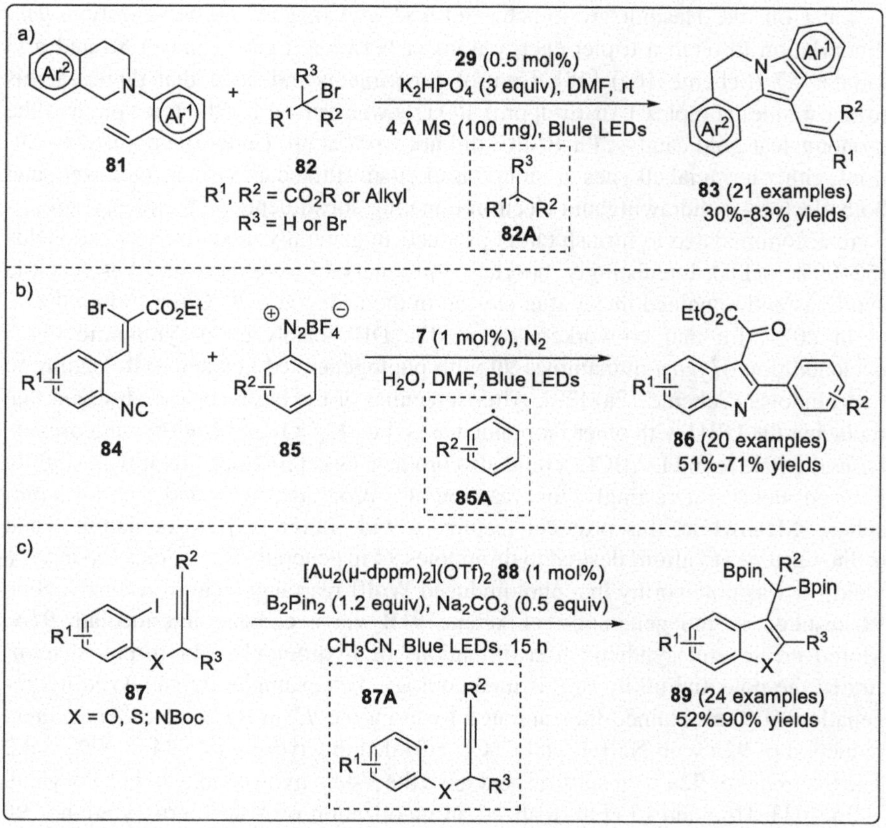

Scheme 16 *C*-radical-mediated cyclization for the synthesis of indoles: a) Relay photocatalytic cascade for the synthesis of indolo[2,1-a]isoquinoline derivatives; b) Tandem cyclization of *ortho*-isocyano-α-bromo cinnamates for the synthesis of 2-substituted indole-3-glyoxylates; c) Energy transfer-mediated carbo-cyclization/*gem*-diborylation of aryl iodides.

dibromides **82** mainly served as the carbon radical precursors, which would undergo single-electron reduction of C–Br bond to produce the key radical intermediate **82A**.

In the same year, the Reiser group disclosed a radical tandem cyclization of *ortho*-isocyano-α-bromo cinnamates **84** to 2-substituted indole-3-glyoxylates **86** under photocatalytic conditions (Scheme 16b, up to 71% yield) [37]. In this reaction, aryl diazonium salts **85** were utilized as aryl radical precursors, with H_2O acting as the terminal nucleophile. Notably, attempts to increase the reaction yield were not successful through changing the loading of diazonium salts and water or replacing the latter by alcohols to avoid hydrolysis of the isonitrile. It was noteworthy that the reaction also proceeded without photocatalyst **7**, albeit with significantly reducing yield, indicating that a radical chain process might be involved. An ^{18}O-labeled experiment using H_2O^{18} supported that H_2O functioned as oxygen source, installing into product **86** via the final nucleophilic attack.

Later on, the Hashmi group achieved a phenyl radical carbo-cyclization/*gem*-diborylation through a triplet energy transfer between a gold catalyst **88** and aryl iodides **87** (Scheme 16c) [38]. Control experiments indicated that dinuclear bis (diphosphine) complex $[Au_2(\mu\text{-dppm})_2](OTf)_2$ was critical for the reaction, and the mononuclear gold catalyst Ph_3PAuCl did not work at all. Under the optimal conditions, either terminal alkynes or mono/bisalkyl-substituted alkynes reacted well, and both electron-withdrawing and electron-donating substituents on the phenyl groups were accommodated to furnish target products in generally moderate to good yields (52–90% yields). Interestingly, borylation products **89** were recognized as versatile synthons and remained intact after storing in the fridge at −20°C for 10 months.

In 2021, Lu and co-workers reported a DBU-catalyzed desymmetric [3+2] cycloaddition of *para*-quinamines **90** with photogenerated ketenes **91B**, leading to hydroindoles (Scheme 17a) [39]. After screening of the bases, it was observed that replacing the DBU with other inorganic bases, i.e., K_2CO_3 and NaOAc, and organic bases, i.e., NEt_3 and DABCO, could also produce target product, albeit with slightly reduced yields. Interestingly, this reaction also proceeded well under an air atmosphere. Meanwhile, this protocol features a wide range of *para*-quinamines and α-diazoketones to afford desired hydroindoles **94** in generally high yields (up to 99% yield). It was noteworthy that photoinduced Wolff rearrangement of α-diazoketone **91** resulted in the generation of ketene **91B** via a carbene intermediate **91A**. Moreover, several synthetic transformations of hydroindole **92a** further demonstrated the potential utility of this methodology. For example, Pd-catalyzed hydrogenation of **92a** obtained the saturated hydroindole **93** in 95% yield. And, direct reduction of **92a** with $NaBH_4$ and $CeCl_3$ afforded the hydroindole **94** in 99% yield. The treatment of **92a** with sulfuric acid gave the *N*-free hydroindole **95** in 51% yield.

In 2023, Hou and Li et al. realized an indolization of *N*-aryl tertiary amines **96** with diazoacetates **97** catalyzed by organophotocatalyst **98** under metal-free conditions (Scheme 17b) [40]. Due to self-coupling of α-aminoalkyl radical **96A** or *C*-radical **97A**, the target indoles were obtained in generally moderate yields (up to 68% yield). Prolonging the reaction time or increasing the intensity of light could not promote this transformation effectively. It was worth mentioning that this reaction exhibits a wide substrate scope and was suitable for late-stage functionalization of complex molecules. The combined mechanistic studies and DFT calculations disclosed a relay visible-light photoredox catalytic process that might involve several discrete photoredox catalytic cycles in a single operation with 4CzIPN as sole organic dye photocatalyst.

Traditionally, *N*-arylacrylamides have been recognized as one class of radical acceptors for straightforward access to indole derivatives. However, hydroarylation of olefinic moiety of *N*-arylacrylamides via the PCET process has been exploited rarely. In this context, Ji and Huang et al. disclosed an intramolecular hydroarylation of alkenes **96** enabling the construction of oxindoles **97** via a visible light-mediated PCET process (Scheme 18a) [41]. Under optimal conditions, it was observed that TfOH served as the proton source, and LiBr acted as a good electron mediator, both of which were essential for the success of this protocol. And this reaction exhibited a broad substrate scope and tolerates the substrates attached with some drug molecules

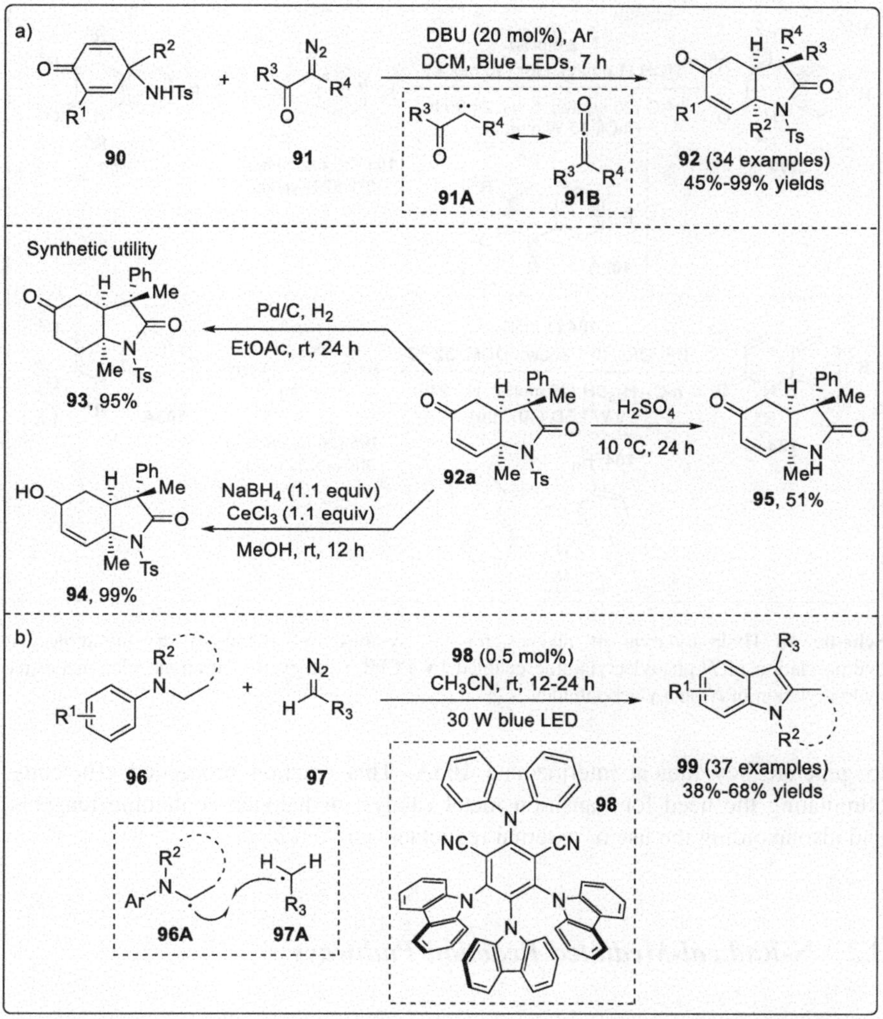

Scheme 17 *C*-radical-mediated cyclization cascade for the synthesis of indoles: a) Desymmetric [3+2] cycloaddition for the synthesis of hydroindoles; b) Indolization of *N*-aryl tertiary amines with diazoacetates.

including ibuprofen, naproxen, salicylic acid, and estrone. Notably, the existence of 3,4-dihydroquinolinone obtained via a 6-*endo*-trig cyclization process was exemplified.

In 2023, Qu and Kang introduced a metal-free intramolecular hydroarylation reaction of *N*-arylacrylamides **103** to obtain oxindoles **105**, employing Lewis acid as an activation factor (Scheme 18b) [42]. In this reaction, organic molecule **104** functioned as photocatalyst, and the coordination between $BF_3{\cdot}OEt_2$ and *N*-arylacrylamides **103** was expected to promote the further single-electron reduction

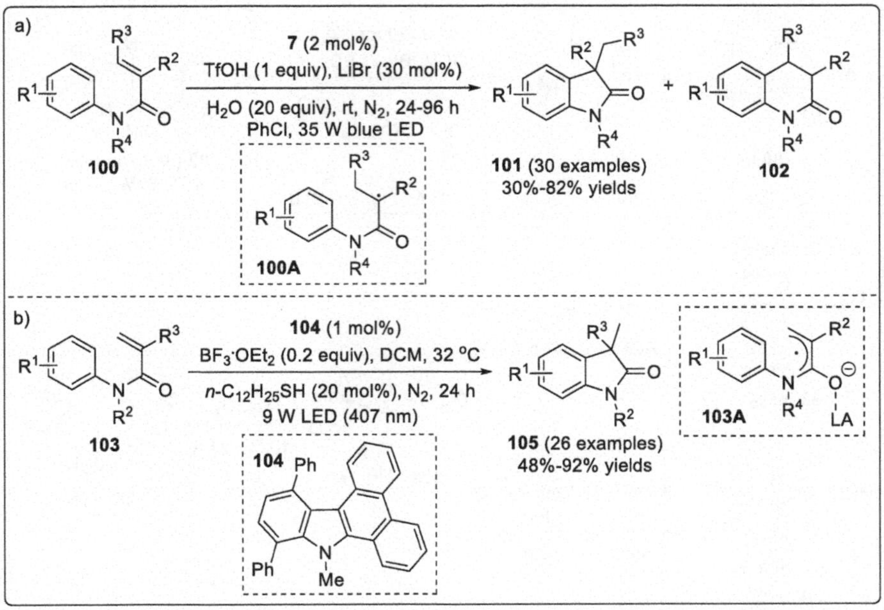

Scheme 18 Hydroarylation of alkenes for the synthesis of oxindoles: a) Intramolecular hydroarylation of *N*-phenylacrylamide enabled by PCET strategy; b) Intramolecular metal-free hydroarylation of *N*-phenylacrylamides.

to generate key radical intermediate **103A**. This reaction proceeded efficiently, eliminating the need for transition metal catalyst or halogen-containing reagents, and also avoiding the use of external reductants and oxidants.

3.2 N-*Radical-Mediated Reaction Pathways*

In contrast to the well-developed 5-*exo*-dig cyclization pathways, the 5-*endo*-dig cyclization of alkynes via *N*-radical-mediated mechanisms, as an alternative route to the synthesis of five-membered nitrogen-containing heterocycles, remains less explored. Along this line, Qian, Sun, and Ye et al. reported a visible light-promoted 5-*endo*-dig *N*-radical cyclization cascade of 2-ethynylanilines **106** in 2021 (Scheme 19a) [43]. This protocol featured a wide range of substituted 2-ethynylanilines, enabling the efficient and controllable synthesis of hydroxy-3-indolinones **107** in moderate to good yields under mild conditions (52–75% yields). The key *N*-radical intermediate **106A** is generated via a deprotonation coupled with a simultaneous single-electron oxidation of N–H bond. And the ambient oxygen serves as the terminal oxidant to complete the Ir^{III}-based catalytic cycle. Notably, this reaction also represents the first *N*-radical 5-*endo*-dig cyclization cascade of 2-ethynylanilines.

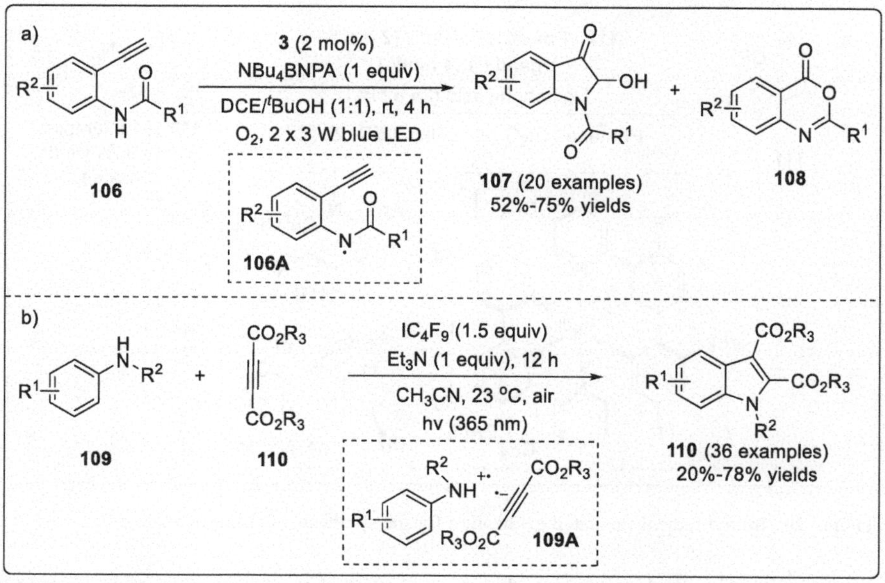

Scheme 19 *N*-radical-mediated cyclization for the synthesis of indolinones and indoles: a) *N*-radical cyclization cascade for the synthesis of 2-hydroxy-3-indolinones; b) The synthesis of indoles via EDA complexes.

Different from the above *N*-radical-mediated methodology, Guo, Fan, and Zheng et al. described an *N*-radical cation-mediated [3+2] annulation of secondary arylamines **109** and alkynes **110** under UV-light irradiation (Scheme 19b) [44]. This reaction proceeded well in the absence of photoredox catalysts and metals, employing IC_4F_9 as the terminal oxidant. The replacement of UV light with visible light sources (465 or 540 nm) was successful, albeit with slightly reduced yield (<45% yield). For the mechanism, it was proposed that the coordination of arylamines **109** with alkynes **110** obtained EDA complex, which was characterized by UV–vis absorption spectrometry and NMR titration experiments. Further internal single-electron transfer of EDA complex led to the formation of key *N*-radical cation intermediate **109A**, which can trigger the sequential annulation reactions.

4 Oxazoles, Thiazoles, and Imidazoles and Their Derivatives

Asymmetric photoredox catalysis has emerged as a powerful route for the synthesis of 5-membered *N*-heterocycles. In this context, an enantioselective radical cyclization mediated by the merger of photocatalysis and chiral copper complex was described by the Nagib group in 2020, affording β-amino alcohols derived from oxazolines **113** (Scheme 20) [45]. In this work, it was proposed that triplet energy

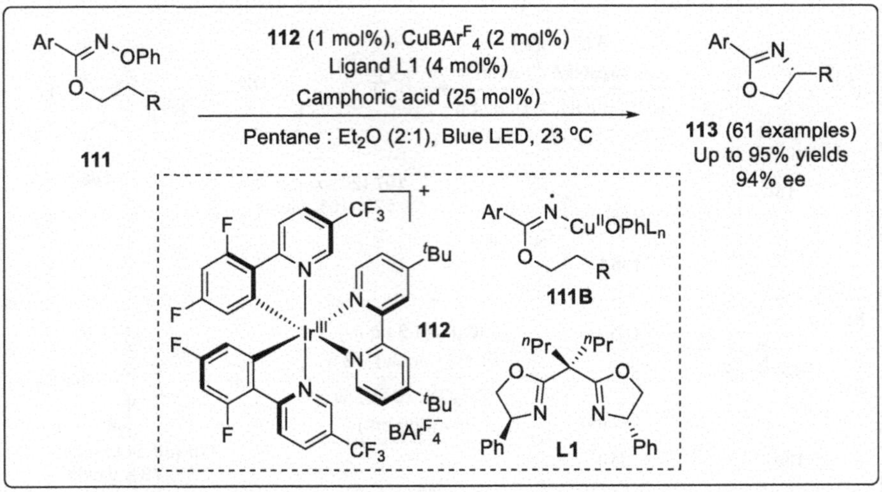

Scheme 20 Iminyl radical-mediated cyclization for the synthesis of chiral oxazolines

sensitization of a Cu-bound radical precursor with the excited state of Ir complex would promote the catalyst-mediated HAT stereoselectivity of the generation of imidate radical. In addition, this reaction showed a broad substrate scope and good functional group tolerance, including the alcohol containing the alkyl, allyl, propargyl, and benzyl C–H bonds, enabling the synthesis of diverse chiral β-amine alcohols with good to excellent yields and up to 94% ee value. Furthermore, the enantioenriched oxazoline intermediates were successfully converted into the phosphinooxazoline (PHOX) ligands and a new bisoxazoline ligand.

Detailed mechanistic experiments were performed to gain insight into the mechanism. The cyclic voltammetry (CV) and density functional theory (DFT) data collectively supported an energy transfer (EnT) pathway rather than the single-electron transfer (SET) process for the formation of the key Cu-bound *N*-centered radical **111B**. A large kinetic isotope effect (KIE) indicated that the HAT process is the product-determining step. Therefore, a credible mechanism was depicted in Scheme 21. Initially, the alcohol reacts with a bench-stable imidoyl chloride chaperone to generate an oxime imidate **111**. Then, the coordination of oxime imidate **111** by the Cu^{I} catalyst, which was illustrated through the photo-quenching experiments, generates the Cu-imidate complex **111A**. Subsequently, complex **111A** is sensitized by Ir* via an energy transfer process to generate the highly reactive chiral Cu-bound imidate radical **111B**. Finally, a regio- and enantioselective HAT process and stereoselective amination occur to obtain the oxazoline **113**, along with the generation of PhOH and Cu^{I} complex. Further acidic hydrolysis of oxazolines **113** would afford chiral β-amino alcohols.

After that, the same group further developed a radical *aza*-Heck cyclization through a multicatalytic mode that integrates energy transfer, electron transfer, and cobalt catalysis, for the synthesis of vinyl oxazolines and oxazoles (Scheme 22) [46]. Notice that the three distinct catalytic cycles cooperate to enable (1) imidate

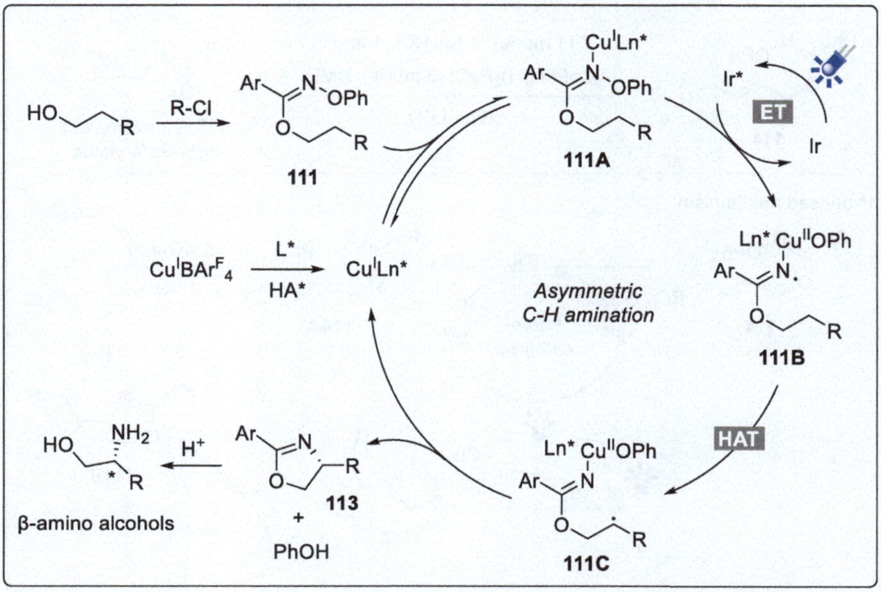

Scheme 21 The proposed mechanism

radical generation via an energy transfer process, (2) dehydrogenation by Co catalysis, and (3) Co and Ir catalyst turnover by electron transfer process. And the use of an ammonium salt ($Et_3N{\cdot}HCl$) could accelerate two separate catalytic turnover events, namely Et_3N reduces $^*Ir^{III}$ complex, and HCl protonates Co^{III}-H species to release H_2.

The plausible mechanism was outlined in Scheme 22. Initially, the reaction commences with an energy transfer process between the allyl imidate **114** and the excited state of Ir photocatalyst. This process produces the imidate radical **114A** through the homolysis of N–O bond, which is more favorable than single-electron transfer process. Then, 5-*exo*-trig cyclization of imidate radical happens to obtain the *C*-radical **114B**, which is quickly trapped by a Co^{II} catalyst to give an alkyl-Co^{III} complex **114C**. At the same time, the Co-catalyzed dehydrogenation cycle is raised. Upon visible-light irradiation, the alkyl-Co^{III} complex **114C** is excited via LMCT to form Co/alkyl radical pair **114D**, which undergoes an MHAT process to give the final vinyl oxazoline product, along with the Co^{III}-H species. Then, under the acidic condition, Co^{III}-H species is protonated to form Co^{III} complex, with the concomitant releasing of H_2. Final single-electron transfer between Co^{III} and reductive state of Ir photocatalyst enables the regeneration of Co^{II}, completing the Co-catalytic cycle.

Then, Liu and co-workers applied the dual-catalytic strategy to realize the dehydrogenative amination of alkenes **116**, mediated by photoredox catalysis in conjunction with Co catalysis, thereby enabling the preparation of *N*-heterocycles **117** (Scheme 23) [47]. This protocol featured a redox-neutral process and an array of

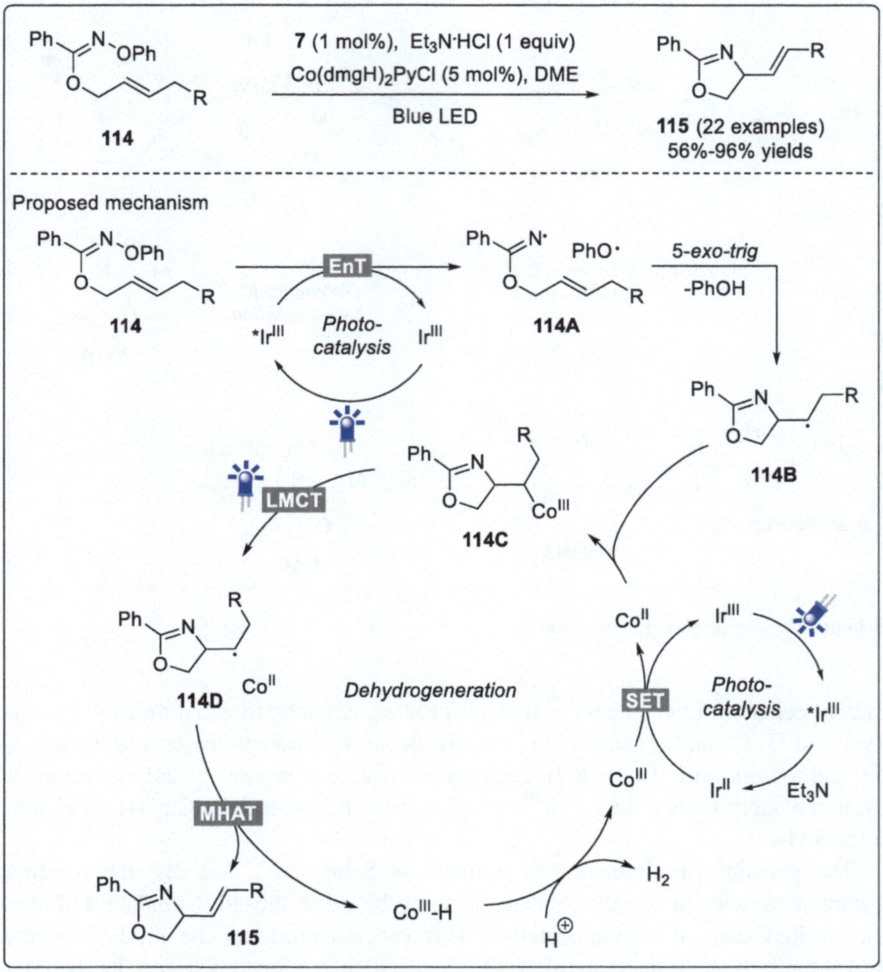

Scheme 22 Imidate radical-mediated cyclization for the synthesis of oxazolines

substrate scope to furnish spiro- and bicyclic compounds. Notably, both *Z*- and *E*-alkenes containing *N*-heterocycles are accessible. Besides, this strategy can be applied into the synthesis of benzoindolizidinones via radical cascade cyclization. For the mechanism, it was proposed that *N*-radical intermediate is generated from amide through the PCET process in the presence of the Brønsted base. Subsequently, a radical addition cyclization takes place, which is followed by a cobalt-catalyzed dehydrogenation step, ultimately leading to the formation of the desired *N*-heterocyclic products **117**.

Researches showed that compounds containing *N*-CF_3 moiety typically exhibit higher lipophilicity, increased metabolic stability, and decreased basicity. But the incorporation of *N*-CF_3 moiety into molecules is highly desirable and challenging. In

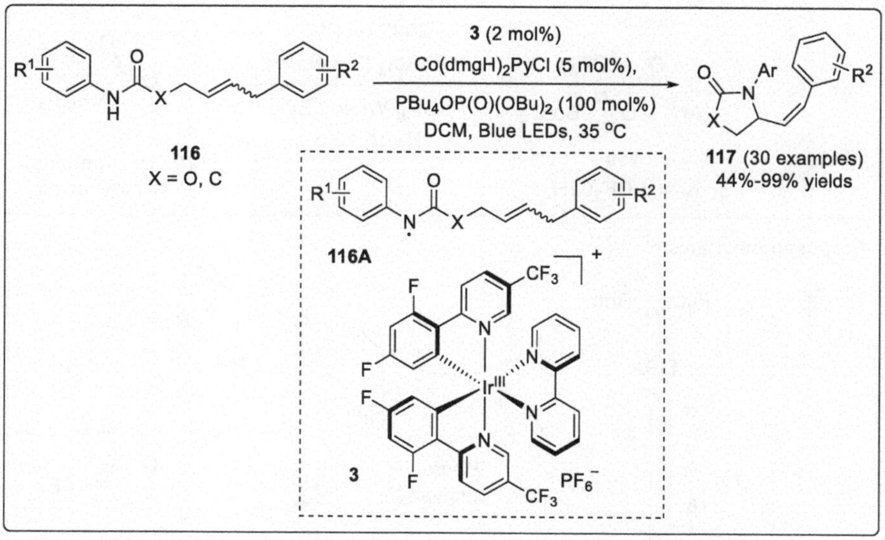

Scheme 23 Amidyl radical-mediated dehydrogenative amination of alkenes

2022, the Xu and Huang group designed and synthesized a novel *N*-Boc-*N*-CF_3 hydroxylamine reagent **119** by a silver-mediated oxidative trifluoromethylation (Scheme 24) [48]. This new reagent, using as *N*-CF_3 *N*-radical precursor, enables the preparation of previously unknown cyclic *N*-CF_3 compounds **120**, encompassing *N*-CF_3 oxazolidinones and oxazolones, through a photoinduced trifluoromethylamination/cyclization of styrenes **118**. In this reaction, the zinc salt was utilized as an additive to enhance the reaction efficiency.

A feasible mechanism was outlined in Scheme 24. Initially, upon visible-light irradiation, the photocatalyst Ir^{III} complex is excited to give its excited state $^*Ir^{III}$, which would be oxidized by *N*-Boc-*N*-CF_3 hydroxylamine reagent **119** to its oxidizing state Ir^{II} species, along with the generation of *N*-radical **119A** and $ArCO_2^-$. Then, a radical addition of *N*-radical **119A** to styrene takes place to obtain the benzyl radical **119B**. This intermediate is then oxidized by Ir^{II} species to form benzyl cation **119C**, with the concomitant release of ground-state Ir^{III} species, thereby completing the photocatalytic cycle. Finally, an intramolecular nucleophilic attack of **119C** gives the oxazolidinones **120**. In contrast to the aforementioned studies, a heterogeneous modified carbon nitride photocatalyst was employed for the reductive amination of alkenes to produce similar oxazolidinones in the existence of disulfide (MesSSMes) as HAT reagent, which was accomplished by the Nocera group in 2021 [49].

In 2022, Schoenebeck and colleagues reported an efficient approach to the formation of *N*-trifluoromethyl imidazolidinone*s* **123** and benzimidazolones **124** through visible-light photoinduced activation of carbamoyl azides (Scheme 25) [50]. These novel scaffolds were demonstrated to be highly robust, allowing their

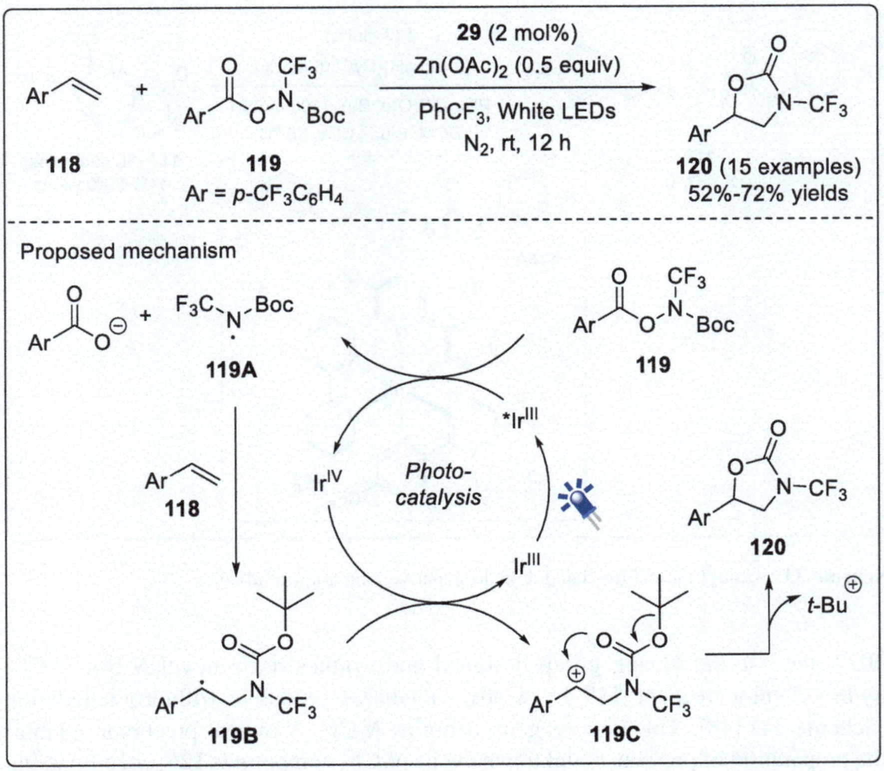

Scheme 24 *N*-radical-mediated cyclization of styrenes for the synthesis of oxazolidinones

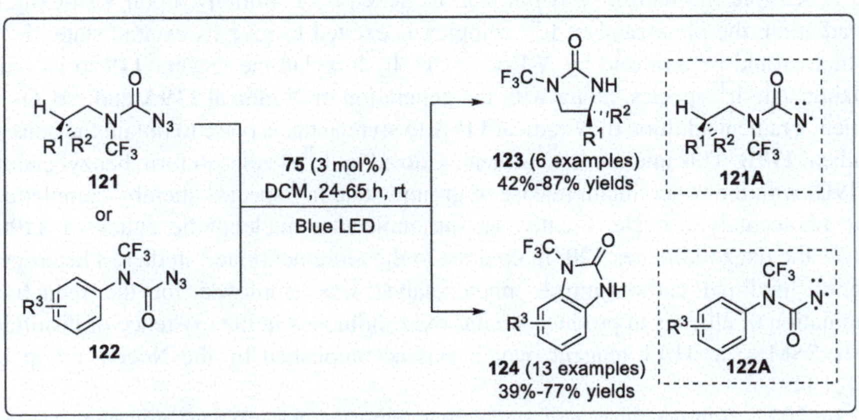

Scheme 25 Nitrene-mediated cyclization for the synthesis of *N*-trifluoromethyl ureas

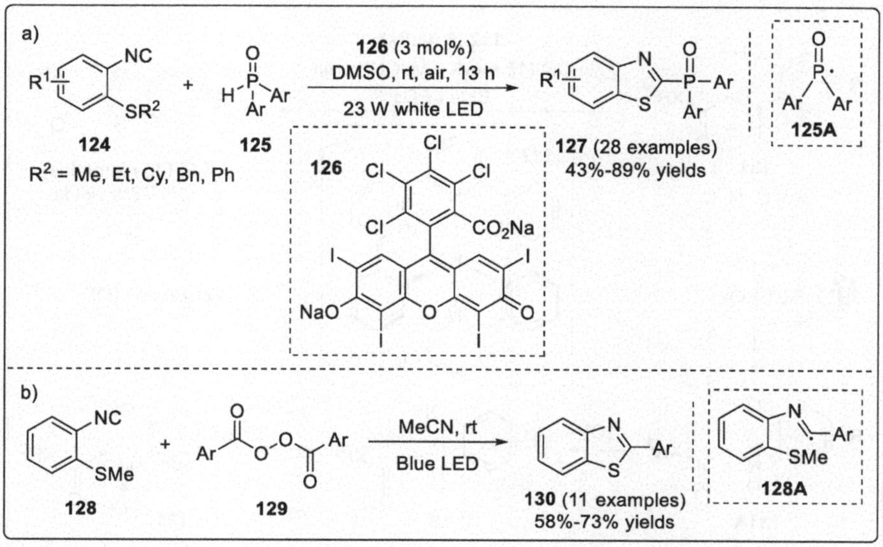

Scheme 26 Radical cyclization of 2-isocyanoaryl thioethers for the synthesis of benzothiazoles: a) Radical C-P and C-S bond formation for synthesis of 2-phosphoryl benzothiazoles; b) Photoinduced facile synthesis of benzothiazoles.

downstream diverse modification. It was proposed that an electron rebound with the Ir^{III} photocatalyst is involved, including a reduction-triggered nitrogen loss, followed by oxidation to the corresponding carbamoyl nitrene and subsequent C–H insertion, which was supported by the results of both computational and experimental studies.

Organophosphorus compounds have found extensive applications in organic synthesis, biochemistry, and medicinal chemistry. A concise synthesis of 2-phosphoryl benzothiazoles **127** via a photocatalytic radical addition cyclization of isocyanides **124** with H-phosphorus oxides **125** was developed by Feng and Yang et al. in 2019 (Scheme 26a) [51]. This reaction exhibited a broad substrate scope under metal-free and mild conditions, furnishing 2-phosphoryl benzothiazoles in generally good to excellent yields (up to 89% yield). And the phosphinoyl radical **125A** is regarded as the key intermediate. After that, the Moorthy group employed the similar strategy to synthesize 2-aryl benzothiazoles **130** in the absence of a photocatalyst (Scheme 26b) [52]. In this reaction, the acyl peroxides proved to be effective aryl radical precursors under the visible-light irradiation.

Besides, the radical ionic-type reactions offer a viable alternative for the preparation of these *N*-heterocyclic compounds. In 2021, Le and Zhu et al. presented a photocatalytic aerobic oxidative cyclization of glycine derivatives **131**, resulting in the formation of 2-substituted benzothiazole **133** (Scheme 27) [53]. In this case, both oxygen, acting as a terminal oxidant, and copper salt, functioning as the Lewis acid to chelate the imine intermediate **131B**, efficiently promote the generation of imine intermediate, intramolecular cyclization, and final oxidative dehydrogenation. This

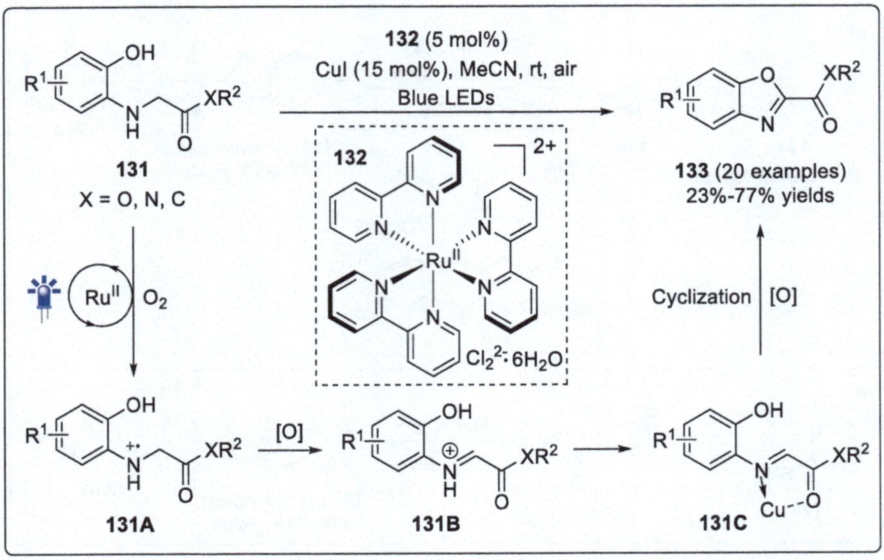

Scheme 27 *N*-radical cation-mediated cyclization for the synthesis of benzoxazoles

protocol exhibited a broad substrate scope and excellent functional group tolerance, including various ester, *N*-substituted, and *N,N*-disubstituted amides.

5 Miscellaneous *N*-Heterocycles

Pyrazole scaffolds are ubiquitous in natural products, pharmaceuticals, and other biologically active compounds. Therefore, continuous attention has been paid to the exploitation of fine synthetic methods for the generation of these core structures. In 2018, Zhu and Zhang et al. presented a visible light-promoted one-pot synthesis of pyrazoles **136** from alkynes **134** and hydrazines, employing air as the terminal oxidant (Scheme 28) [54]. The success of this reaction was attributed to the dual-catalytic strategy, which involves a light-promoted cascade of Glaser coupling/annulation driven by copper catalysis and a photoredox catalysis. Detailed mechanistic studies indicated that diyne is regarded as the key intermediate, and the oxidant is essential for this reaction.

The mechanism, which mainly encompasses two parts, was depicted in Scheme 28. The first part involves the generation of diyne **134E** through copper-catalyzed Glaser reaction. The second part is the annulation of diyne **134E** with hydrazine mediated by photocatalytic cycle. The hydrazine, serving as a reductant, is oxidized by $*Ru^{II}$ to afford the radical ion intermediate and Ru^{I}. Upon the deprotonation of radical ion, *N*-centered radical intermediate is generated, followed by a radical addition to diyne **134E** to form radical intermediate **134F**. Subsequently, HAT

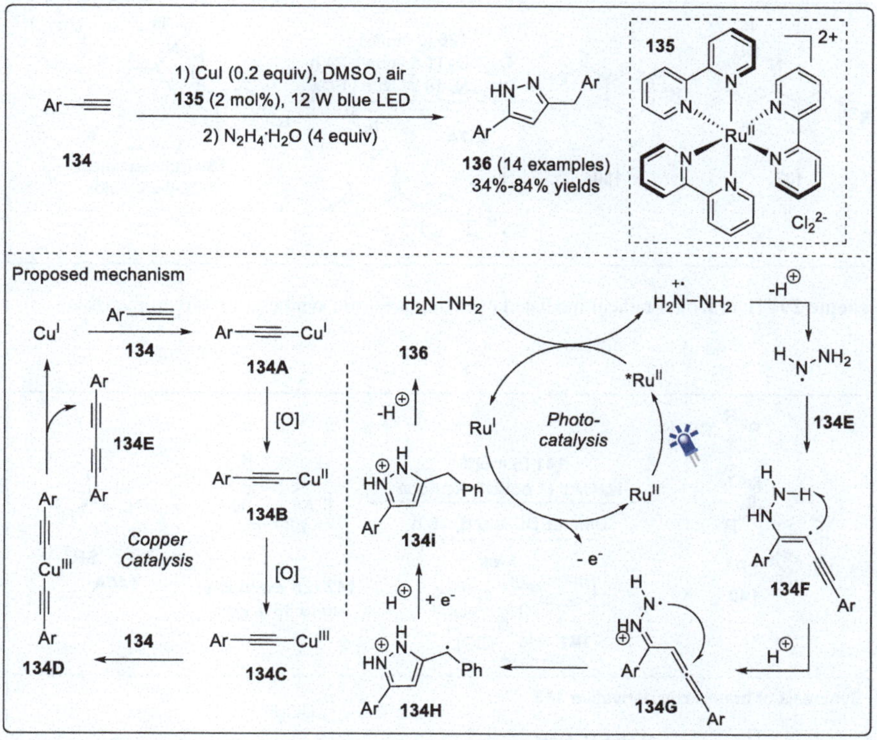

Scheme 28 One-pot photochemical synthesis of pyrazoles from alkynes and hydrazines

process and intramolecular radical addition happen to yield the benzyl radical **134H**. Finally, an SET process between benzyl radical and Ru^I complex together with a deprotonation process occurs, providing pyrazoles **136** and Ru^{II}, accomplishing the photocatalytic cycle.

N-centered hydrazonyl radicals have also been proven to be a highly useful radical intermediate for the construction of *N*-heterocycles. In this field, Xiao and Chen et al. have made a significant and continuous contributions [4a, b]. In 2021, Zhou, Yu, and Sun et al. described a photocatalytic generation of hydrazonyl radical and tandem radical cyclization/defluorinated alkylation of β,γ-unsaturated hydrazones **137** and α-trifluoromethyl alkenes **138** (Scheme 29) [55a]. This protocol provided a general and effective route to synthesize various dihydropyrazole-fused *gem*-difluoroalkenes **139** in moderate to excellent yields under the mild and redox-neutral conditions. And the key hydrazonyl radical **137A** is formed via oxidative deprotonation of N–H bond. Afterwards, employing urea-derived carbon nitride as optimal photocatalyst, The Cheng and Wang group achieved a similar *N*-radical cyclization of β,γ-unsaturated hydrazones, thereby yielding a variety of dihydropyrazole derivatives [55b].

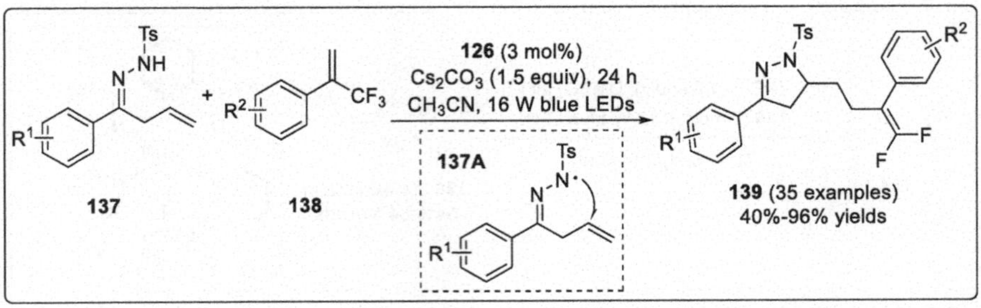

Scheme 29 Hydrazonyl radical-mediated cyclization for the synthesis of dihydropyrazoles

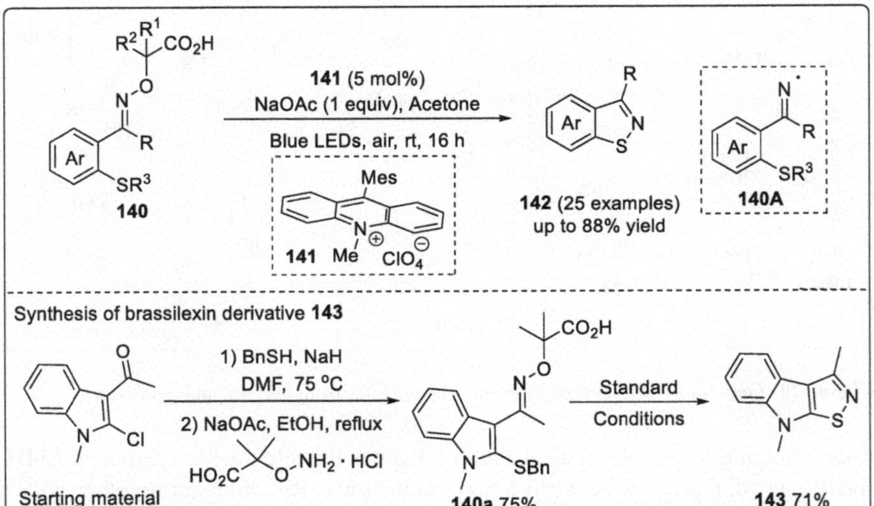

Scheme 30 Iminyl radical-mediated cyclization for the synthesis of isothiazoles

Isothiazoles are regarded as one important class of heterocyclic scaffolds in medicinal chemistry and the agriculture industry due to their two electronegative heteroatoms in a 1,2 relationship. In 2020, Alemán and Maestro et al. achieved the iminyl radical-mediated cyclization enabling the synthesis of isothiazoles **142** via the formation of N–S bond (Scheme 30) [56]. This reaction employed the α-amino-oxy acid auxiliary as the iminyl radical precursor and oxygen as the terminal oxidant under mild and metal-free reaction conditions, providing various isothiazoles in moderate to excellent yields. And it is worthy of note that this method could be further applied into the synthesis of natural product brassilexin derivative **143** through three steps in total 53% yield, showcasing the versatile utilities of this strategy.

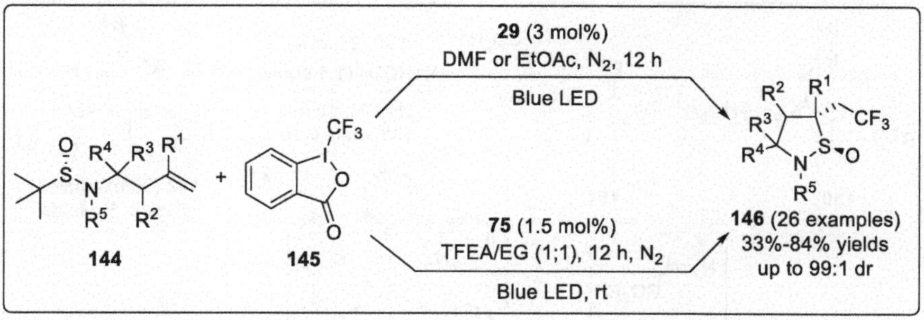

Scheme 31 Asymmetric radical cyclization of alkenes for the synthesis of cyclic sulfinamides

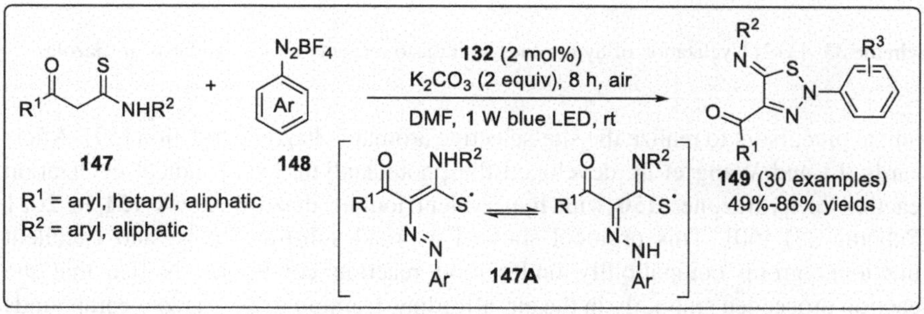

Scheme 32 Formal [3+2] heterocyclization of thioamides to thiadiazoles

The assembly of the chiral sulfinamides remains relatively challenging and less developed. Accordingly, in 2022, the Zhu group described an asymmetric radical cyclization of alkenes via stereospecific homolytic substitution of sulfinamides **146** (Scheme 31) [57]. This reaction utilized the readily available feedstocks to synthesize a variety of cyclic sulfinamides **146** with high diastereoselectivities and reaction efficiency.

Meanwhile, a visible light-induced aerobic formal [3+2] heterocyclization of thioamides **147** with aryldiazonium salts **148** to thiadiazoles **149** was documented by Singh and co-workers in 2021 (Scheme 32) [58]. This reaction showed a broad substrate scope under mild conditions, resulting in the synthesis of 2,4-disubstituted 5-imino-1,2,3-thiadiazoles **149** in moderate to good yields. Mechanistic studies have shown that α-phenylhydrazone adduct of thioamides to aryl diazonium salt can lead to the formation of *S*-centered radical intermediate **147A** through a sequential deprotonation and oxidation process.

Diazo compounds are widely recognized as highly valuable building blocks owing to their exceptional reactivity and versatility. However, reports of diazo compounds serving as diazomethyl radical precursors are relatively rare. In 2018, the Suero group used the hypervalent iodine diazo reagents as the diazomethyl

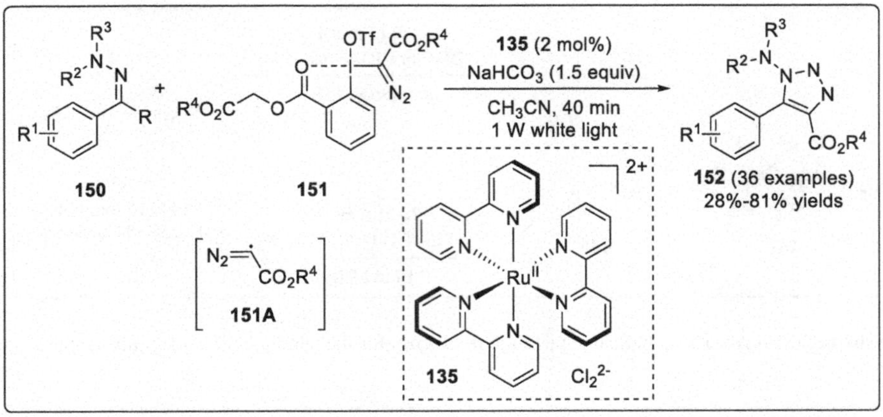

Scheme 33 [3+2] Cyclization of hydrazones with diazo reagents for the synthesis of triazoles

radical precursors to realize the site-selective aromatic diazomethylation [59]. Afterwards, Li and Wang et al. developed the photocatalytic [3+2] radical cyclization reaction of hydrazones **150** with hypervalent iodone diazo reagents **151** in 2021 (Scheme 33) [60]. This protocol showed a broad substrate scope and excellent functional group compatibility under mild reaction conditions. Notice that the reaction proceeded smoothly in the air, affording 1-amino-1,2,3-triazole compounds **152** in 28–81% yields. And this strategy holds potention for the late-stage functionalization of natural products.

In the same year, Wang, Zhou, and Yang et al. documented a visible light-induced decarboxylation/radical addition cyclization of *N*-aryl glycines **154** with azobenzenes **153** (Scheme 34) [61]. This reaction employed the organic dye methylene blue **21** (MB) as photocatalyst, and *N*-aryl glycines **154** as α-amino *C*-centered radical precursors, delivering the 1,2,4-triaryl 1,2,4-triazolidines **155** in good to excellent yields (up to 95% yield). A plausible mechanism was proposed based on detailed mechanistic studies. Firstly, MB is excited to its excited-state MB* under visible-light irradiation. Then, the *N*-aryl glycine **154** undergoes an MB*-mediated single-electron oxidation, followed by a decarboxylation process to generate the key α-amino *C*-centered radical **154A**. The addition of radical **154A** to azobenzene **153** occurs to afford the *N*-centered radical **154B**, which undergoes a radical-radical cross-coupling with another α-amino *C*-centered radical to give hydrazine **154C**. Ultimately, the protonated hydrazine **154D** undergoes the release of an aniline molecule, leading to the formation of the iminium cation **154E**, which then undergoes an intramolecular cyclization, thus culminating in the synthesis of the final products **155**.

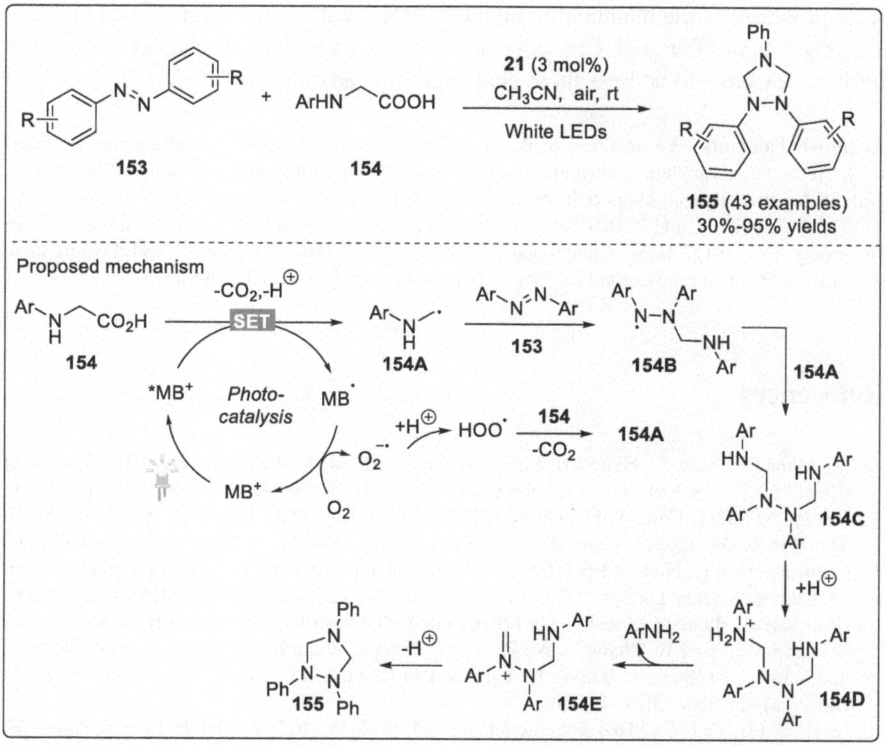

Scheme 34 Cyclization of *N*-aryl glycines and azobenzenes for the synthesis 1,2,4-triazolidines

6 Conclusion

Over the past 5 years, the domain of visible light-induced radical cyclization cascades has witnessed remarkable strides, especially in the assembly of five-membered *N*-heterocycles. This research has been marked by the generation of diverse radical species, which act as key reactive intermediates. These radical species can be generated either in conjunction with photocatalysts or through alternative pathways, thereby significantly expanding the range of radical cyclization methodologies. Therefore, this chapter focuses on the notable advancements in this field over the past 5 years, highlighting the development of catalytic systems, expansion of substrate scope, diversity of various radical species, and elucidation of reaction mechanisms.

Despite the considerable advances outlined, several challenges persist in this domain. On the one hand, the highly reactive and transient nature of radical species poses a significant hurdle in controling reactivities and selectivities. Particularly, developing novel and efficient asymmetric methods to suppress competing background reactions and achieve enantioselective transformations remains a formidable task. On the other hand, scaling up visible light-induced reactions to large-scale processes while maintaining high efficiency and avoiding diminished yields is a highly sought-after goal. Consequently, we anticipate that this chapter will inspire further endeavors to address these challenges and advance the field.

Acknowledgements We sincerely thank our collaborators and co-workers, whose names are listed in the references, for their contributions to the research presented in this chapter. We are also grateful for the financial support from the National Natural Science Foundation of China (Nos. 22301109, 32202334, and 22161047), the Research Funds for Talent Introduction of Jiangsu Ocean University (No. KQ23065), Lianyungang Haiyan project (No. KK24005), and Postgraduate Research & Practice Innovation Program of Jiangsu Province (SJCX24_2095).

References

1. (a) Ahmad S, Alam O, Naim MJ, Shaquiquzzaman M, Alam MM, Iqbal M (2018) Eur J Med Chem 157:527–561. (b) Hatai J, Schmuck C (2019) Acc Chem Res 52:1709–1720. (c) Reddy GS, Pal M (2021) Curr Med Chem 28:4531–4568. (d) Bajad NG, Singh SK, Singh SK, Singh TD, Singh M (2022) Curr Res Pharmacol Drug Discov 3:100119. (e) Anthoni U, Christophersen C, Nielsen PH (1999) In: Alkaloids: chemical and biological perspectives, vol 13. Elsevier Science Ltd., pp 163–236. (f) Acharya BN, Saraswat D, Tiwari M, Shrivastava AK, Ghorpade R, Bapna S, Kaushik MP (2010) Eur J Med Chem 45:430–438. (g) Ke S, Zhang Z, Shi L, Liu M, Fang W, Zhang Y, Wu Z, Wan Z, Long T, Wang K (2019) Org Biomol Chem 17: 3635–3639. (h) Zhang C, Zhong B, Yang S, Pan L, Yu S, Li Z, Li S, Su B, Meng X (2015) Bioorg Med Chem 23:3774–3780
2. (a) Hong FL, Ye LW (2020) Acc Chem Res 53:2003–2019. (b) Li H, Guo H, Fang Z, Aida TM, Smith RL (2020) Green Chem 22:582–611. (c) Kaur N, Verma Y, Grewal P, Ahlawat N, Bhardwaj P, Jangid NK (2020) Synth Commun 50:1286–1334. (d) Si YF, Lv QY, Yu B (2021) Adv Synth Catal 363:4640–4666. (e) Dehaen W, Pulikkal Veettil S, Pookkandam Parambil S (2021) Synthesis 54:910–924. (f) Hawkins BC, Bissember AC (2022) Chem Catal 2:1849–1850
3. For selected reviews on visible light photoredox catalysis, see: (a) Wei Y, Zhou QQ, Tan F, Lu LQ, Xiao WJ (2019) Synthesis 51:3021–3054. (b) Cannalire R, Pelliccia S, Sancineto L, Novellino E, Tron GC, Giustiniano M (2021) Chem Soc Rev 50:766–897. (c) Chan AY, Perry IB, Bissonnette NB, Buksh BF, Edwards GA, Frye LI, Garry OL, Lavagnino MN, Li BX, Liang Y, Mao E, Millet A, Oakley JV, Reed NL, Sakai HA, Seath CP, MacMillan DWC (2022) Chem Rev 122:1485–1542. (d) Rivas M, Palchykov V, Jia X, Gevorgyan V (2022) Nat Rev Chem 6:544–561
4. (a) Yu XY, Zhao QQ, Chen J, Xiao WJ, Chen JR (2020) Acc Chem Res 53:1066–1083. (b) Wang P, Zhao Q, Xiao W, Chen J (2020) Green Synth Catal 1:42–51. (c) Wang HZ, Li JZ, Guo Z, Zheng H, Wei WT (2021) ChemSusChem 14:4658–4670. (d) Murray PRD, Cox JH, Chiappini ND, Roos CB, McLoughlin EA, Hejna BG, Nguyen ST, Ripberger HH, Ganley JM, Tsui E, Shin NY, Koronkiewicz B, Qiu G, Knowles RR (2022) Chem Rev 122:2017–2291. (e) Pratley C, Fenner S, Murphy JA (2022) Chem Rev 122:8181–8260
5. (a) Xiao W, Wu J (2023) ChemCatChem 15: e202300541. (b) Zhao QQ, Gu M, Chen JR (2024) Chin J Chem 42:2412–2416
6. Pantaine LRE, Milligan JA, Matsui JK, Kelly CB, Molander GA (2019) Org Lett 21:2317–2321
7. Zhang W, Zou Z, Wang Y, Liang Y, Wu Z, Zheng Y, Pan Y (2019) Angew Chem Int Ed 58: 624–627
8. Reed NL, Lutovsky GA, Yoon TP (2021) J Am Chem Soc 143:6065–6070
9. Venditto NJ, Liang YS, El Mokadem RK, Nicewicz DA (2022) J Am Chem Soc 144:11888–11896

10. Walker JCL, Werrel S, Donohoe TJ (2019) Chem Eur J 25:13114–13118
11. von Zuben TW, Cariello Silva G, Salles AG (2021) Green Chem 23:6361–6365
12. Cybularczyk-Cecotka M, Predygier J, Crespi S, Szczepanik J, Giedyk M (2022) ACS Catal 12: 3543–3549
13. Shen Q, Zheng X, Li L, Zhong T, Yin C, Yu C (2022) Org Lett 24:2556–2561
14. Huang P, Yan Z, Ling J, Li P, Wang J, Li J, Sun B, Jin C (2023) Green Chem 25:3989–3994
15. Cai S-H, Wang D-X, Ye L, Liu Z-Y, Feng C, Loh T-P (2018) Adv Synth Catal 360:1262–1266
16. Shen X, Huang C, Yuan XA, Yu S (2021) Angew Chem Int Ed 60:9672–9679
17. Yuan Y, Dong WH, Gao XS, Xie XM, Zhang ZG (2019) Chem Commun 55:11900–11903
18. Usami K, Yamaguchi E, Tada N, Itoh A (2018) Org Lett 20:5714–5717
19. Han Y, Jin Y, Jiang M, Yang H, Fu H (2019) Org Lett 21:1799–1803
20. Wu K, Du Y, Wei Z, Wang T (2018) Chem Commun 54:7443–7446
21. Ward RM, Schomaker JM (2021) J Org Chem 86:8891–8899
22. Zhu Q, Graff DE, Knowles RR (2018) J Am Chem Soc 140:741–747
23. Roos CB, Demaerel J, Graff DE, Knowles RR (2020) J Am Chem Soc 142:5974–5979
24. Ji YX, Li J, Li CM, Qu S (2021) Org Lett 23:207–212
25. (a) McAtee RC, Noten EA, Stephenson CRJ (2020) Nat Commun 11:2528. (b) Liu XY, Qin Y (2019) Acc Chem Res 52:1877–1891
26. Zheng S, Gutierrez-Bonet A, Molander GA (2019) Chem 5:339–352
27. (a) Lei Z-Y, Hu K, He Y-X, Geng S, Chen L-N, Zou S, Pan L, Ding Z-J, Huang F (2022) Org Biomol Chem 20:2397–2401. (b) Dahiya A, Das B, Sahoo AK, Patel BK (2022) Adv Synth Catal 364:966–973
28. Lee W, Koo Y, Jung H, Chang S, Hong S (2023) Nat Chem 15:1091–1099
29. Shi C, Guo L, Gao H, Luo M, Yang C, Xia W (2022) Org Lett 24:4365–4370
30. (a) Liu Y, Parodi A, Battaglioli S, Monari M, Protti S, Bandini M (2019) Org Lett 21:7782–7786. (b) Kurpil B, Otte K, Mishchenko A, Lamagni P, Lipinski W, Lock N, Antonietti M, Savateev A (2019) Nat Commun 10:945. (c) Borra S, Borkotoky L, Newar UD, Das B, Maurya RA (2020) Adv Synth Catal 362:3364–3368
31. Karki BS, Devi L, Pokhriyal A, Kant R, Rastogi N (2019) Chem Asian J 14:4793–4797
32. Han WJ, Zhang JW, Yan CX, Wang JW, Zhou PP, Han B (2022) Org Lett 24:542–547
33. Liang YQ, Xu YX, Cai ZJ, Ji SJ (2022) Chem Commun 58:10206–10209
34. Sahoo AK, Rakshit A, Dahiya A, Pan A, Patel BK (2022) Org Lett 24:1918–1923
35. Festa AA, Voskressensky LG, Van der Eycken EV (2019) Chem Soc Rev 48:4401–4423
36. Cheng J, Xie J, Zhu C (2018) Chem Commun 54:1655–1658
37. Vidyasagar A, Shi J, Kreitmeier P, Reiser O (2018) Org Lett 20:6984–6989
38. Zhang L, Si X, Rominger F, Hashmi ASK (2020) J Am Chem Soc 142:10485–10493
39. Liu D, Shi B, Jiang H, Cheng Y, Xiao WJ, Lu LQ (2021) Chem Commun 57:8496–8499
40. Li W, Qiu J, Li H, Chen W, Hou C, Li H (2023) Org Lett 25:3778–3783
41. Liu Z, Zhong S, Ji X, Deng G-J (2021) ACS Catal 11:4422–4429
42. Cai X-P, Han B-H, Cen F-T, Qu J-P, Kang Y-B (2023) Org Lett 25:2863–2867
43. Tan T-D, Zhai T-Y, Liu B-Y, Li L, Qian P-C, Sun Q, Zhou J-M, Ye L-W (2021) Cell Rep Phys Sci 2:10057
44. Guo W, Tao K, Xie Z, Cai L, Zhao M, Tan W, Liu G, Mei W, Deng L, Fan X, Zheng L (2019) J Org Chem 84:14168–14178
45. Nakafuku KM, Zhang Z, Wappes EA, Stateman LM, Chen AD, Nagib DA (2020) Nat Chem 12:697–704
46. Prusinowski AF, Sise HC, Bednar TN, Nagib DA (2022) ACS Catal 12:4327–4332
47. Tu J-L, Tang W, He S-H, Su M, Liu F (2022) Sci China Chem 65:1330–1337
48. Liu S, Huang Y, Wang J, Qing FL, Xu XH (2022) J Am Chem Soc 144:1962–1970
49. Rieth AJ, Qin Y, Martindale BCM, Nocera DG (2021) J Am Chem Soc 143:4646–4652
50. Bouayad-Gervais S, Nielsen CD, Turksoy A, Sperger T, Deckers K, Schoenebeck F (2022) J Am Chem Soc 144:6100–6106

51. Yang W, Li B, Zhang M, Wang S, Ji Y, Dong S, Feng J, Yuan S (2020) Chin Chem Lett 31: 1313–1316
52. Yadav N, Bhatta SR, Moorthy JN (2023) J Org Chem 88:5431–5439
53. Zhu ZQ, Liu S, Hu ZY, Xie ZB, Tang J, Le ZG (2021) Adv Synth Catal 363:2568–2572
54. Meng Y, Zhang T, Gong X, Zhang M, Zhu C (2019) Tetrahedron Lett 60:171–174
55. (a) Gao QS, Niu Z, Chen Y, Sun J, Han WY, Wang JY, Yu M, Zhou MD (2021) Org Lett 23: 6153–6157. (b) Yang M, Lian R, Zhang X, Wang C, Cheng J, Wang X (2022) Nat Commun 13: 4900
56. Cabrera-Afonso MJ, Cembellín S, Halima-Salem A, Berton M, Marzo L, Miloudi A, Maestro MC, Alemán J (2020) Green Chem 22:6792–6797
57. Chen Y, Wu X, Yang S, Zhu C (2022) Angew Chem Int Ed 61:e202201027
58. Pali P, Shukla G, Saha P, Singh MS (2021) Org Lett 23:3809–3813
59. Wang Z, Herraiz AG, Del Hoyo AM, Suero MG (2018) Nature 554:86–91
60. Dong JY, Wang H, Mao S, Wang X, Zhou MD, Li L (2021) Adv Synth Catal 363:2133–2139
61. Yang J, Song M, Zhou H, Qi Y, Ma B, Wang X-C (2021) Green Chem 23:5806–5811

Top Heterocycl Chem (2025) 62: 41–66
https://doi.org/10.1007/7081_2025_89

Published online: 3 June 2025

Photoredox Catalytic Asymmetric Functionalization of Azaarene Derivatives

Tianju Shao, Xu Ban, and Zhiyong Jiang

Contents

Abstract The enantioenriched azaarene derivatives have garnered significant interest in pharmaceutical and synthetic chemistry, prompting the development of efficient catalytic methodologies. Recent advancements have leveraged the high reactivity of radicals in conjunction with the hydrogen-bonding interactions characteristic of azaarenes. Notably, the integration of visible light-driven photocatalysis with chiral Brønsted acid catalysis has emerged as a powerful platform for synthesizing these valuable compounds. This chapter presents an overview of the state-of-the-art strategies established for the enantioselective synthesis of diverse and complex azaarene derivatives, emphasizing sustainable and mild reaction conditions. The discussion is structured according to the stereocenter positions within azaarenes – specifically, α-, β-, γ-, and various other positions. Ultimately, this chapter aims to elucidate current achievements while identifying existing challenges, thereby inspiring chemists to explore novel or more efficient approaches for accessing enantiomerically pure azaarene derivatives, whether their structures are known or yet to be discovered, thus fostering rapid advancements within the pharmaceutical sector.

T. Shao, X. Ban, and Z. Jiang (✉)
School of Chemistry and Chemical Engineering, Pingyuan Laboratory, Henan Normal University, Xinxiang, Henan, People's Republic of China
e-mail: jiangzhiyong@htu.edu.cn

Keywords Asymmetric catalysis · Azaarenes · Chiral hydrogen bond · Photocatalysis

1 Introduction

Imine-containing azaarenes, such as pyridines and quinolines, represent critical structural motifs prevalent in a vast array of bioactive natural and nonnatural products, including agrochemicals, pharmaceuticals, functional materials, catalysts, and ligands [1] (Fig. 1). The synthesis of azaarene derivatives has long captivated chemists, leading to the development of numerous efficient methodologies. Among these, protocols that exploit the electron-deficient nature of azaarenes have emerged as essential tools, enabling the direct utilization of simple azaarenes as substrates. This approach not only simplifies the synthetic process but also facilitates subsequent modifications to yield valuable and complex compounds [2]. However, the relatively weak basicity of these azaarenes may inhibit sufficient enantioselective interactions between substrates and products, resulting in diminished enantio-control during asymmetric functionalization.

In recent years, photocatalysis has been recognized as a promising and versatile platform to address the inherent challenges associated with the "harsh conditions"

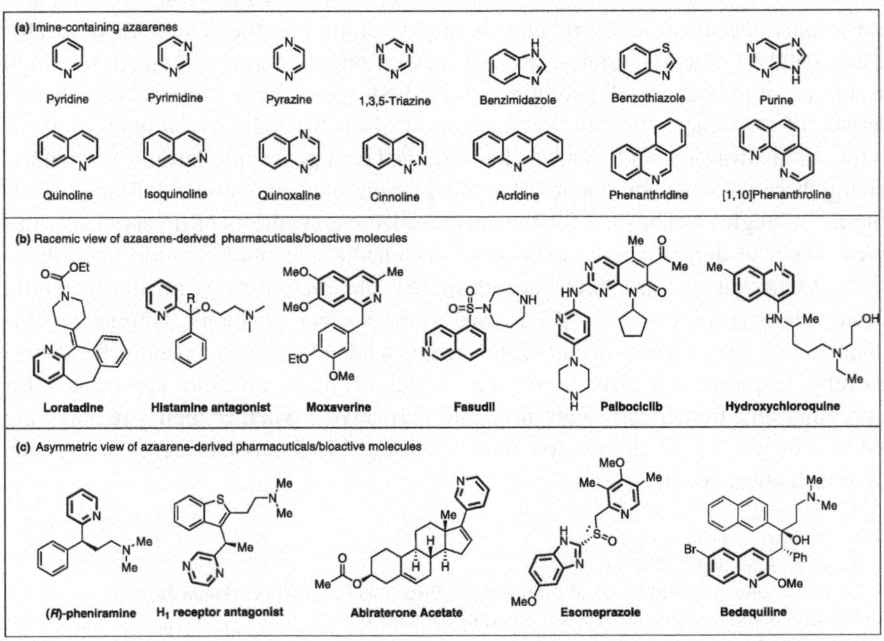

Fig. 1 Structures of imine-containing azaarenes

often required for the functionalization of chiral azaarene derivatives. Notably, a dual chiral hydrogen bond (CHB) strategy combined with photocatalysis has emerged as one of the most effective protocols for synthesizing functionalized azaarenes. Traditionally, the synthesis of such compounds has relied on tenuous hydrogen bonding interactions between the nitrogen atom of azaarenes and the catalysts [3]. To achieve robust stereocontrol under these delicate interactions, it is typically necessary to employ exceedingly mild reaction conditions and catalysts with high selectivity. The gentle and adaptable nature of photocatalytic systems aligns seamlessly with these requirements, promising significant advancements in the asymmetric functionalization of azaarenes and the assembly of intricate chiral azaarene compounds.

Over the past two decades, researchers have developed a variety of eco-friendly reaction systems for the functionalization of chiral azaarene derivatives through photocatalysis [4]. Notably, the Jiang research group has made substantial contributions in this field, advancing the green and efficient synthesis of chiral azaarenes. However, the creation of site-versatile azaarene derivatives remains a significant challenge. To date, the majority of asymmetrically functionalized azaarenes have focused on the C-2 position of nitrogen-containing aromatic rings, with limited documentation on the C-4 position, particularly in pyridines, and even fewer reports on chiral construction at the C-3 or other sites.

This chapter aims to review and consolidate progress over the past decade, focusing on the stereocenter positions within azaarenes, including α-, β-, γ-, and various other locations. The goal is to provide adaptable guidelines for the future evolution of chiral azaarenes. The chapter concludes with a comprehensive overview of the most recent published research as of April 2, 2024.

2 Methods to the Construction of α-Stereocenters

2.1 From Simple Azaarenes

The Minisci-type reaction has long been recognized as one of the most effective methodologies for synthesizing functionalized azaarenes directly from simple azaarenes via radical addition processes [5]. Over the past century, a vast array of protocols have been developed, predominantly employing stoichiometric amounts of strong protonic acids (e.g., trifluoroacetic acid, sulfuric acid) or utilizing aggressive reagents (e.g., persulfate, hydrogen peroxide) as terminal oxidants [6]. However, with the rapid advancement of photocatalysis, chemists have unveiled a significantly milder, environmentally friendly, and more efficient catalytic Minisci reaction system. Despite these advancements, achieving an asymmetric version of the Minisci reaction remained a formidable challenge due to the inherently harsh conditions typically required.

It was not until 2018 that Phipps and colleagues reported the first asymmetric Minisci reaction under photoredox catalysis, employing an iridium complex as the

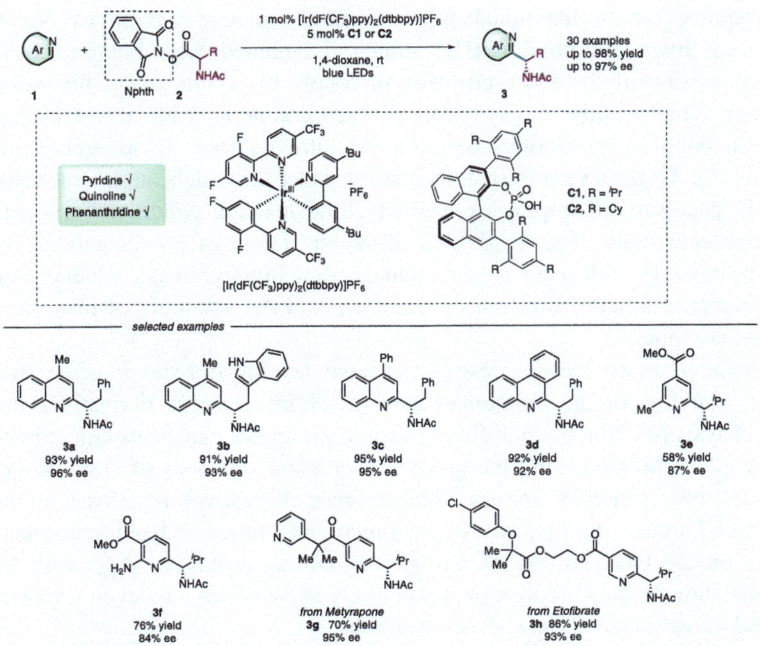

Scheme 1 Photoredox-catalyzed asymmetric Minisci-type reactions

photocatalyst [7] (Scheme 1). In this groundbreaking work, an α-amino acid-derived redox-active ester (RAE) was utilized as a radical precursor, which could be reduced by the generated Ir(II) species to release an aminoalkyl radical intermediate. This intermediate was subsequently added to azaarenes with the activation of chiral phosphoric acids (CPAs). This asymmetric platform demonstrated a versatile substrate scope, wherein a series of activated α-amino acid derivatives were smoothly reacted with azaarenes to yield a wide range of chiral, functionalized azaarenes (**3**) with high yields (up to 98%) and exceptional regioselectivity and enantioselectivity (up to 97% ee). Remarkably, the authors achieved the late-stage asymmetric functionalization of two commercially available pyridine-derived drugs, metyrapone (**3g**) and etofibrate (**3h**), which exhibited commendable regio- and enantioselectivities as well as high yields.

Almost concurrently, Jiang and co-workers illustrated the utility of an enantioselective Minisci reaction for synthesizing isoquinoline derivatives [8] (Scheme 2). Utilizing a dual-organocatalytic system featuring a dicyanopyrazine-derived chromophore as the photosensitizer alongside a chiral SPINOL-derived CPA, they successfully introduced a variety of prochiral α-aminoalkyl radicals onto isoquinolines. This process led to the formation of novel α-isoquinoline-substituted chiral secondary amines (**5**), achieving high yields (up to 95%) and excellent enantioselectivities (up to 93% ee).

Building on these significant advancements, Phipps and his colleagues undertook a comprehensive examination of the complex mechanisms underlying chiral-

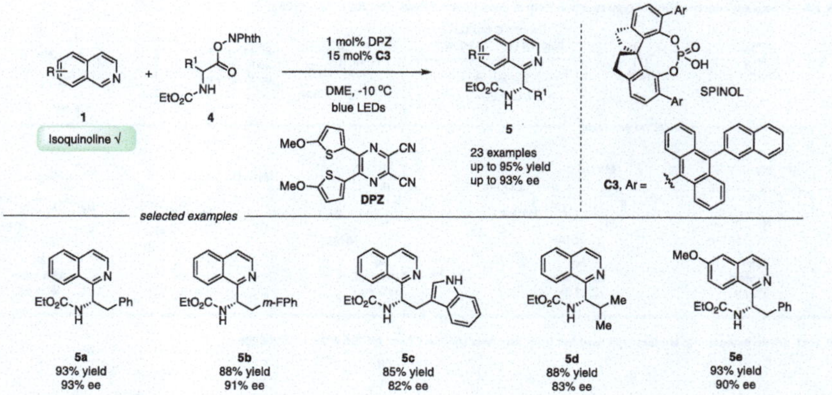

Scheme 2 Asymmetric Minisci-type reaction on isoquinolines

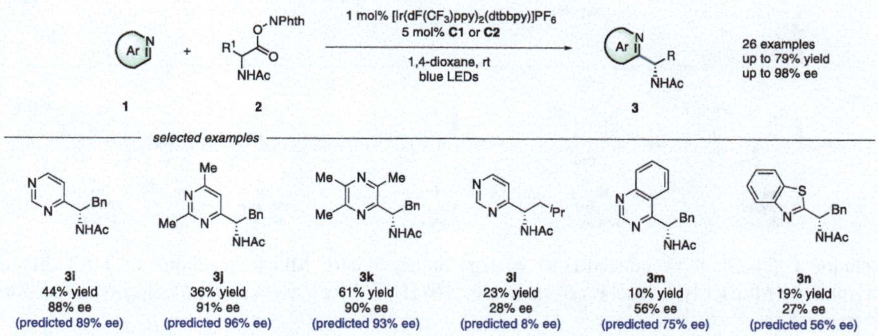

Scheme 3 Predictive MLR analysis strategies on Minisci reactions

selective control in photoredox-catalyzed asymmetric Minisci reactions. This integrative computational and experimental study not only refined the platform but also transformed it into a highly compelling approach [9a]. In 2019, they introduced a predictive multivariate linear regression (MLR) analysis strategy to facilitate the optimization of azaarenes to pyrimidines and pyrazines, building on their prior success with quinolines and pyridines [9b] (Scheme 3). Fortunately, most predictions regarding enantioselectivity were remarkably consistent with observed results, with predicted ee values closely aligning ($\leq$5%) with actual observations. This data-driven predictive approach for asymmetric synthesis holds considerable practical value, enabling significant savings in time and resources.

In 2021, Phipps and co-workers advanced a cooperative hydrogen atom transfer (HAT) and CHB catalysis strategy for the enantioselective Minisci reaction [10] (Scheme 4a). By incorporating a stoichiometric amount of 2,3-butanedione (diacetyl), the conversions between azaarenes and *N*-acetylalkylamines proceeded smoothly, yielding a series of chiral α-aminoalkyl pyridine and quinoline derivatives. Notably, a chiral α-amino acid-derived 1,5-diamine was effectively utilized in this transformation, resulting in the corresponding 1,2,6-triamine quinoline product

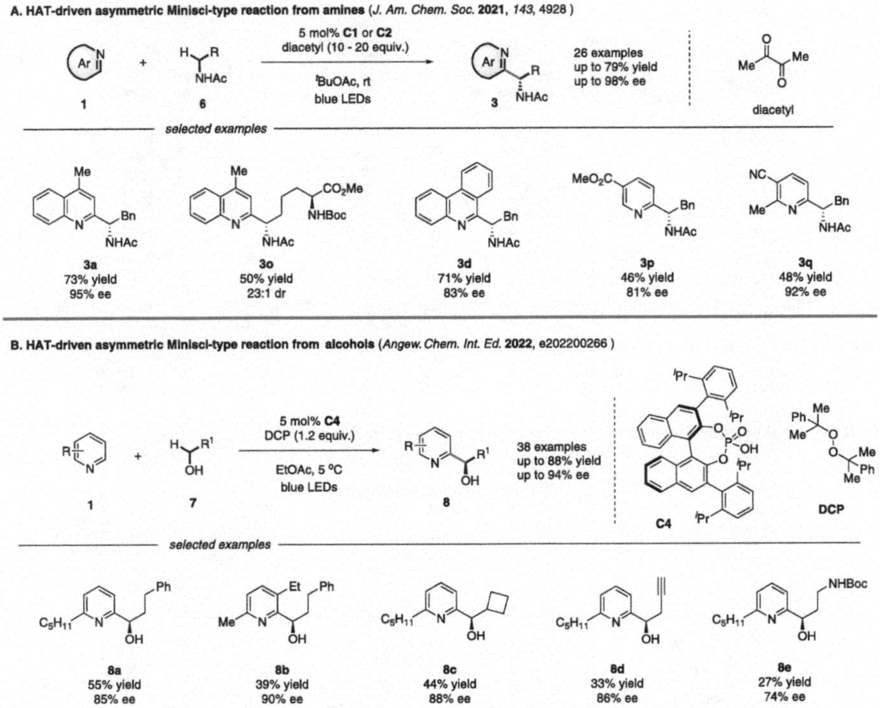

Scheme 4 Visible light-induced HAT strategy on asymmetric Minisci reactions. (**a**) HAT-driven asymmetric Minisci-type reaction from amines; (**b**) HAT-driven asymmetric Minisci-type reaction from alcohols

(**3o**) being obtained in a moderate yield (50%) with an excellent diastereomeric ratio (dr) of 23:1. In 2022, they further achieved the construction of α-chiral tertiary azaarene alcohols from linear alcohols and heteroarenes using a similar HAT-driven approach [11] (Scheme 4b). While the scope of azaarene application in pyridines was somewhat limited, there was notable tolerance for alkyl alcohols containing functional groups, effectively processing cycloalkyl alcohols and branched alkenes or alkynes under standard conditions. However, challenges in terms of lower yields and enantioselectivities emerged with more complex substrates, such as doubly amine- and alcohol-substituted linear alkanes.

In 2023, Xiao and colleagues achieved an elegant synthesis of axial and central chiral azaarenes by merging pro-axial chiral azaarenes with α-amino acid-derived RAEs [12] (Scheme 5). This innovative reaction employed a dual-photoredox organocatalysis and CHB catalysis strategy, resulting in a diverse array of intriguing 5-arylpyrimidines (**10**). These compounds displayed remarkable regio-, diastereo-, and enantioselectivities, with yields reaching up to 82%, diastereoselectivities exceeding 19:1, and enantioselectivities surpassing 99%. In addition to the model pyrimidine derivatives (**10a–e**), the methodology also facilitated the seamless conversion of a quinoline derivative into the corresponding product (**10f**). Importantly,

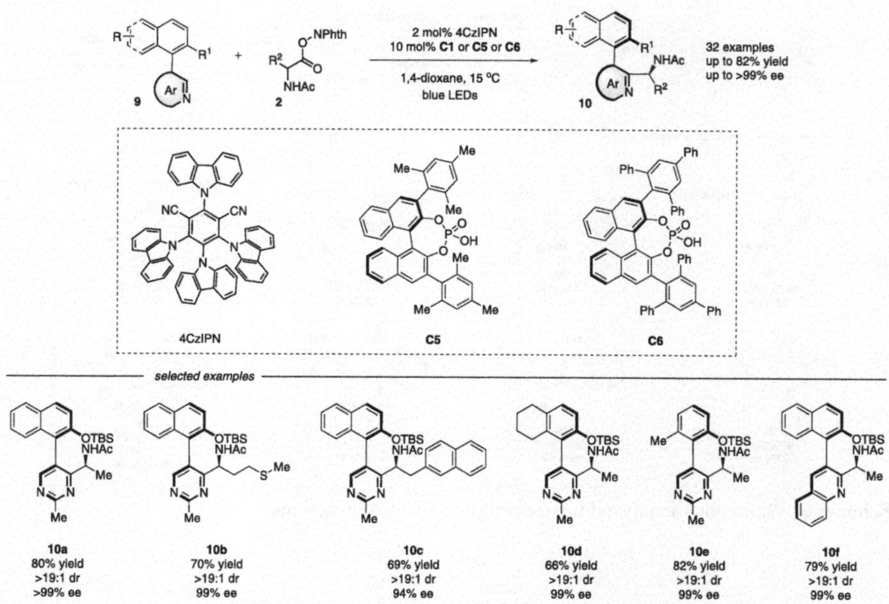

Scheme 5 Catalytic asymmetric construction of axial and central heterobiaryls by Minisci reaction

this study introduced a novel platform for desymmetrization through radical-involved atroposelective functionalization on the heteroarene of prochiral heterobiaryls.

While the aforementioned asymmetric Minisci reactions primarily utilized a "two-component" system, in 2019, Studer and co-workers ingeniously developed a three-component radical relay addition strategy to synthesize α-aminoalkyl stereocentered azaarenes [13] (Scheme 6). This dual photoredox and CHB catalysis approach enabled the transformation of azaarenes (**1**), α-bromo carbonyls (**11**), and enamides (**12**) into the desired products. The three-component mode offered an expanded substrate scope, allowing for the straightforward synthesis of a variety of valuable azaarene-containing chiral γ-amino acids (**3r**–**u**) and 1,2-diamine derivatives (**3v**–**x**) with good to excellent enantiomeric excesses (ees), reaching up to 96% ee. Notably, the introduction of a 1,2-branched enamide yielded a double α- and β-stereocentered quinoline (**3y**) with a 90% yield and enantiomeric excesses of 92% and 90%, albeit with a diastereomeric ratio of 2:1.

In 2023, Jiang and collaborators unveiled a groundbreaking three-component asymmetric Minisci-type transformation, which for the first time combined quinoxalinones, styrenes, and electron-withdrawing alkyl bromides [14] (Scheme 7). This study represents the inaugural instance of achieving chemodivergent synthesis through asymmetric photocatalysis. By employing two inorganic bases (Na_3PO_4 and Na_2HPO_4), two distinct sets of valuable products (**15** and **16**) were synthesized, exhibiting moderate to good yields and impressive enantiomeric excesses (up to 79% yield, 98% ee). The success of this transformation hinges on

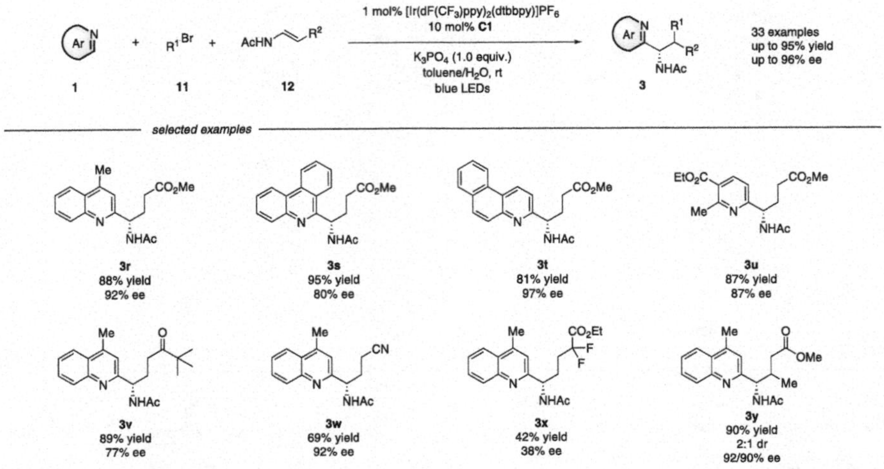

Scheme 6 Photoredox-catalyzed three-component Minisci reactions

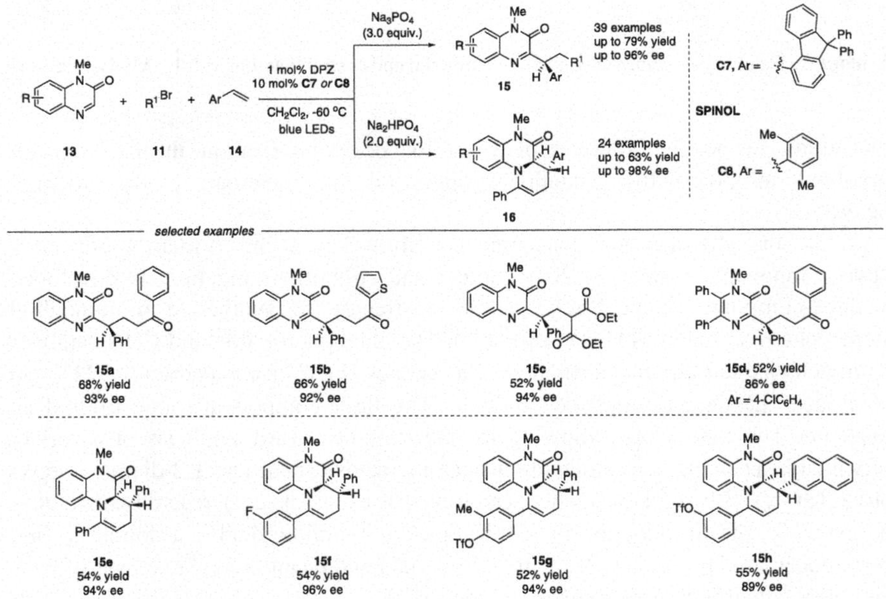

Scheme 7 Photoredox-catalyzed enantioselective chemodivergent three-component tandem reactions

the distinct pK_a environment, where radicals generated on the quinoxalin-2(1*H*)-one rings, following two radical addition processes, are subjected to either single-electron oxidation or single-electron reduction. Notably, this work marks the first-ever utilization of quinoxalin-2(1*H*)-ones in asymmetric photoredox catalysis.

2.2 From Azaarene Derivatives

In contrast to the Minisci-type reactions that typically initiate with simple azaarenes, this section focuses on asymmetric transformations derived from pre-functionalized azaarene derivatives (Fig. 2). Within the framework of cooperative photoredox and CHB catalysis combined with *N*-heterocarbene (NHC) catalysis, enantioselective intermolecular and intramolecular reactions – including radical addition, cross-coupling, and reduction-protonation – have been successfully executed at the C-2 or C-4 positions of azaarenes [15].

In 2018, Jiang and collaborators reported a novel radical conjugate addition-enantioselective protonation of α-branched 2-vinylazaarenes (**17**) with *N*-aryl glycines (**18**) via the synergistic action of dual DPZ and CHB catalysis [16] (Scheme 8). This process yielded a diverse array of enantioenriched 3-(2-pyridine/quinoline)-3-substituted amines (**19**), achieving high yields of up to 97% and remarkable enantiomeric excesses (ees) exceeding 99%. Notably, adduct **19a** could be efficiently

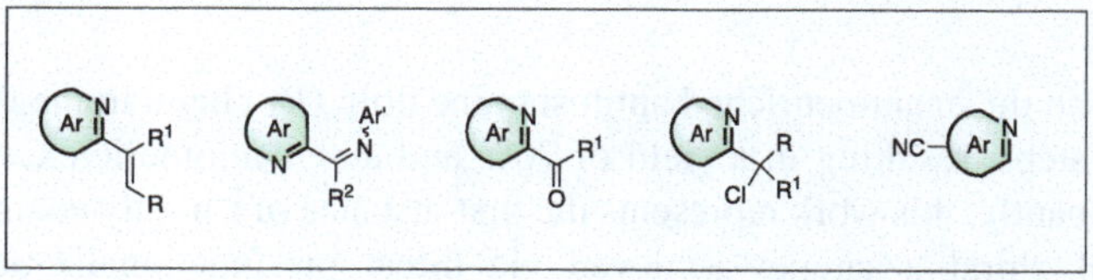

Fig. 2 Some developed active azaarene derivatives

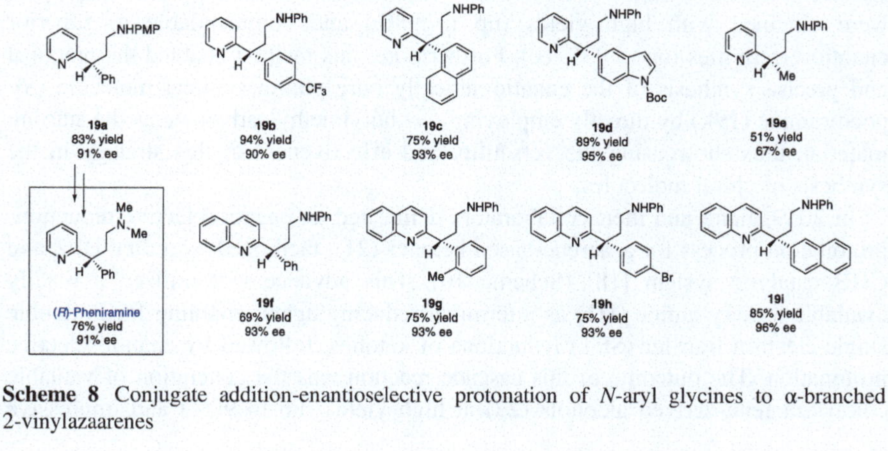

Scheme 8 Conjugate addition-enantioselective protonation of *N*-aryl glycines to α-branched 2-vinylazaarenes

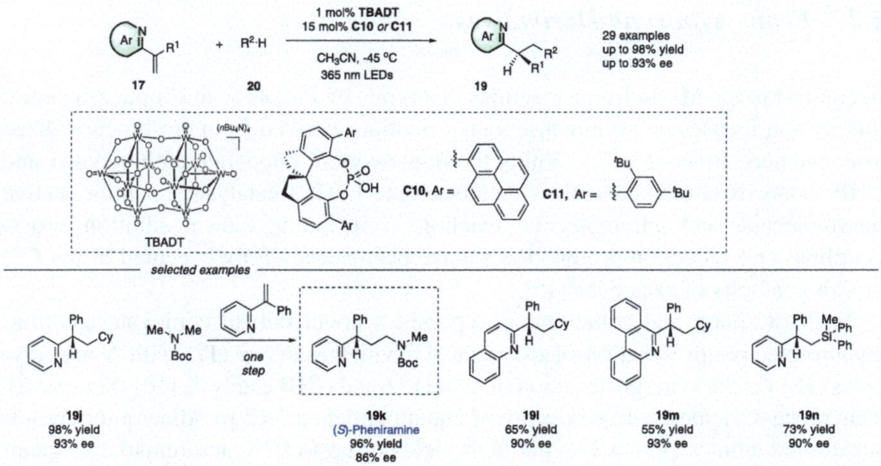

Scheme 9 Cooperative HAT and CHB catalysis on conjugate addition-enantioselective protonation reactions

converted into the enantioenriched antihistamine drug (*R*)-pheniramine through two subsequent steps, resulting in a yield of 76% and an enantiomeric excess of 91%. Most significantly, this work represents the first instance of a highly enantioselective synthesis of chiral azaarenes achieved via direct pre-functionalization through asymmetric photoredox catalysis.

Following the success of the aforementioned enantioselective protonation process, Jiang and co-workers further explored cooperative light-induced HAT and CHB catalysis in 2022, utilizing the same α-branched 2-vinylazaarenes in conjunction with commercially available hydrocarbons and silanes [17] (Scheme 9). Departing from the commonly utilized radical addition-elimination scaffolds, this innovative approach capitalizes on the simplicity and ubiquity of hydrocarbons and silanes, offering a straightforward and atom-economical route to a variety of enantiomerically enriched α-tertiary carbonyl azaarenes (**19**). These compounds were obtained with high yields (up to 98%) and commendable to superior enantioselectivities (up to 93% ee). Furthermore, this method enabled the practical and precise synthesis of the enantiomerically pure pharmaceutical molecule (*S*)-pheniramine (**19k**) by directly employing *tert*-butylmethylcarbamate as the starting material, thus showcasing the versatility and effectiveness of this strategy in the synthesis of chiral molecules.

In 2019, Jiang and their collaborators delineated an enantioselective reduction-protonation process for azaarene-based ketones (**21**), facilitated by a dual DPZ and CHB catalytic system [18] (Scheme 10). This advancement utilized a readily available tertiary amine (**H1**) as a terminal reducing agent, enabling facile double single-electron transfer (SET) reductions of ketones, followed by enantioselective protonation. The outcome of this cascade reaction was the generation of valuable chiral azaarene-derived alcohols (**22**) at high yields (up to 99%) and impressive

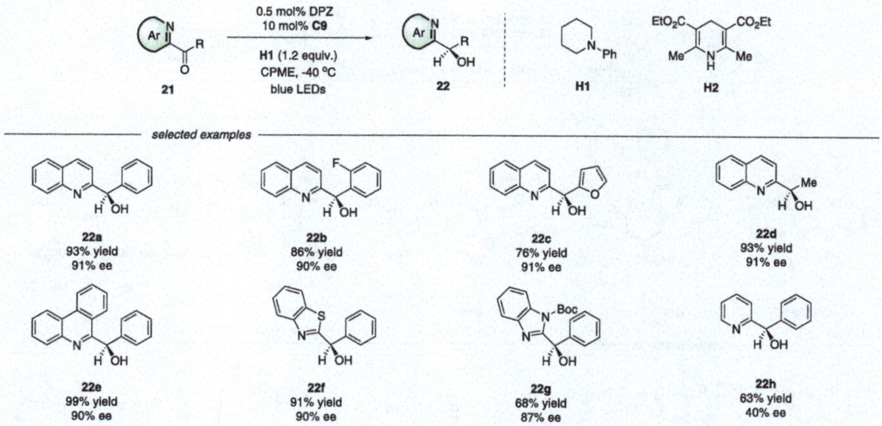

Scheme 10 Photoredox-catalyzed enantioselective reduction of azaarene-based ketones

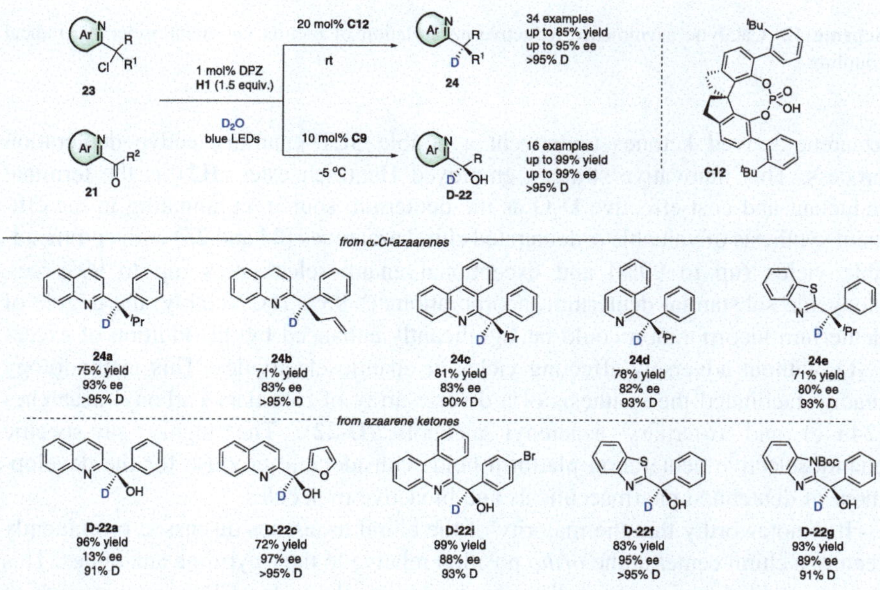

Scheme 11 Photoredox-catalyzed enantioselective α-deuteration of azaarenes with D_2O

enantioselectivities (up to 97% ee). Notably, a wide array of ketone derivatives – including quinolines, phenanthridine, benzothiazole, and benzimidazole – were seamlessly transformed, with the exception of one pyridine-derived alcohol (**22h**), which was obtained in a less desirable manner (63% yield, 40% ee).

Jiang and co-workers have made significant advances in asymmetric deuteration methodologies by expanding the double SET reduction approach into a robust "asymmetric deuteration" strategy [19] (Scheme 11). Utilizing cooperative catalysis involving dual DPZ and CHB catalysts, both racemic α-chloroazaarenes and

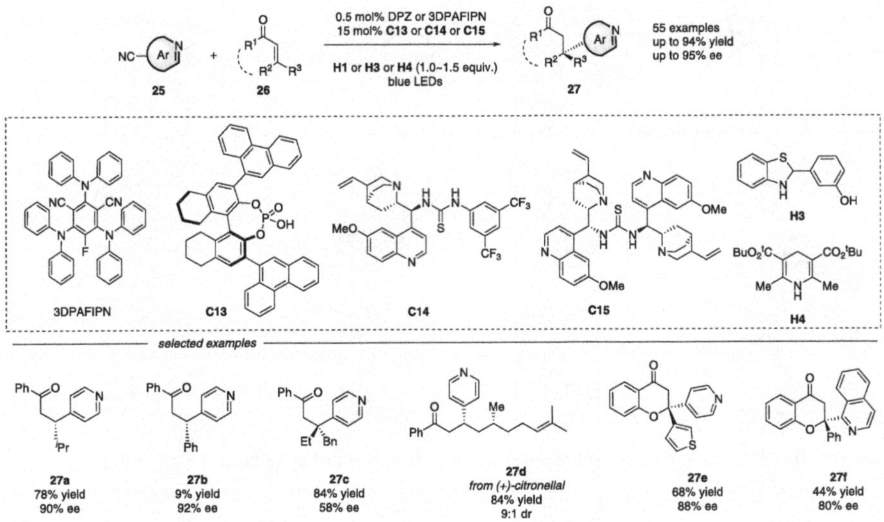

Scheme 12 Catalytic asymmetric reductive azaarylation of olefins via enantioselective radical coupling

azaarene-derived ketones underwent a double SET enantioselective deuteration process. This innovative strategy employed Hantzsch ester (**H2**) as the terminal reductant and cost-effective D_2O as the deuterium source, culminating in the efficient synthesis of valuable α-deuterated chiral azaarenes (**24** and **25**) with commendable yields (up to 99%) and exceptional enantioselectivities (up to 99% ee), alongside substantial deuterium incorporations (>95% D). Notably, the degree of deuterium incorporation could be significantly enhanced by the addition of excess D_2O, without adversely affecting yields or enantioselectivities. This methodology readily facilitated the synthesis of a diverse array of α-tertiary carbonyl azaarenes (**24a–e**) and α-tertiary azaarenyl alcohols (**D-22**). The highly site-specific enantioselective deuteration platform holds considerable promise for the development of deuterated pharmaceuticals and bioactive molecules.

It is noteworthy that the majority of the chiral azaarenes discussed prominently feature a chiral center at the *ortho* position relative to their pyridine analogues. This constraint likely arises from the active sites available for CPA catalysts, which encompass Minisci-type reactions, radical additions, or reduced SET protonations. Consequently, achieving chirality at multiple sites within an azaarene remains an unresolved challenge in this field.

In 2022, Jiang and colleagues introduced cyanoazaarenes (**25**) and enones (**26**) into a radical enantioselective cross-coupling reaction utilizing dual DPZ and CHB catalysis under blue LED irradiation [20] (Scheme 12). This approach enabled the elegant formation of α-stereocentered azaarenes with functionalization at the C-4 position of pyridines. Two classes of hydrogen-bonding catalysts were employed to ensure adequate enantioselective control over different structures of enones. Specifically, a CPA catalyst (**C13**) was utilized for acyclic enones in the transformations of

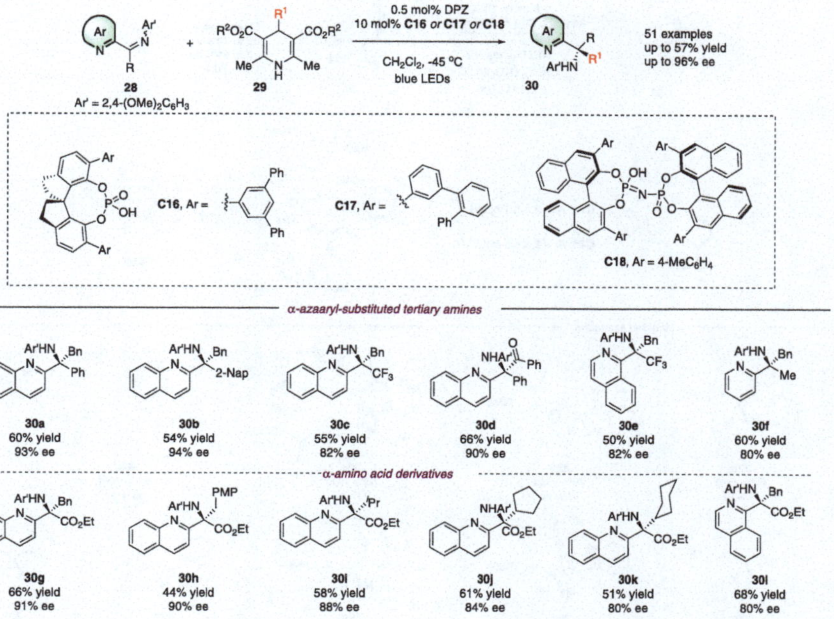

Scheme 13 Catalytic asymmetric synthesis of azaarene-functionalized tertiary amines and α-amino acid derivatives

chiral pyridine-4-functionalizations (**27a–d**), while cyclic enones were transformed using cinchona alkaloid-derived bifunctional catalysts (**C14** and **C15**), yielding valuable enantioenriched azaarene flavone variants (**27e** and **27f**) with high yields and ees. Importantly, the methodology was well tolerated with isonicotinonitriles and isoquinoline-1-carbonitriles, leading to the successful synthesis of 1-quaternary α-stereocentered isoquinoline (**27f**) in 44% yield and 80% ee.

In the subsequent year, Jiang and co-workers demonstrated a radical addition-enantioselective protonation strategy, facilitating the synthesis of valuable α-azaaryl-substituted tertiary amines (**30a–f**) and α-amino acid derivatives (**30g–l**) from azaarene-substituted *E*/*Z*-ketimines and 4-substituted 1,4-dihydropyridines (Scheme 13) [21]. This approach afforded a wide range of aza-quaternary carbonyl α-stereocentered azaarenes with high yields and ees. Notably, in addition to model quinoline-derived ketimines, isoquinoline and pyridine derivatives (**30e**, **30f**, and **30l**) also demonstrated good performance, albeit with slightly lower ees. This method significantly facilitated the efficient synthesis of nonnatural α-amino acids containing an azaarene moiety, highlighting its relevance in the discovery of α-amino acid-like bioactive molecules.

Subsequently, Jiang and colleagues reported a potent asymmetric olefin isomerization strategy that successfully generated a diverse range of α-tertiaryallylic azaarenes (**32**) utilizing cooperative DPZ, *N*-hydroxy-1,8-naphthalimide (NHI), and CHB catalysis (Scheme 14) [22]. As a result, a series of quinoline derivatives (**32a–d**) were synthesized with commendable yields and exceptional ees. Notably,

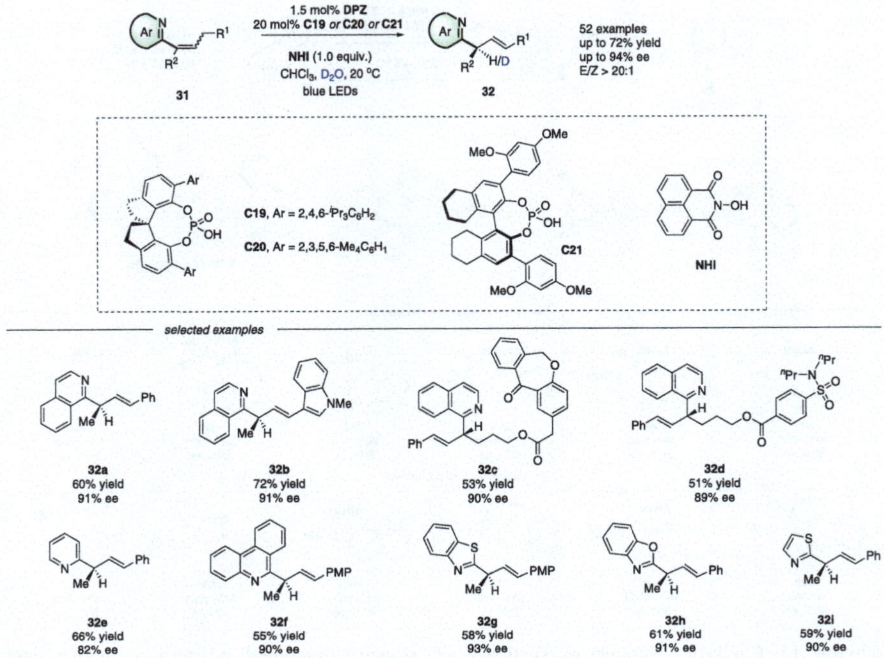

Scheme 14 Asymmetric olefin isomerization via photoredox catalytic hydrogen atom transfer and enantioselective protonation

the substrate scope was broadened to include isoquinoline-derived tertiary alkenes (**32a–b**) and bioactive molecular-derived alkenes (**32c–d**), which were produced in moderate to high yields alongside excellent ees. Additionally, various azaarenes, including pyridine (**32e**), phenanthridine (**32f**), and benzothiazole (**32g**), exhibited good reactivity, yielding satisfactory results. This HAT-driven asymmetric olefin isomerization strategy demonstrated remarkable atom economy and *E*/*Z* selectivity, contributing significantly to the advancement of green and sustainable chemistry.

In a recent study, Jiang and co-workers unveiled an intriguing methodology involving azaarene-involved bicyclo[2.1.1]hexanes (BCHs) through a highly enantioselective [2+2] photocycloaddition of bicyclo[1.1.0]butanes (BCBs) and allylic azaarenes [23] (Scheme 15). This innovative process was catalyzed by a bifunctional chiral photosensitizer, specifically a polycyclic aromatic hydrocarbon (PAH)-containing CPA. A wide array of pharmaceutically relevant BCH derivatives (**35**) were synthesized with high yields (up to 99%), ees (exceeding 99% ee), and impressive diastereoselectivities (greater than 19:1 dr). In addition to the diverse 1-ketocarbonyl-3-substituted BCBs (**35a–b**), α/β-substituted vinylazaarenes (**35c–d**) were shown to be compatible with this groundbreaking photoredox catalytic pathway, enabling the successful formation of an all-carbon quaternary stereocenter (**35e**) or two adjacent tertiary stereocenters in the product (**35f**). Notably, allylic pyrimidine (**35c**) performed well, yielding an impressive 99% ee, albeit with lower yields (24%).

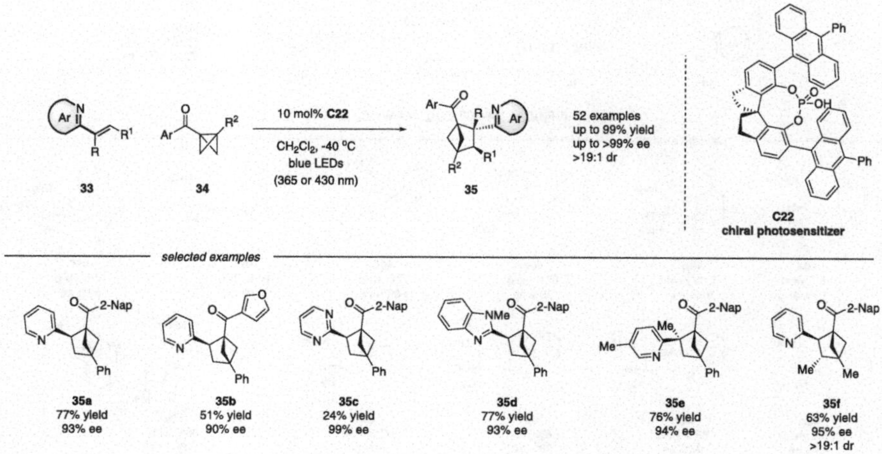

Scheme 15 Photoredox-catalyzed enantioselective [2π+2σ] cycloadditions of BCBs with vinylazaarenes

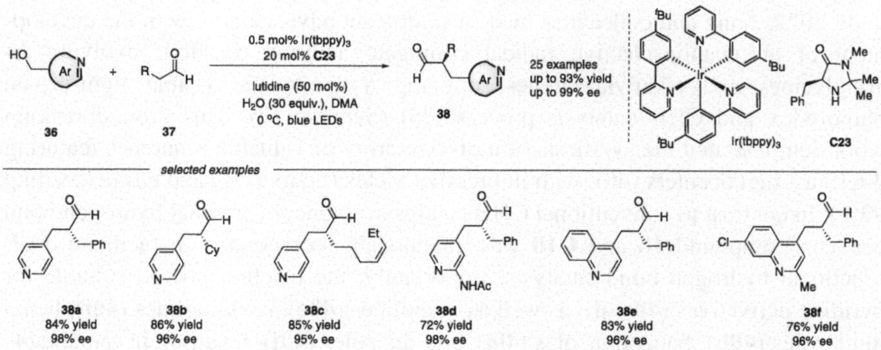

Scheme 16 SCS-enabled enantioselective β-functionalization of azaarenes

3 Methods to the Construction of β-Stereocenters

In 2018, MacMillan and colleagues introduced a dual photoredox and chiral amine catalytic strategy that leverages a radical-mediated "spin-center shift" (SCS) mechanism [24] (Scheme 16). This innovative approach utilized a small molecular catalyst (**C23**) as the chiral source, facilitating the transformation of 4-(hydroxymethyl)-pyridines and aldehydes into β-chiral azaarenes, specifically pyridines (**38a–d**) and quinolines (**38e–f**). The method achieved high yields (up to 93%) and exceptional enantiomeric excesses (ees) (up to 99%). Notably, this covalent bond-driven catalytic platform significantly enhanced site-specific selectivity at the 4-position of both pyridines and quinolines, marking a valuable addition to the toolkit for constructing chiral azaarenes compared to traditional CHB catalysis.

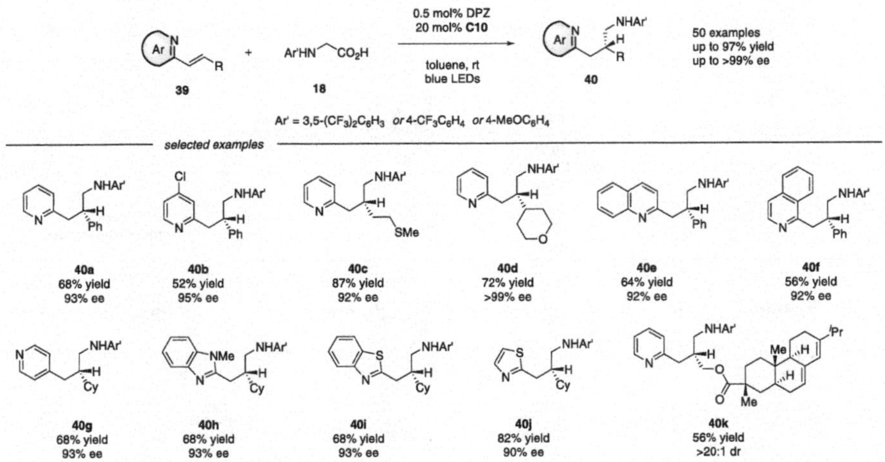

Scheme 17 Asymmetric hydroaminoalkylation of alkenylazaarenes from 1,2-branched vinylazaarenes

In 2022, Jiang and colleagues made a significant advancement with the development of an enantioselective radical conjugate addition reaction involving *N*-arylglycines and alkenylazaarenes, utilizing a synergistic visible light-driven photoredox and CHB catalysis process [25] (Scheme 17). This groundbreaking work demonstrated the synthesis of a diverse array of valuable azaarenes featuring β-tertiary stereocenters (**40**), with impressive yields (up to 97%) and ees (exceeding 99%). In contrast to conventional CHB catalysis, a nonconventional hydrogen bond between compound **18** and **C10** was ingeniously incorporated to facilitate dual-functional hydrogen bond catalysis. Importantly, the reaction proved suitable for pyridine derivatives (**40a–d**), as well as quinolines (**40e**), isoquinolines (**40f**), benzimidazoles (**40h**), benzothiazoles (**40i**), and thiazoles (**40j**), resulting in remarkable outcomes. Notably, the CHB catalysis exhibited versatility by yielding a 4-pyridine product (**40g**) with a β-tertiary stereocenter in 68% yield and 93% ee, further underscoring its efficacy in generating a wide range of chiral azaarenes. Additionally, the biologically active compound abietic acid (**40k**) was synthesized with a yield of 56% and a significant diastereomeric ratio exceeding 20:1, highlighting its potential in bioactive molecule synthesis.

In 2024, Jiang and colleagues reported an innovative radical coupling-enantioselective protonation strategy for synthesizing β-stereocentered azaarenes [26] (Scheme 18). Consistent with their previous studies, this dual catalytic method employed two oxidative substrates (**25** and **41**) and incorporated a sacrificial reductant (**H4**) to initiate the catalytic cycle, resulting in the formation of a diverse range of β-chiral substituted azaarenes (**42**) with good yields and excellent ees. Notably, this reaction module successfully incorporated a commercially available Fasudil analogue, yielding a product with an 81% yield and 91% ee (**42e**).

Recently, Jiang and colleagues introduced a challenging sensitization-initiated electron transfer strategy that utilized DPZ and a pyrenyl-substituted CPA as

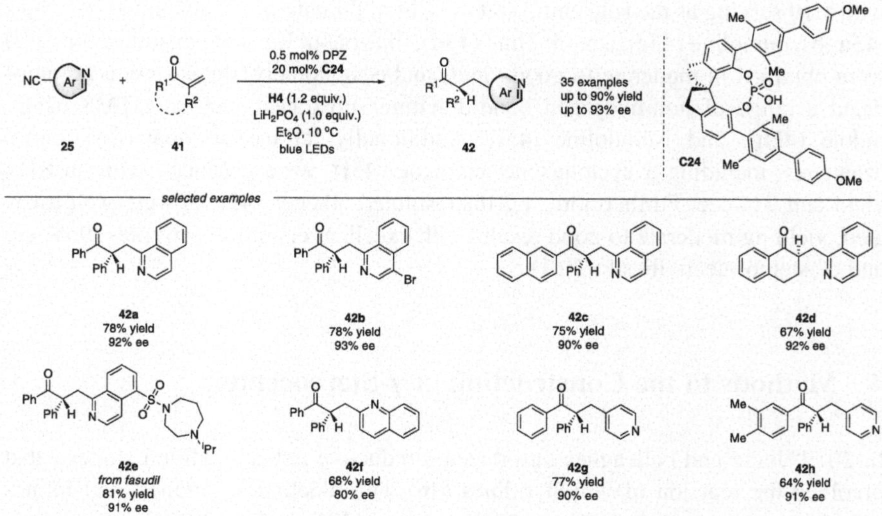

Scheme 18 Radical cross-coupling and enantioselective protonation of cyanoazaarenes and α-substituted enones

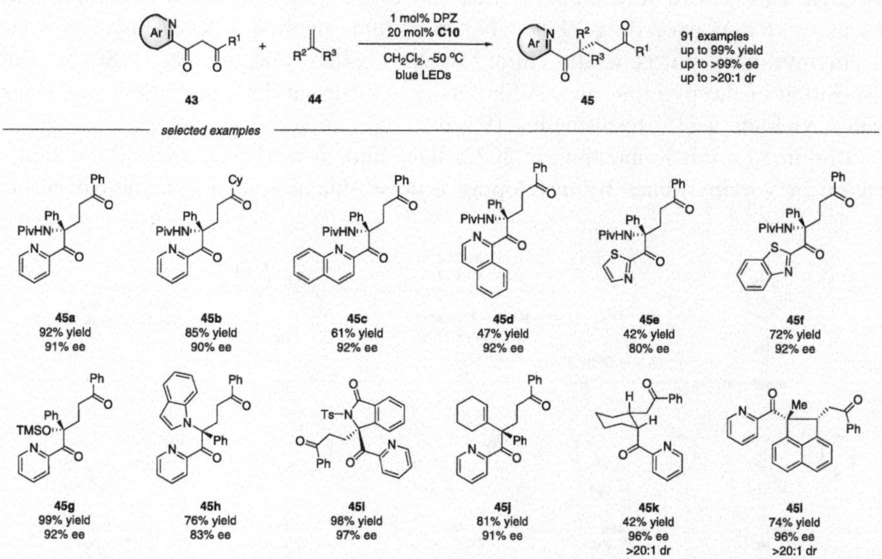

Scheme 19 Asymmetric photoredox catalytic formal de Mayo reaction enabled by sensitization-initiated electron transfer

co-sensitizers [27] (Scheme 19). Through meticulous design of this dual-catalyst system, they successfully executed an asymmetric de Mayo-type reaction, resulting in a wide array of valuable enantioenriched azaarene-containing 1,5-diketones (**45**). Impressively, this platform demonstrated robust tolerance for various substrates, with **C10** serving as the sole chiral source. Chiral β-quaternary substituted pyridines (**45a–b**), quinoline (**45c**), isoquinoline (**45d**), thiazole (**45e**), and benzothiazole (**45f**) were obtained in moderate to good yields and ees. Notably, the system accommodated a range of non-hydrogen bond-donating substrates, such as OTMS (**45g**), indole (**45h**), and isoindoline (**45i**). Additionally, all-carbon quaternary chiral azaarenes, including a cyclohexene analogue (**45j**), were produced with an 81% yield and 91% ee. Furthermore, 1,2-disubstituted alkenes (**45k–l**) were well tolerated, yielding moderate to good results with excellent enantioselectivities (96% ee) and diastereoselectivities (>20:1).

4 Methods to the Construction of γ-Stereocenters

In 2019, Jiang and colleagues introduced a reductive radical-addition strategy that involved the reaction of vinylpyridines (**46**) with aldehydes, ketones, and imines (**47**) through cooperative DPZ and CHB catalysis [28] (Scheme 20). This innovative approach enabled the synthesis of a diverse array of pharmaceutically and biologically significant γ-secondary and tertiary hydroxyl- and amine-substituted chiral azaarenes (**48**) with remarkable yields (up to 99%) and excellent enantiomeric excesses (ees) exceeding 99%. Notably, the method successfully utilized 4-vinylpyridines to generate chiral alcohols (**48e**) and amines (**48i**) at the 4-position of the pyridine ring. Additionally, an estrone derivative (**48f**) was evaluated, yielding a 41% result with a 19:1 dr.

Building on this foundation, in 2022, Jiang and co-workers expanded the application of vinylpyridines by developing a novel enantioselective reductive cross-

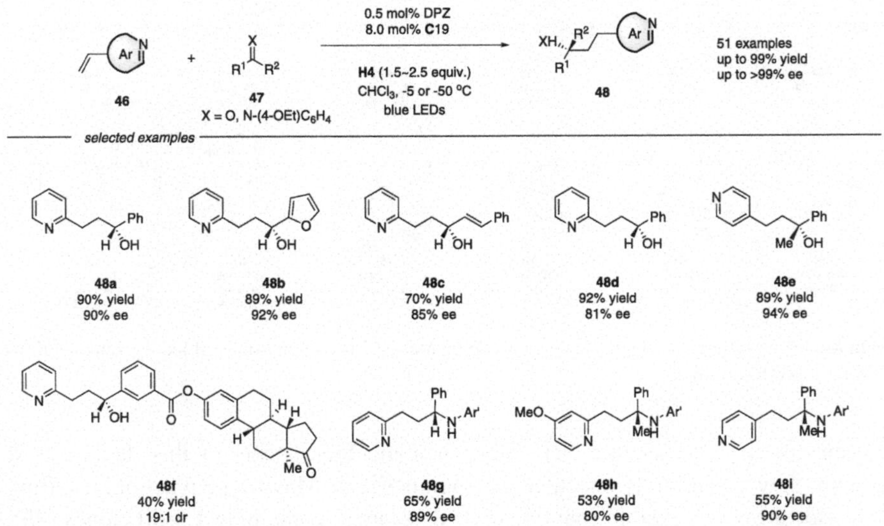

Scheme 20 Photoredox-catalyzed enantioselective addition of prochiral radicals to vinylpyridines

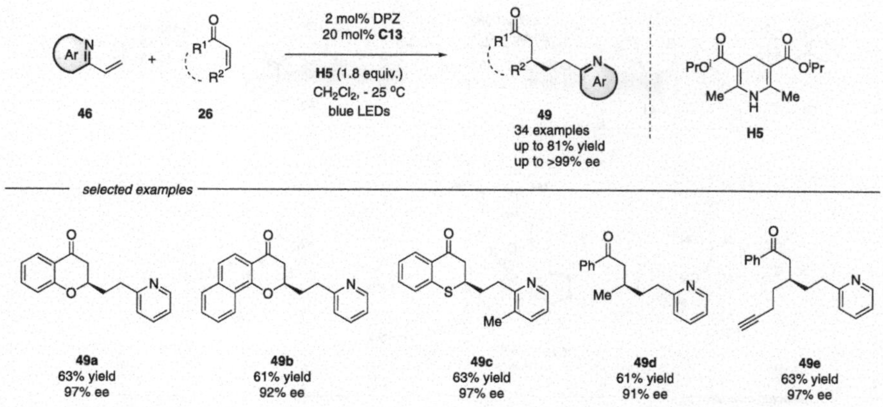

Scheme 21 Photoredox-catalyzed cross-coupling between 2-vinylpyridines and enones

coupling strategy for synthesizing γ-tertiary carbon stereocentered azaarenes from cyclic and acyclic enones [29] (Scheme 21). This innovative method, enabled by cooperative DPZ and CHB catalysis, utilized Hantzsch ester (**H5**) as a terminal reductant, facilitating the direct synthesis of various biologically relevant chiral γ-stereocentered pyridines (**49**) with yields of up to 93% and ees reaching >99%. Importantly, a wide range of cyclic and acyclic enones demonstrated compatibility under the optimized conditions, although the azaarenes were limited to vinylpyridines.

In subsequent investigations, Jiang and co-workers explored a compelling synthetic route for γ-chiral-centered pyridines using 2-(chloromethyl)-azaaren-1-ium chlorides (**50**) and α-substituted enones [26] (Scheme 22). In this process, a novel organic photosensitizer TDPZ was employed in place of DPZ to facilitate the smooth reaction of various chloromethylazaarene salts with both cyclic (**51a–c**) and acyclic α-branched (**51d**) enones. Although the study presented only nine examples, these results convincingly highlighted the success of enantioselective synthesis of remote-stereocentered azaarenes through the application of chloromethylazaarenes.

Most recently, Jiang and co-workers unveiled a direct enantioselective reduction of pyridine-substituted β-polyfluoroalkylated enones, aimed at constructing γ-fluoro-containing stereocentered azaarenes [30] (Scheme 23). Utilizing a dual 3DPAFIPN and CHB catalysis approach, they successfully synthesized a diverse range of derivatives including pyridines (**53a–c**), quinolines (**53d**), and 7,8-benzoquinolines (**53e**) with high yields and ees. Notably, a stoichiometric amount of H_2O was employed as the H^+ source, without detriment to yield or ee. When D_2O was used under standard conditions, γ-chiral D-azaarenes (**53f**) were produced with remarkable deuteration (>99%), achieving a 70% yield and 98% ee. Furthermore, to address the synthesis of more complex multi-deuterium-containing azaarenes, the researchers cleverly designed tandem reaction sequences that facilitated the straightforward access to chiral azaarenes with single- (**53h**), double- (**53i**), and triple-deuterium (**53g, 53j**) incorporation.

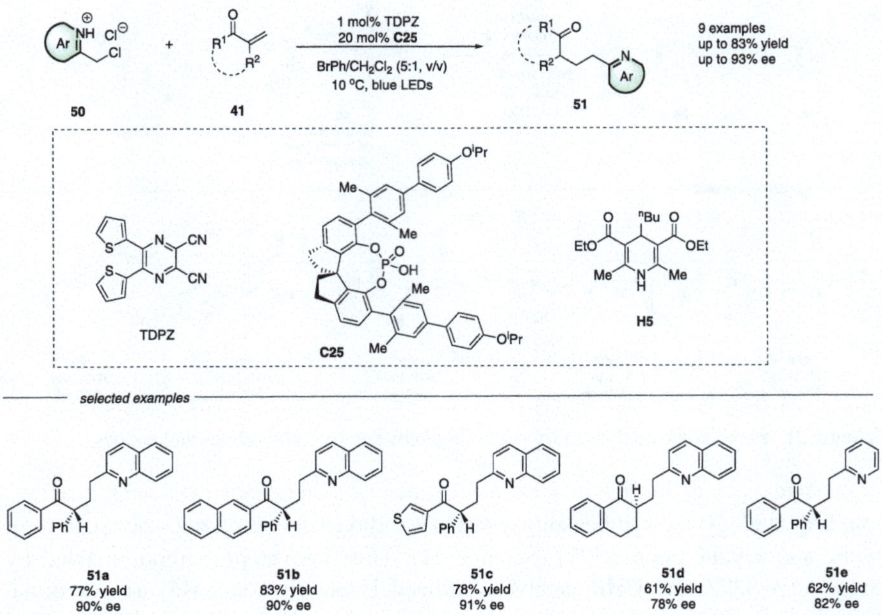

Scheme 22 Photoredox-catalyzed radical cross-coupling between 2-(chloromethyl)-azaaren-1-ium chlorides and α-substituted enones

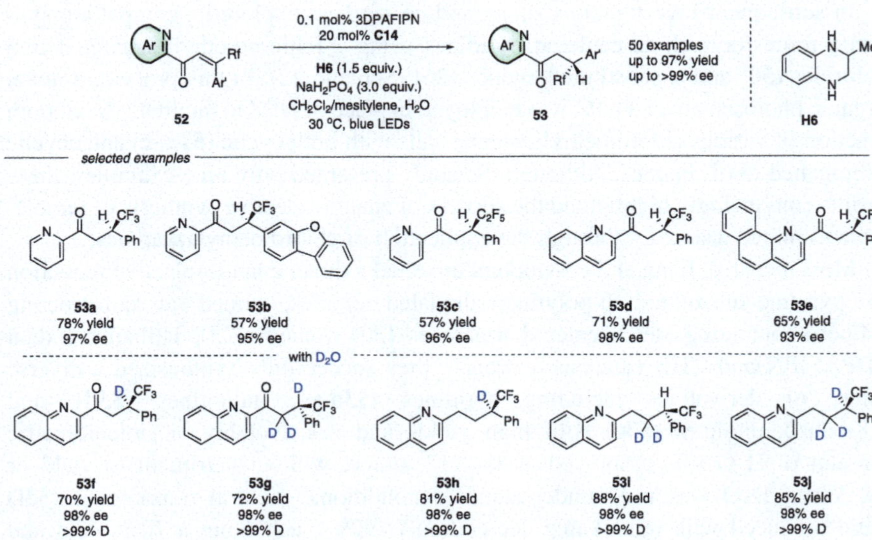

Scheme 23 Photoredox-catalyzed enantioselective reduction of C=C bond of β-polyfluoro-alkylated enones

5 Miscellaneous

With advancements in α-, β-, and γ-stereocentered azaarenes, researchers have reported several multi-stereocentered examples utilizing a similar dual-catalytic strategy. In 2020, Jiang and co-workers developed a photoredox catalytic [3+2] cycloaddition strategy for constructing α,β-distereocentered azaarenes by combining α-branched 2-vinylazaarenes with cyclopropylamines [31] (Scheme 24). This redox-neutral approach showcased the synthesis of a diverse array of highly valuable azaarenes featuring α-all-carbon quaternary and β-tertiary stereocenters (**55**). The method achieved impressive yields (up to 97%), diastereoselectivities (up to >19:1 dr), and enantioselectivities (up to >99%). Notably, the reaction was effective not only with pyridine analogues (**55a**) but also extended to benzimidazoles (**55b**), quinolines (**55c**), isoquinolines (**55d**), and thiazoles (**55e**), all yielding outstanding results. Crucially, this platform demonstrated exceptional versatility, synthesizing a 3-pyridine product (**55f**) with 83% yield and 91/92% ee at a 1.5:1 dr. The use of a triple-solvent system (Et_2O/CPME/benzene) and ultralow temperatures (−60°C) were ingeniously employed in these transformations, leading to a more stereocontrolled and efficient process.

In 2021, Jiang and co-workers reported a divergent asymmetric synthesis of azaarene-functionalized cyclic alcohols through a cooperative DPZ- and CHB-catalyzed Beckwith-Enholm cyclization [32] (Scheme 25). This approach yielded a variety of azaarene-functionalized carbocyclic and heterocyclic alcohols, bearing adjacent β,γ-stereocenters (**57a–h**) and nonadjacent α,γ-stereocenters (**58a–b**), with high yields and excellent enantio- and diastereoselectivities. By exploring various skeletal modifications of these azaarene-based olefins, a series of CPAs (**C8**,

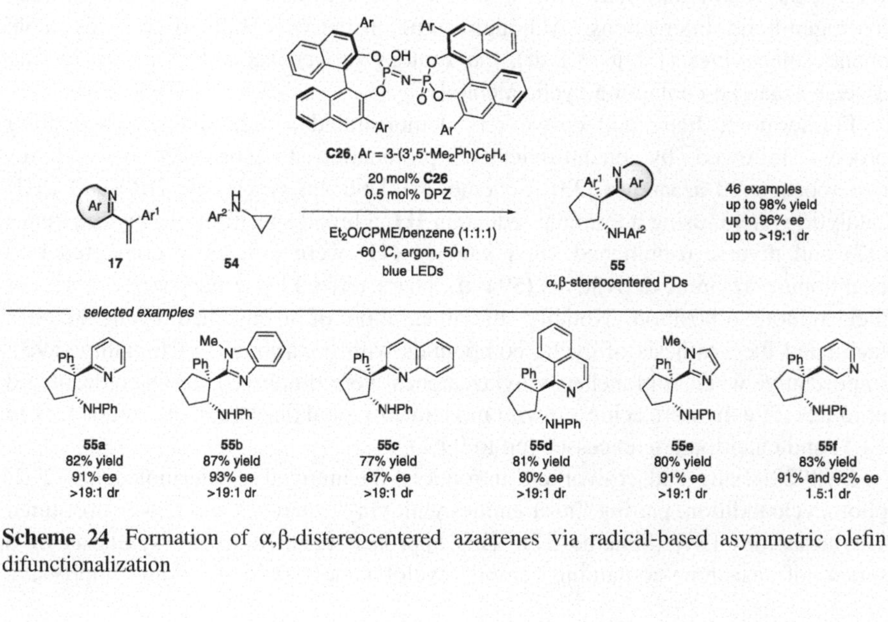

Scheme 24 Formation of α,β-distereocentered azaarenes via radical-based asymmetric olefin difunctionalization

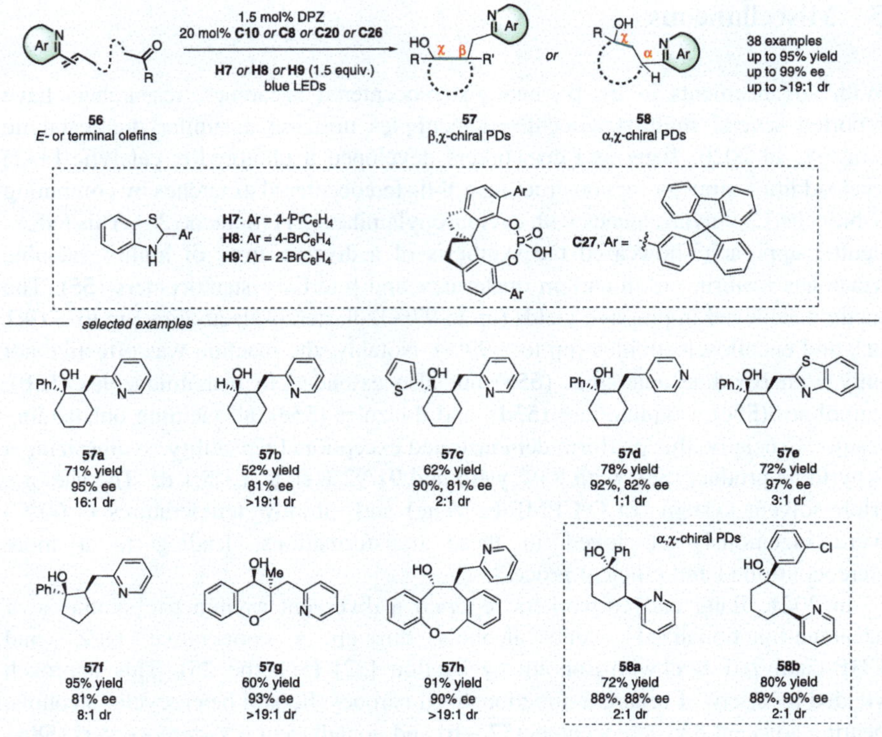

Scheme 25 Photoredox-catalyzed divergent synthesis of azaarene-functionalized cyclic alcohols

C10, **C20**, **C27**) and reductants (**H7–9**) were introduced to enhance efficient hydrogen-bond interactions. Although some instances exhibited less favorable enantioselectivities (1:1 to 3:1 dr), the strategy proved versatile for constructing diverse azaarene-containing cyclic alcohols.

Furthermore, Jiang and co-workers demonstrated a reductive cross-coupling process followed by enantioselective protonation to construct δ- and α,-δ-stereocentered azaarenes [33] (Scheme 26). Under a synergistic DPZ and CHB catalytic system, using a terminal reductant **H1**, a broad spectrum of vinylazaarenes (**17**) and diverse α-branched vinyl ketones (**41**) were efficiently converted into enantiopure azaarene derivatives (**59a–d**), characterized by tertiary stereocenters at their remote δ-position. Notably, the integration of α-branched cyclic ketones facilitated the synthesis of cyclic compounds with δ-carbonyl functionality (**59d**). Importantly, when α-branched vinylazaarenes were employed, α,δ-stereocentered products (**59e–h**) were achieved with moderate to good diastereomeric ratios (3:1 to 12:1) and enantiomeric excesses (up to 99%).

In 2022, Jiang and co-workers introduced an innovative intermolecular [2+2] photocycloaddition, pairing linear enones with vinylazaarenes and (*E*)-2-substituted vinylazaarenes [34] (Scheme 27). This approach resulted in the synthesis of a series of azaarene-containing chiral cyclobutanes (**61a–e**) with impressive

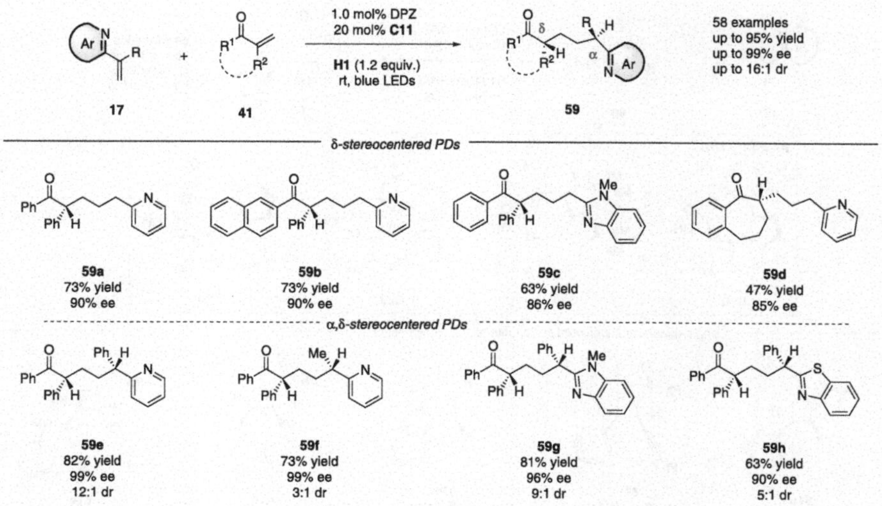

Scheme 26 Photoredox-catalyzed radical cross-coupling and enantioselective protonation of olefins

chemoselectivity and high enantioselectivity (up to 99% ee). Furthermore, this catalytic method demonstrated remarkable adaptability, successfully applied to complex combinations of (*E*)-2-substituted vinylazaarenes and vinylazaarenes, leading to the generation of a diverse range of chiral bi-azaarenes (**61f–j**) with moderate to high yields and good enantioenrichment. Notably, this platform facilitated the construction of intricate all-carbon quaternary stereocenters (**61e**, **61i**) and four consecutive stereocenters (**61c–d**), underscoring its versatility in stereochemical synthesis.

6 Conclusion and Future Perspective

In conclusion, we have compiled recent advancements in the enantioselective functionalization of azaarene derivatives over the past decade. Our review highlights a consistent stream of novel enantioenriched azaarene derivatives, which are poised to facilitate broader applications of chiral azaarenes in pharmaceuticals, life sciences, and materials science. The expanding structural diversity of chiral azaarenes has significantly enhanced the discovery of potential active molecules, unique pharmacological activities, and strategic design of compounds featuring chiral azaarene moieties.

However, several limitations persist in these advancements: (1) The incorporation of chiral moieties remains largely confined to specific positions within azaarene scaffolds. For instance, asymmetric functionalization of pyridine and quinoline is predominantly restricted to the C-2 and C-4 positions, respectively, while

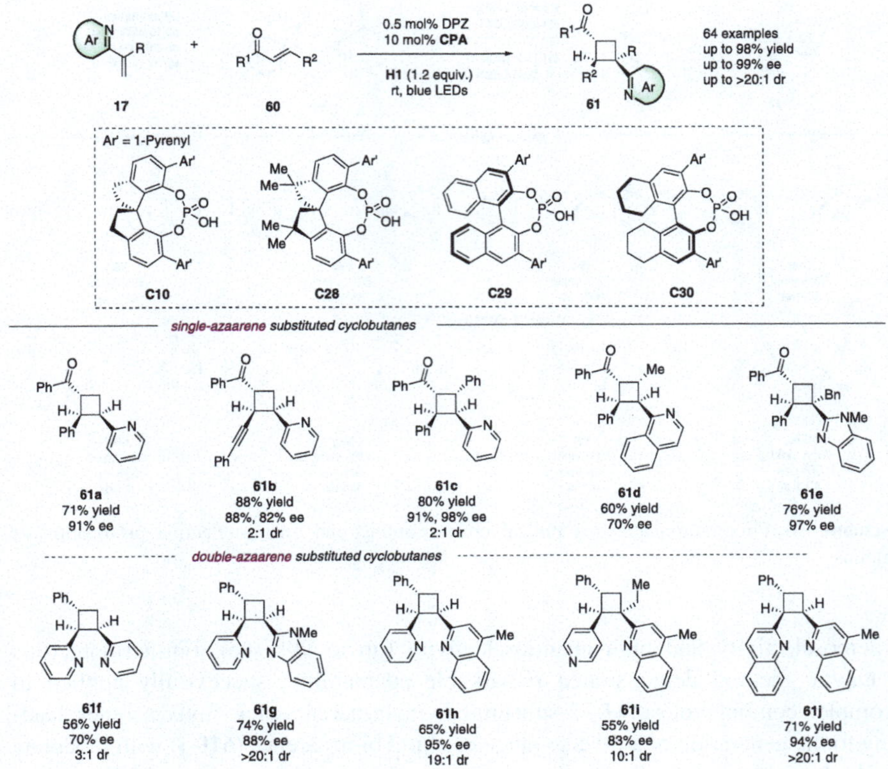

Scheme 27 Visible light-induced enantioselective intermolecular [2+2] photocycloadditions

isoquinoline is limited to the C-1 position. (2) The proximity of chiral centers to azaarenes is constrained, with current strategies primarily targeting α-, β-, γ-, and δ-positions, leaving the incorporation of chiral centers at greater distances and the establishment of multiple chiral centers as significant challenges. (3) The application of cooperative asymmetric and photoredox catalysis strategies for the asymmetric functionalization of azaarenes remains notably limited. To date, research has focused primarily on synergistic catalytic strategies that involve photoredox and CHB catalysis. This approach relies heavily on hydrogen bonding interactions between azaarenes and CHB catalysts, which restricts the catalytically active sites to specific positions within the pyridine rings, thereby limiting the spatial arrangement of chiral centers.

Consequently, the exploration of novel synergistic catalytic modes, alongside the development of azaarenes with diverse functional group sites and enhanced chiral group diversity under the guidance of photoredox catalysis, promises to be of significant research importance moving forward.

Acknowledgements We acknowledge the financial support from NSFC (21925103, 22301066), Henan Normal University, and Pingyuan Laboratory.

References

1. (a) Pozharskii AF, Soldatenkov AT, Katritzky AR (2011) Heterocycles in life and society: an introduction to heterocyclic chemistry, biochemistry and applications. John Wiley & Sons, Chichester. (b) Li JJ (ed) (2013) Heterocyclic chemistry in drug discovery. John Wiley & Sons, Hoboken. (c) Guan A-Y, Liu C-L, Sun X-F, Xie Y, Wang M-A (2016) Bioorg Med Chem 24: 342–353
2. (a) Best D, Lam HW (2014) J Org Chem 79:831–845. (b) Evano G, Theunissen C (2019) Angew Chem Int Ed 58:7558–7598
3. (a) Yin Y, Zhao X, Jiang Z (2020) ChemCatChem 12:4471–4489. (b) Yin Y, Zhao X, Jiang Z (2022) Chin J Org Chem 42:1609–1625. (c) Yin Y, Zhao X, Qiao B, Jiang Z (2020) Org Chem Front 7:1283–1296
4. (a) Prier CK, Rankic DA, MacMillan DWC (2013) Chem Rev 113:5322. (b) Shaw MH, Twilton J, MacMillan DWC (2016) J Org Chem 81:6898–6926. (c) Sun A, McAtee R, McClain E, Stephenson C (2019) Synthesis 51:1063–1072
5. For reviews see: (a) Duncton MAJ (2011) Med Chem Commun 2:1135–1161. (b) Tauber J, Imbri D, Opatz T (2014) Molecules 19:16190–16222. (c) Proctor RSJ, Phipps RJ (2019) Angew Chem Int Ed 58:2–36. (d) Sun AC, McAtee RC, McClain EJ, Stephenson CRJ (2019) Synthesis 51:1063–1072. (e) Li W, Wang Y, Chen J, Shi D, Yu X, Wang Z-J (2024) Chin J Org Chem 44:2110–2123
6. Selected examples see: (a) Zhao W-M, Chen X-L, Yuan J-W, Qu L-B, Duan L-K, Zhao Y-F (2014) Chem Commun 50:2018–2020. (b) Mai DN, Baxter RD (2016) Org Lett 18:3738–3741. (c) O'Brien AG, Maruyama A, Inokuma Y, Fujita M, Baran PS, Blackmond DG (2014) Angew Chem Int Ed 53:11868–11871. (d) Xuan J, Zhang Z-G, Xiao W-J (2015) Angew Chem Int Ed 54:15632–15641
7. Proctor RSJ, Davis HJ, Phipps RJ (2018) Science 360:419–422
8. Liu X, Liu Y, Chai G, Qiao B, Zhao X, Jiang Z (2018) Org Lett 20:6298–6301
9. (a) Reid JP, Proctor RSJ, Sigman MS, Robert RJ (2019) J Am Chem Soc 141:19178–19185. (b) Ermanis K, Colgan AC, Proctor RSJ, Hadrys BW, Phipps RJ, Goodman JM (2020) J Am Chem Soc 142:21091–21101
10. Proctor RSJ, Chuentragool P, Colgan AC, Phipps RJ (2021) J Am Chem Soc 143:4928–4934
11. Colgan AC, Proctor RSJ, Gibson DC, Chuentragool P, Lahdenperä ASK, Ermanis K, Phipps RJ (2022) Angew Chem Int Ed 61:e202200266
12. Liang D, Chen J-R, Tan L-P, He Z-W, Xiao W-J (2022) J Am Chem Soc 144:6040–6049
13. Zheng D, Studer A (2019) Angew Chem Int Ed 58:15803–15807
14. Ma C, Shen J, Qu C, Shao T, Cao S, Yin Y, Zhao X, Jiang Z (2023) J Am Chem Soc 145: 20141–20148
15. For selected reviews see: (a) Yin Y, Zhao X, Qiao B, Jiang Z (2020) Org Chem Front 7:1283–1296. (b) Inamdar SM, Shinde VS, Patil NT (2015) Org Biomol Chem 13:8116–8162. (c) Jiang C, Chen W, Zheng W-H, Lu H (2019) Org Biomol Chem 17:8673–8689. (d) Holmberg-Douglas N, Nicewicz DA (2022) Chem Rev 122:1925–2016
16. Yin Y, Dai Y, Jia H, Li J, Bu L, Qiao B, Zhao X, Jiang Z (2018) J Am Chem Soc 140:6083–6087
17. Tan Y, Yin Y, Cao S, Zhao X, Qu G, Jiang Z (2022) Chin J Catal 43:558–563
18. Qiao B, Li C, Zhao X, Yin Y, Jiang Z (2019) Chem Commun 55:7534–7537
19. Shao T, Li Y, Ma N, Li C, Chai G, Zhao X, Qiao B, Jiang Z (2019) iScience 16:410–419
20. Li Y, Han C, Huang X, Zhao X, Qiao B, Jiang Z (2022) J Am Chem Soc 144:7805–7814
21. Song X, Zhang Y, Li Y, Zhao X, Yin Y, Ban X, Jiang Z (2023) ACS Catal 13:6396–6402
22. Liu Y, Zhang L, Zhang Y, Cao S, Ban X, Yin Y, Zhao X, Jiang Z (2023) J Am Chem Soc 145: 18307–18315
23. Fu Q, Cao S, Wang J, Lv X, Wang H, Zhao X, Jiang Z (2024) J Am Chem Soc 146:8372–8380
24. Nacsa ED, Macmillan DW (2018) J Am Chem Soc 140:3322–3330
25. Chai X, Hu X, Zhao X, Yin Y, Cao S, Jiang Z (2022) Angew Chem Int Ed 61:e202115110

26. Kong M, Wang Z, Ban X, Zhao X, Yin Y, Zhang J, Jiang Z (2024) Adv Sci 11:2307773
27. Sun X, Liu Y, Yin Y, Ban X, Zhao X, Jiang Z (2024) Nat Chem 16:1169–1176
28. Cao K, Tan SM, Lee R, Yang S, Jia H, Zhao X, Qiao B, Jiang Z (2019) J Am Chem Soc 141: 5437–5443
29. Cao K, Li C, Tian D, Zhao X, Yin Y, Jiang Z (2022) Org Lett 24:4788–4792
30. Zhang L, Ma J, Ban X, Zhao X, Yin Y, Jiang Z (2024) Sci China Chem 67:2016–2021
31. Yin Y, Li Y, Gonçalves TP, Zhan Q, Wang G, Zhao X, Qiao B, Huang K-W, Jiang Z (2020) J Am Chem Soc 142:19451–19456
32. Guo Z, Chen X, Fang H, Zhao X, Jiang Z (2021) Sci China Chem 64:1522–1529
33. Kong M, Tan Y, Zhao X, Qiao B, Tan C-H, Cao S, Jiang Z (2021) J Am Chem Soc 143:4024–4031
34. Tian D, Sun X, Cao S, Wang E-M, Yin Y, Zhao X, Jiang Z (2022) Chin J Catal 43:2732–2742

Top Heterocycl Chem (2025) 62: 67–100
https://doi.org/10.1007/7081_2025_88

Published online: 17 April 2025

Photoinduced C–H Functionalization of Pyridines with *N*-Functionalized Pyridinium Salts

Hyewon Ju and Sungwoo Hong

Contents

Abstract The radical-mediated C–H functionalization of pyridines, catalyzed by visible light, has emerged as a powerful and versatile strategy in organic chemistry. This approach enables direct and selective modification of the pyridine framework, significantly expanding the toolkit available for organic synthesis. In recent years, *N*-functionalized pyridinium salts have gained considerable attention in this field, serving dual roles as both radical precursors and pyridine surrogates. *N*-functionalized pyridinium salts offer several key advantages over traditional pyridine reactants. They enhance reactivity and selectivity in synthetically valuable transformations; perform excellently under mild, acid-free conditions; and provide superior regiocontrol for nontraditional Minisci-type reactions. These properties

H. Ju and S. Hong (✉)
Department of Chemistry, Korea Advanced Institute of Science and Technology (KAIST) and Center for Catalytic Hydrocarbon Functionalizations, Institute for Basic Science (IBS), Daejeon, South Korea
e-mail: hongorg@kaist.ac.kr

have made them increasingly valuable in modern organic synthesis, particularly in the context of late-stage functionalization and complex molecule synthesis. Recent advancements in this area have been substantial and diverse. Researchers have developed a wide array of synthetic applications using pyridinium salts under visible-light conditions, ranging from C–H functionalization to heterocycle formation. Notably, *N*-substituted pyridinium salts have been ingeniously utilized as bifunctional reagents for alkene difunctionalization, opening new avenues for molecular complexity generation. Innovative approaches involving light-absorbing electron donor-acceptor (EDA) complexes between pyridinium salts and electron-rich donors have been introduced, enabling new reactivity patterns in photocatalyst-free conditions. This strategy not only simplifies reaction setups but also expands the scope of accessible transformations. Furthermore, the establishment of enantioselective reactions using pyridinium salts represents a significant breakthrough, further broadening their synthetic utility and addressing long-standing challenges in asymmetric synthesis. These developments collectively demonstrate the versatility and potential of *N*-functionalized pyridinium salts in advancing the frontiers of organic synthesis. This comprehensive chapter provides an in-depth overview of these developments, meticulously organized based on the *N*-substituent of pyridinium salts and their distinct reactivity patterns. It delves into recent advancements in *N*-functionalized pyridinium salt chemistry, discussing a wide range of organic reactions utilizing these compounds under visible-light conditions. The authors explore crucial structure-reactivity relationships based on *N*-substituents as well as innovative activation modes and their mechanistic implications, providing insights that are vital for reaction design and optimization.

Keywords C–H functionalization · Photochemistry · Pyridinium salts · Radical · Selectivity

1 Introduction

The pyridine moiety stands as a cornerstone in the realm of organic chemistry, permeating the structures of numerous natural products, pharmaceuticals, and industrial chemicals [1–3]. Its ubiquity and versatility have long captivated the attention of synthetic chemists, spurring the development of a myriad of methodologies aimed at modifying and functionalizing this essential heterocycle. Among these approaches, radical-based strategies have emerged as particularly powerful tools, offering unique pathways to manipulate the pyridine core – a process of paramount importance in medicinal and materials chemistry [4–10].

Traditionally, the functionalization of pyridines has relied heavily on Minisci-type reactions, often necessitating harsh acidic conditions [11]. However, a persistent challenge in this field has been achieving site-specific functionalization, as the inherent reactivity of pyridine often leads to competition between multiple reactive sites, particularly at the C2 and C4 positions [12]. This limitation has driven an intense research focus towards developing more selective methods for pyridine functionalization, with the ultimate goal of facilitating the synthesis of valuable pyridine derivatives.

In recent years, a paradigm shift has occurred with the emergence of pyridinium salts as versatile alternatives to pyridines. These compounds exhibit enhanced reactivity and undergo a wide array of transformations under the influence of transitio*n*-metal or photoredox catalysis. Notably, pyridinium salts have proven to be effective precursors for generating useful radicals through single-electron transfer (SET) processes, particularly when activated by visible light. This discovery has opened new avenues for site-selective functionalization strategies, leveraging the unique electronic properties of *N*-substituted pyridinium salts.

The advent of visible light-induced transformations has been particularly transformative, enabling milder reaction conditions for pyridine functionalization – a domain previously dominated by transition-metal catalysis. This shift towards photochemical methods has not only expanded the toolkit available to synthetic chemists but also aligned with the growing emphasis on sustainable and environmentally friendly synthetic practices. Furthermore, the design of *N*-substituted pyridinium salts has evolved to include bifunctional reagents capable of facilitating alkene difunctionalization, showcasing the remarkable versatility of these compounds [13–16]. This development highlights the potential for pyridinium salts to serve as key building blocks in the assembly of complex molecular architectures.

As we delve into the recent advances in this field, our focus will be on the photochemical, radical-mediated synthetic applications of pyridinium salts, with particular emphasis on visible light-induced functionalizations. We will explore the critical role of *N*-substituents in influencing reactivity and regioselectivity – a crucial consideration in designing efficient and selective functionalization strategies. By organizing our discussion around the various activation modes of *N*-pyridinium salts and the types of synthetic strategies employed, we aim to provide a comprehensive overview that will not only elucidate the current state of the art but also inspire future innovations in this dynamic area of organic synthesis. This introduction sets the stage for an exploration of a field that is rapidly reshaping our approach to pyridine chemistry, promising to yield new tools and methodologies that will significantly impact drug discovery, materials science, and beyond (Fig. 1).

2 Reactions with N–O Pyridinium Derivatives

2.1 *Alkylation of* N-*Methoxypyridinium Salts*

The field of Minisci-type reactions using *N*-methoxypyridinium salts was pioneered by Mitchell et al. in 1989 [17]. They developed a hydroxymethylation method using *N*-methoxypyridinium salts (**1a**) with substoichiometric ammonium persulfate as an oxidizing initiator. While this method successfully demonstrated monoalkylation of

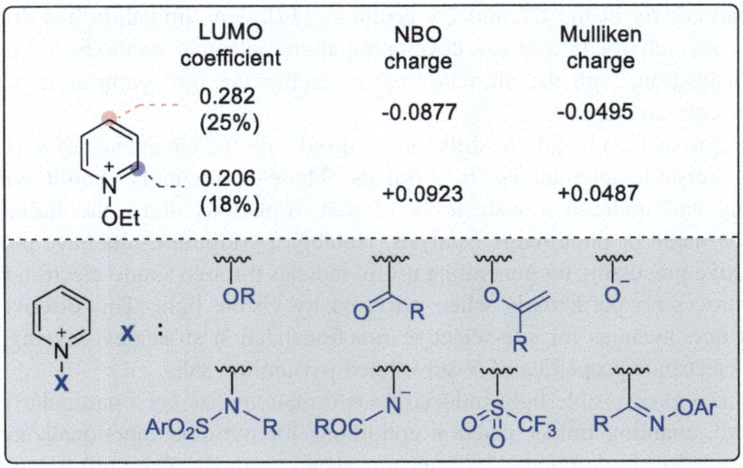

Fig. 1 Examples of *N*-activated pyridinium salts

pyridines, it was constrained by solvent limitations and moderate C2 vs. C4 selectivity. In 2016, Herzon's group made significant advancements in this area by reporting the hydroxypyridylation of unactivated alkenes (**1c**) [18]. Their method utilized *N*-methoxypyridinium salts (**1a**) under stoichiometric cobalt-mediated hydrogen atom transfer (HAT) conditions to produce alkylpyridine products (**1d**). The proposed mechanism involves a cobalt(III) hydride complex transferring to a highly substituted alkene, generating an alkyl radical.

Interestingly, Herzon's group observed distinct selectivity patterns in their study. Harder secondary radicals favored C2 addition with modest selectivity, while softer tertiary radicals preferred C4 positions. Assuming irreversible radical addition, they hypothesized that electrostatic interactions guide secondary radicals to C2 positions, whereas frontier orbital interactions explain C4 addition by tertiary radicals with higher singly occupied molecular orbital (SOMO) energies. This work significantly expanded the scope and understanding of pyridinium salt reactivity in organic synthesis. It represents an important advancement in heterocyclic chemistry, providing new methods for functionalizing pyridine rings and deepening our comprehension of radical addition reactions. The research improved upon previous limitations in solvent choice and selectivity while also elucidating the mechanisms behind the observed reactivity patterns. Minisci-type reactions, which involve the addition of alkyl radicals to heterocyclic bases, have been further refined through this research. The use of *N*-methoxypyridinium salts, activated forms of pyridine, in conjunction with hydrogen atom transfer (HAT) processes, has opened up new possibilities in synthetic organic chemistry. The insights gained into the role of SOMO energies and electrostatic interactions in determining reaction outcomes provide valuable tools for predicting and controlling the selectivity of these reactions (Scheme 1).

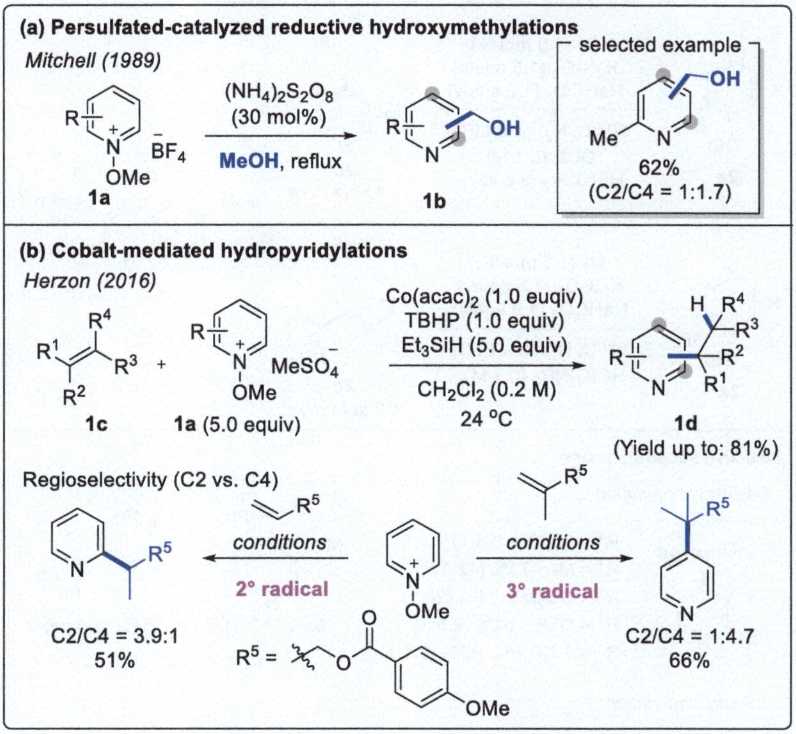

Scheme 1 Radical-mediated alkylation of *N*-methoxypyridinium salts

2.2 *Photochemical Reactions of N–O Pyridinium Derivatives*

Visible-light photocatalysis has emerged as a powerful synthetic tool, enabling novel transformations under mild conditions. This approach leverages photocatalysts' ability to convert visible light into electrochemical potential, facilitating the synthesis of valuable compounds. Our research group developed a visible light-induced, site-selective C–H functionalization of *N*-ethoxypyridinium salts **2a** (Scheme 2) [12], using $\mathbf{Q_1}$ as an organic photocatalyst. This method achieved phosphonylation and carbamoylation under mild, transition metal-free conditions and proved applicable to other heteroarenium substrates like quinolinium and pyridazinium salts. The proposed mechanism involves excited photocatalyst $\mathbf{Q_1}$* undergoing single-electron transfer (SET) with **2a**, generating ethoxy radical.

High quantum yields for phosphonylation ($\Phi = 9.9$) and carbamoylation ($\Phi = 8.4$) support a radical chain pathway. Notably, phosphinoyl radicals selectively produced C4 products **2b**, likely due to steric interactions from larger phosphorus atoms. Conversely, carbamoyl radicals yielded C2 products **2c**, possibly favored by electrostatic interactions between the pyridinium substrate's nitrogen center and the carbamoyl radical's oxo moiety. This work demonstrates the potential of visible-

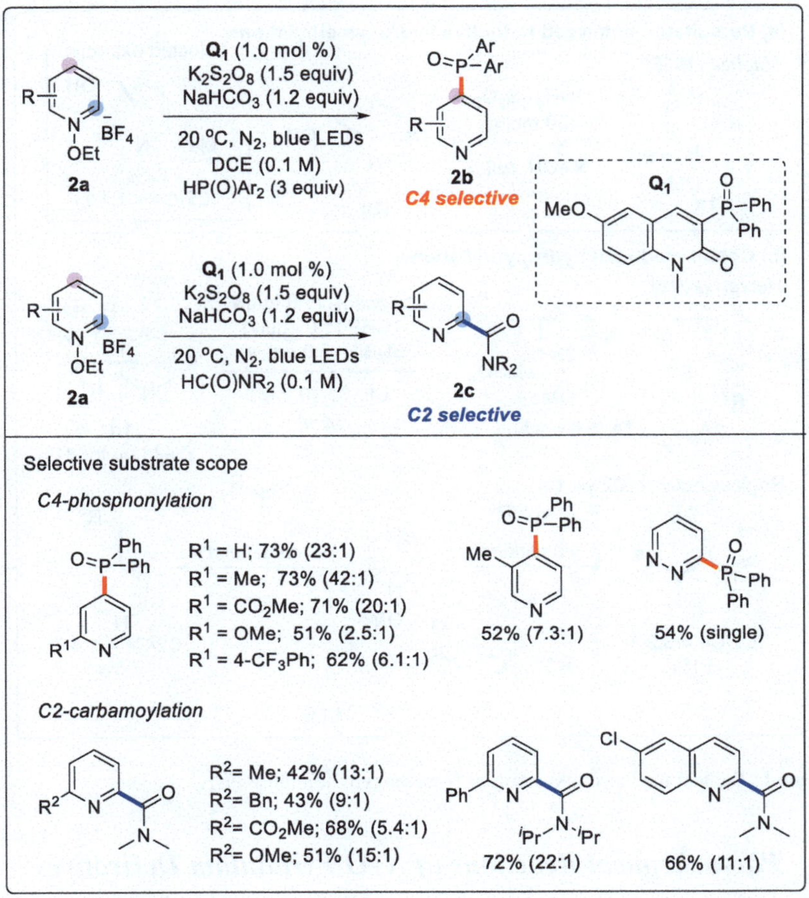

Scheme 2 Visible light-induced C4-selective phosphonylation and C2-selective carbamoylation of *N*-ethoxypyridinium salts

light photocatalysis in achieving selective functionalization of heterocyclic compounds under mild conditions.

Our group developed a versatile strategy using *N*-alkoxypyridinium salts **3a** for visible light-mediated heteroarylation of remote C(sp^3)–H bonds and tandem radical processes, employing $\mathbf{Q_1}$ as a photocatalyst (Scheme 3) [19, 20]. These reactions follow an oxidative quenching photoredox cycle, with the following key steps: Photoexcited $\mathbf{Q_1}$* undergoes single-electron transfer (SET) with **3a**, generating alkoxy radical. The generated alkoxy radical initiates a 1,5-hydrogen atom transfer (HAT) process, forming carbon-centered radical, which adds to the pyridine core, followed by deprotonation to form product **3c**.

We extended this approach to a tandem radical process involving N–O bond cleavage, intramolecular alkoxy radical cyclization, and pyridylation, yielding

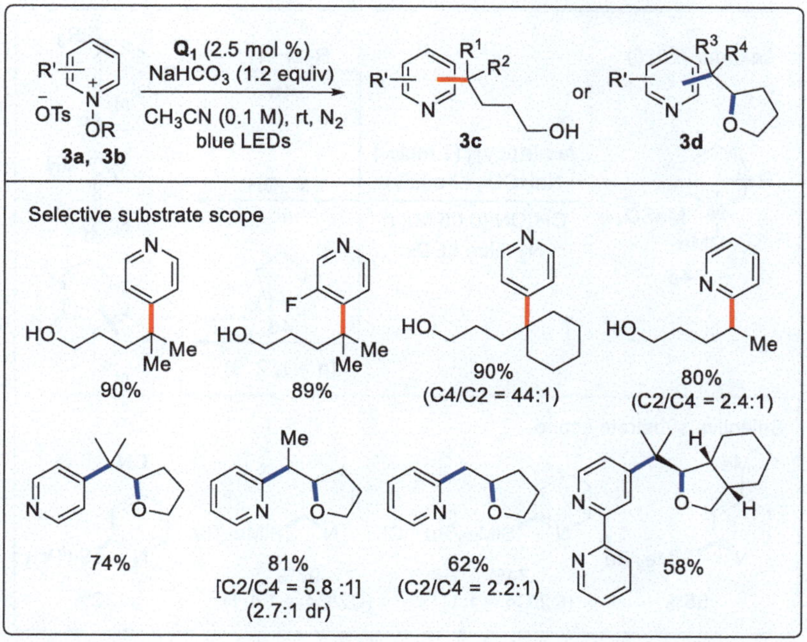

Scheme 3 Remote C(sp^3)–H pyridylation by radical translocation or radical cascade

pyridine-tethered tetrahydrofurans **3d**. This process begins with photoexcitation of **Q**$_1$, followed by N–O bond cleavage in *N*-alkenyloxypyridinium salt **3b** via SET reduction, producing alkoxy radical. The alkoxy radical is then trapped by the olefin, forming a tetrahydrofuran ring through 5-exo-trig cyclization and generating alkyl radical. Finally, the alkyl radical undergoes addition to form product **3d**. Notably, pyridinium salts with diene chains underwent smooth radical cascade reactions, resulting in bicyclic systems. These methodologies demonstrate the versatility of *N*-alkoxypyridinium salts in visible light-mediated transformations, enabling both remote C–H functionalization and complex ring-forming reactions.

Lakhdar and colleagues developed a photocatalytic method for silylation and hydroxymethylation of pyridinium salts using *fac*-Ir(ppy)$_3$ as a photocatalyst (Scheme 4) [21]. Key features of their approach include the use of *N*-methoxypyridinium salt **4a** as a dual-purpose reagent, serving both as a heteroaryl source and a precursor for the methoxy radical, an effective hydrogen atom abstractor. Blue-light irradiation was used to obtain silylated (**4e**) and hydroxymethylated (**4f**) products. The method demonstrated broad applicability across various salts and tolerated electronic variations in heteroarene sources. The proposed mechanism involves photocatalytic N–O bond homolysis of **4a** via SET from Ir*, generating a methoxy radical and pyridine. Methoxy radicals then participate in HAT reactions with silane **4b** or methanol **4c**, producing silyl and hydroxymethyl radicals. The nucleophilic silyl radical preferentially attacks the C2 position of the pyridine ring. Similarly, the less nucleophilic hydroxymethyl radical also favors addition at the C2

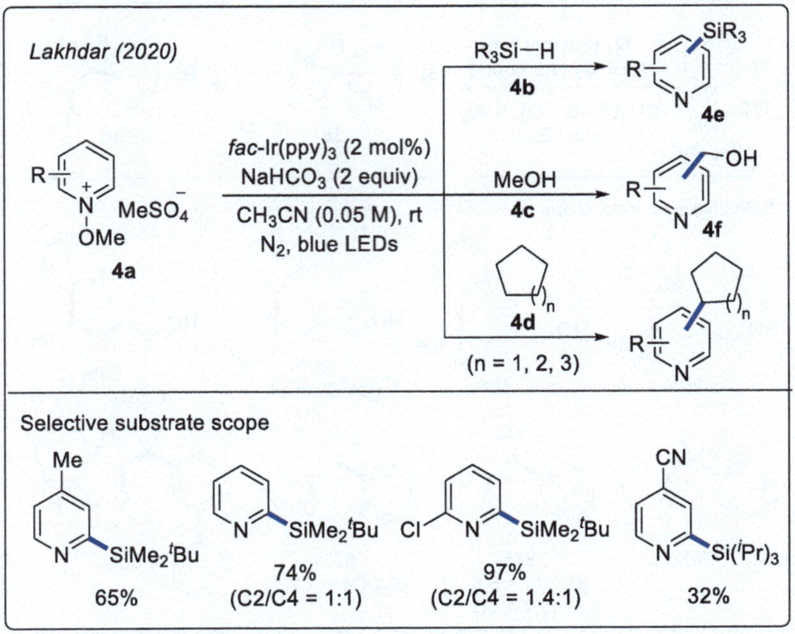

Scheme 4 Visible light-induced silylation, hydroxymethylation, and alkylation

position. This sequence of events explains the observed regioselectivity and product formation in the reaction. The modest regioselectivity observed is attributed to a slight energy difference (0.2–0.6 kcal/mol) between the transition states of these radical additions. Lakhdar's group extended this approach to visible light-induced C–H alkylation of **4a** with cycloalkanes **4d** using an Ir photocatalyst [22]. They successfully demonstrated monoalkylation of various pyridines and quinolines. While this HAT strategy effectively generated alkyl radicals through direct activation of aliphatic C–H bonds, the challenge of modest regioselectivity remained.

Niu and colleagues recently advanced the field of visible light-enabled C–H alkylation reactions, focusing on *N*-methoxypyridinium salt **5b** and utilizing alkyl sulfoxide **5c** as a radical precursor (Scheme 5) [23]. A key application of this method was the synthesis of various pyridyl C-glycosides from glycosyl sulfoxides. The mechanism proceeds through photoexcitation of an electron donor-acceptor (EDA) complex formed among the three reactants under visible-light irradiation. This is followed by homolytic cleavage of the N–O bond, generating a methoxy radical. The methoxy radical then adds to the sulfoxide substrate, producing an alkyl radical. This newly formed alkyl radical attacks the pyridinium salt, forming radical cation. Finally, deprotonation and rearomatization of the intermediate yield the final product. This sequence of events explains the observed reactivity and product formation in the absence of an external photocatalyst. Despite the innovative approach, the reaction exhibits only modest regioselectivity. Nevertheless, this method represents a significant advancement in visible light-mediated C–H functionalization,

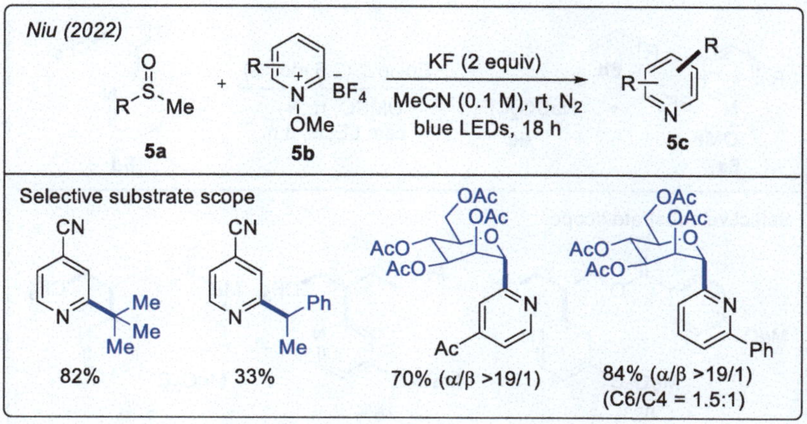

Scheme 5 Alkylation of *N*-methoxypyridinium salts using alkyl sulfoxides

particularly in the context of synthesizing complex heterocyclic compounds such as pyridyl C-glycosides. This work demonstrates the potential of EDA complexes in organic synthesis, offering a more sustainable and efficient alternative to traditional methods for pyridinium salt activation and functionalization.

2.3 *Difunctionalization of Alkenes Using N–O Pyridinium Salts*

A strategy that simultaneously introduces distinct functional groups on each side of a double bond is highly desirable, as it rapidly increases molecular complexity [24]. Particularly appealing are site-selective multicomponent couplings based on nonclassical Minisci-type reactivity using *N*-alkoxypyridinium salts, which can be achieved by controlling the characteristics of the generated radical. This section explores recent advancements in this approach. In 2018, our group introduced a protocol enabling visible light-mediated trifluoromethylative heteroarylation of unactivated alkenes (Scheme 6). This method employs *N*-methoxypyridinium salts **6a** as heteroarene sources. The reaction is initiated by the generation of CF_3 radicals from $NaSO_2CF_3$ (**6c**) using eosin Y as a photocatalyst. The reaction mechanism proceeds as follows: The CF_3 radical is trapped by alkene **6b**, generating alkyl radical. This alkyl radical then reacts with **6a** to form radical cation. The radical cation is subsequently reduced by the radical anion form of eosin Y. Finally, elimination of methanol yields β-trifluoromethyl alkylpyridine derivatives **6d**. This sequence of events elucidates the formation of the desired product through a series of radical-mediated transformations and electron transfer processes. Alternatively, radical cation may undergo deprotonation followed by N–O bond homolytic

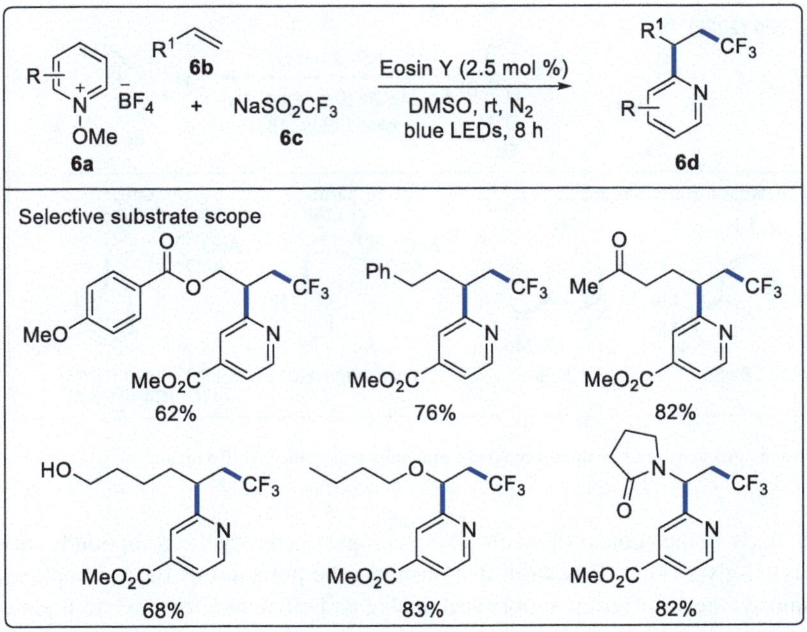

Scheme 6 Multicomponent trifluoromethylative heteroarylation of alkenes

cleavage to directly form the final product **6d**. This approach demonstrates the potential of visible light-mediated multicomponent couplings.

The regioselectivity of alkyl radical additions to *N*-alkoxypyridinium salts is heavily influenced by the radical substituents' order, with secondary radicals typically showing only modest selectivity between C2 and C4 positions. To address this regioselectivity challenge, we developed a remote pyridyl *ortho*-migration strategy for trifluoromethylative pyridylation of olefins in 2020 (Scheme 7) [25]. This approach successfully engaged both secondary and tertiary alkyl radicals in C2-selective additions, yielding C2-fluoroalkyl heteroarenes. We also extended the method to reactions involving P-centered radicals. The reaction mechanism proceeds as follows: Single-electron transfer (SET) occurs from the Langlois reagent to the excited eosin Y*, generating a CF_3 radical. This radical adds to the olefin of substrate **7a**, forming nucleophilic radical. This radical then undergoes intramolecular addition to the C2 position of the pyridine. Notably, substrates with varying tether lengths ($n = 2$, 3, and 4) successfully yielded the corresponding products **7c** through six-, seven-, and eight-endo cyclizations, respectively. This strategy represents a significant advancement in controlling regioselectivity in radical additions to pyridinium salts, offering a powerful tool for the synthesis of complex fluoroalkyl heteroarenes.

Our group demonstrated the versatility of well-designed *N*-alkenoxypyridinium salts as bifunctional reagents in a photocatalytic carbopyridylation of alkenes (Scheme 8) [26]. This method leverages the photocatalytic generation of relatively

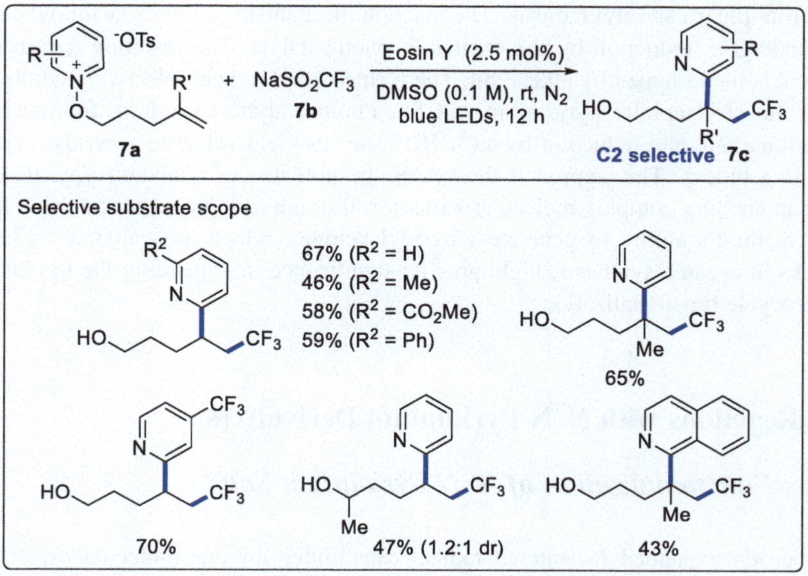

Scheme 7 Trifluoromethyl pyridylation of alkenes by remote ortho-migration

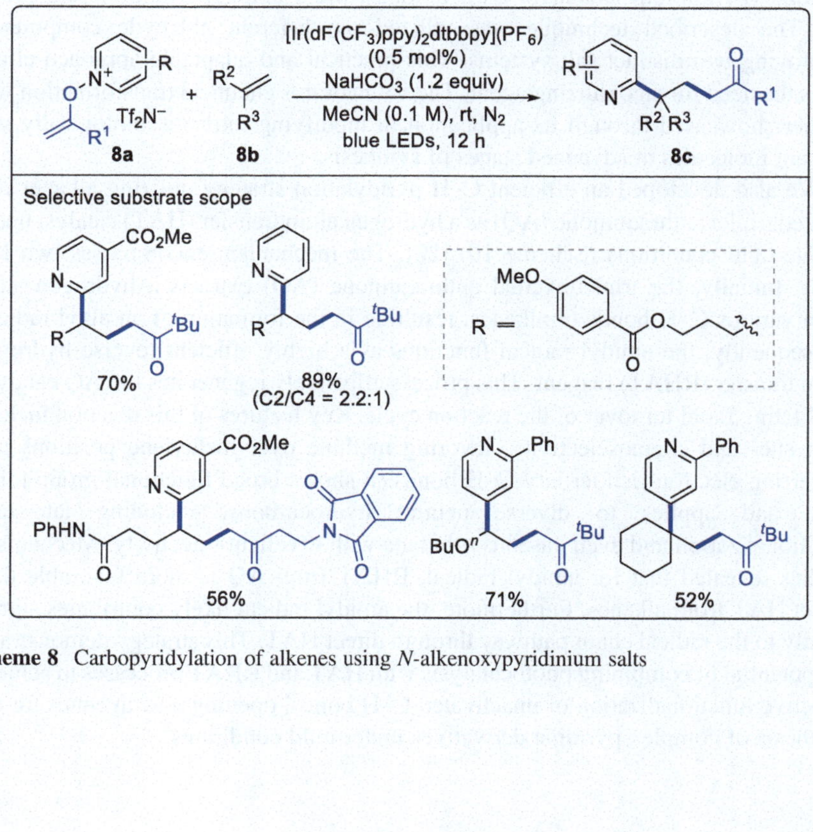

Scheme 8 Carbopyridylation of alkenes using *N*-alkenoxypyridinium salts

electrophilic α-carbonyl radicals. The reaction mechanism proceeds as follows: Salt **8a** undergoes reduction by the excited Ir photocatalyst. The resulting α-carbonyl radical is then trapped by alkene **8b**. The formed alkyl radical selectively combines with the electrophilic pyridinium salt **8a**. Photocatalytic oxidation, followed by deprotonation, and reduction by an Ir(III)* species yield valuable γ-pyridyl ketone building blocks. This approach showcases the potential of *N*-alkenoxypyridinium salts in creating complex molecular structures through a single, photocatalytic step. The method's ability to generate γ-pyridyl ketones, which are valuable building blocks in organic synthesis, highlights its significance in expanding the toolkit for heterocycle functionalization.

3 Reactions with N–N Pyridinium Derivatives

3.1 Functionalization of N–N Pyridinium Salts

Our group expanded *N*-centered radical capabilities for intermolecular hydrogen extraction (Scheme 9) [27]. Alkoxy or amidyl radicals from the *N*-alkoxy- or *N*-amino-pyridinium salts extract hydrogen from aldehydes, forming acyl radical, which is then captured by the pyridinium salt. Regioselectivity differed: the *N*-aminoypyridinium salts favor C4 addition, while *N*-alkoxypyridinium salts prefer C2. The described technique was utilized on different aldehyde compounds, employing two distinct salt systems. This practical and adaptable approach eliminates the need for an oxidizing agent. The value of this chemical transformation was further showcased through its application in modifying intricate, biologically significant molecules at advanced stages of synthesis.

We also developed an efficient C–H pyridylation strategy utilizing alkanes and excited triplet anthraquinone (AQ) as a hydrogen atom transfer (HAT) catalyst under visible-light conditions (Scheme 10) [28]. The mechanism encompasses two key steps. Initially, the triplet-excited anthraquinone (AQ) extracts a hydrogen atom from various C–H bonds of alkanes, resulting in the formation of an alkyl radical. Subsequently, the amidyl radical functions as a highly efficient reverse hydrogen atom transfer (RHAT) reagent. This process effectively regenerates the AQ catalyst, facilitating rapid turnover of the reaction cycle. Key features of this reaction include high site- and chemoselectivity, favoring methine over methylene positions and preferring electron-rich tertiary C–H bonds. It shows broad functional group tolerance and applies to diverse neutral hydrocarbons, including late-stage functionalization and even gaseous *n*-butane with excellent selectivity. Mechanistic studies revealed that for amidyl radical, RHAT from AQ is more favorable than direct HAT from alkanes. Furthermore, the amidyl radical likely contributes significantly to the radical chain pathway through direct HAT. This strategy demonstrates the potential of combining photocatalysis with HAT and RHAT processes to achieve selective functionalization of unactivated C–H bonds, opening new avenues for the synthesis of complex pyridine derivatives under mild conditions.

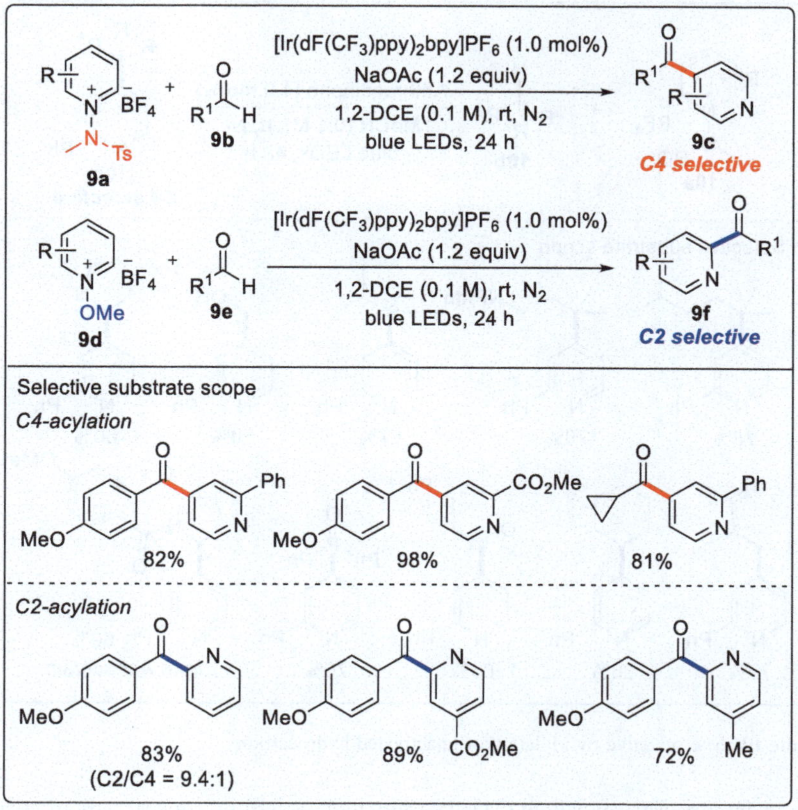

Scheme 9 Site-selective C–H acylation of pyridines

A unified strategy for deoxygenative or desulfurative pyridylation of various alcohols and thiols was developed using a single-electron transfer (SET) process involving frustrated Lewis pairs (FLPs) derived from pyridinium salts and $PtBu_3$ (Scheme 11). Mechanistic studies unveiled a detailed reaction pathway. The process begins with the formation of a frustrated Lewis pair (FLP), where *N*-amidopyridinium salts function as effective Lewis acids, pairing with $PtBu_3$ acting as a Lewis base. This FLP interaction triggers a single-electron transfer, leading to the generation of phosphine radical cations. These radical cations then undergo ionic coupling with xanthates, which are formed in situ from alcohol or thiol substrates. The resulting intermediate subsequently experiences β-scission, producing alkyl radicals without requiring photocatalysts. These alkyl radicals then add specifically to the C4 position of the pyridinium salt, creating a new carbon-carbon bond. Finally, the released amidyl radical acts as a single-electron transfer (SET) oxidant, facilitating a radical chain pathway.

This innovative method overcomes traditional challenges in cleaving strong C–OH and C–SH bonds, offering selective heteroarylation under mild conditions. The

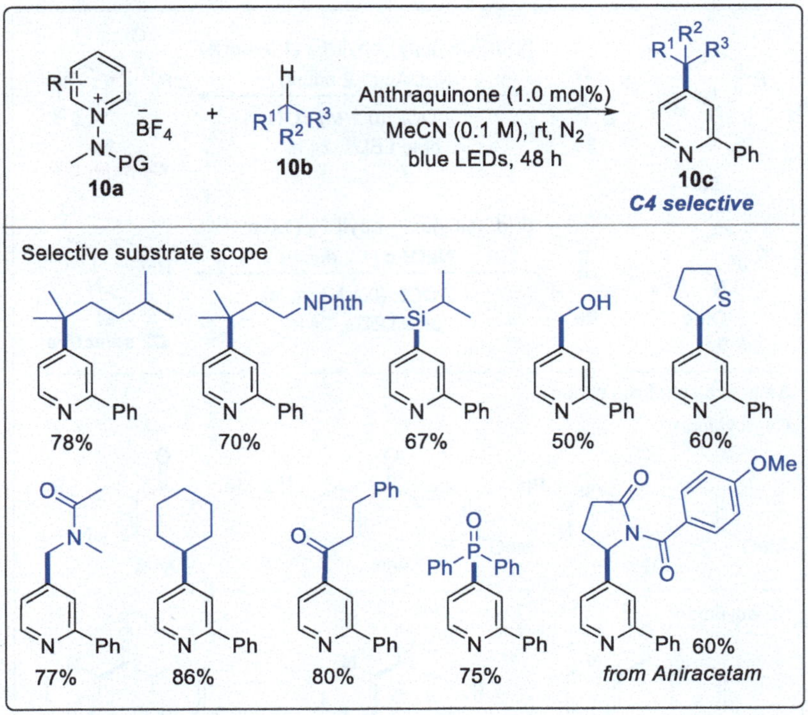

Scheme 10 Site-selective pyridylation of unactivated hydrocarbons

method demonstrates excellent C4 selectivity and versatility, successfully modifying pharmaceutically relevant molecules. It provides facile access to high-value pyridine-containing compounds from readily available alcohol and thiol feedstocks.

An efficient photocatalytic method for C4-selective C–H alkylation of pyridines was developed using alkyl carboxylic acids under visible-light irradiation (Scheme 12) [29]. This approach operates under external oxidant-free and metal-free conditions. Mechanistically, the process generates alkyl radicals from aliphatic carboxylic acids via SET oxidation, followed by CO_2 extrusion. The released *N*-centered radical serves as an effective SET oxidant, efficiently regenerating the photocatalyst. The system demonstrates broad substrate compatibility, accommodating carboxylic acids and pyridines with oxidatively sensitive functionalities. It enables C4-selective pyridylation of diverse primary, secondary, and tertiary carboxylic acids and amino acids. From a sustainability perspective, this method allows direct conversion of renewable feedstocks into valuable pharmacophores and functional materials. The process begins with an initiation step where, under basic conditions, the carboxylate undergoes single-electron transfer (SET) with the excited state of Mes-Acr+ upon absorption of visible light at 440 nm. This leads to radical formation, where the resulting carboxylic radical undergoes decarboxylation to form an alkyl radical. The alkyl radical then adds to the C4 position of the *N*-amidopyridinium salt. This process results in the formation of the desired product and an *N*-centered amidyl

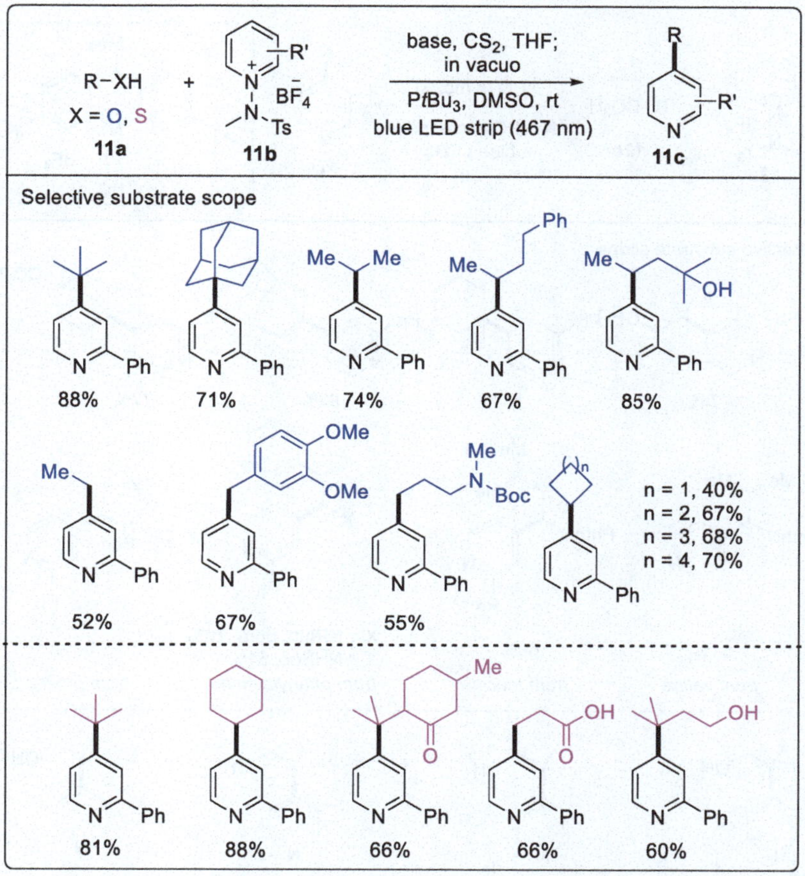

Scheme 11 C4-selective C–H alkylation of pyridines involving frustrated Lewis pairs (FLPs)

radical. Finally, the mechanism concludes with catalyst regeneration, where the in situ-generated amidyl radical regenerates the photocatalyst via direct SET.

An efficient method to synthesize β-pyridylated carbonyl compounds **13c** was developed using cyclopropanols **13b** as starting materials (Scheme 13) [30]. Key features of this method include its efficiency attributed to the BDE difference and the successful extension to cyclopropanone hemiketals and secondary cyclopropanols under mild reaction conditions, yielding β-pyridylated esters and aldehydes, respectively. The process begins with the formation of β-keto radicals through hydrogen atom transfer (HAT). This HAT is initiated by an amidyl radical, which is derived from N-aminopyridinium salts **13a**. The HAT process is facilitated by the difference in bond dissociation energies (BDEs) between the amidyl radical (102 kcal/mol) and cyclopropanols (O–H bond, approximately 90 kcal/mol). Following this, β-scission of an alkoxy radical occurs, which is thermodynamically driven by the release of ring strain. This approach demonstrates an innovative use of cyclopropanol

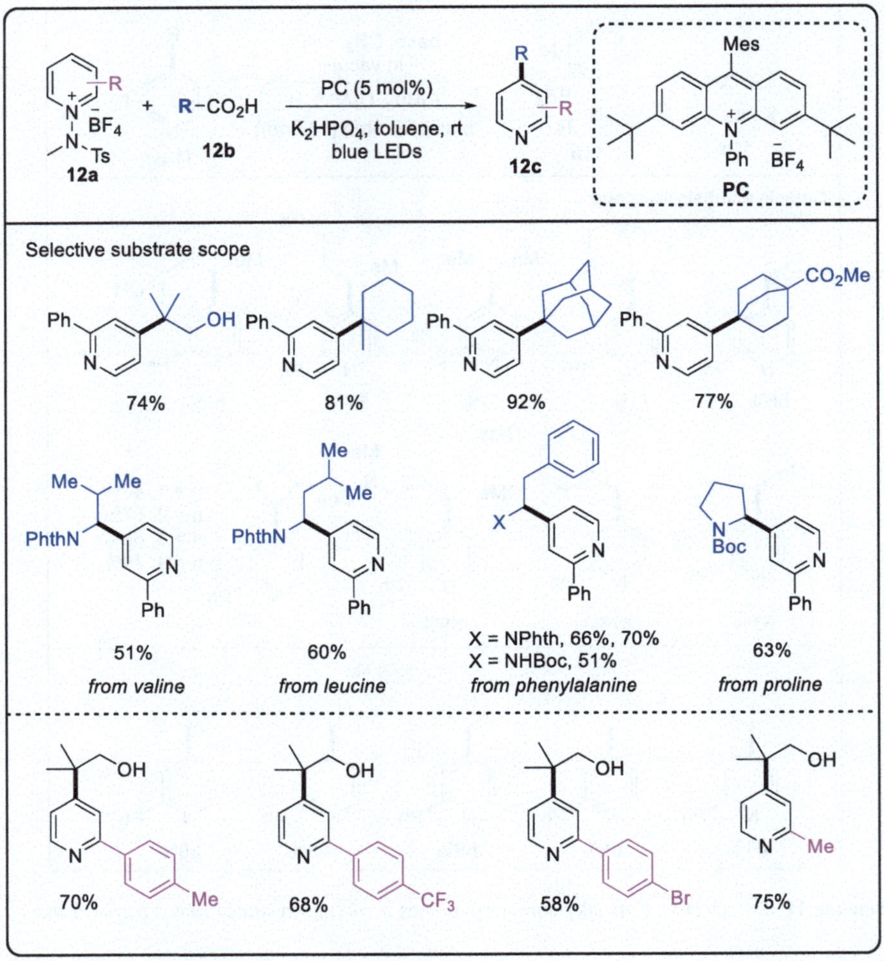

Scheme 12 Decarboxylative pyridylation of carboxylic acid

derivatives in radical chemistry, leveraging their unique structural properties to achieve selective functionalization. It showcases the potential of utilizing strain release as a driving force in organic synthesis, opening new avenues for the preparation of complex pyridine-containing compounds.

N-aminopyridinium derivatives generate nitrogen radicals via single-electron transfer (SET) and N–N bond homolysis, typically through photocatalyst oxidative quenching or electron donor-acceptor (EDA) complex formation. These processes often enable efficient chain pathways. The resulting nitrogen-centered radical [31] serves as a hydrogen-atom abstractor, single-electron oxidant, or aminating source. Based on these observations, our group developed a visible light-mediated intermolecular alkene aminopyridylation using *N*-amidopyridinium salts **14a** as

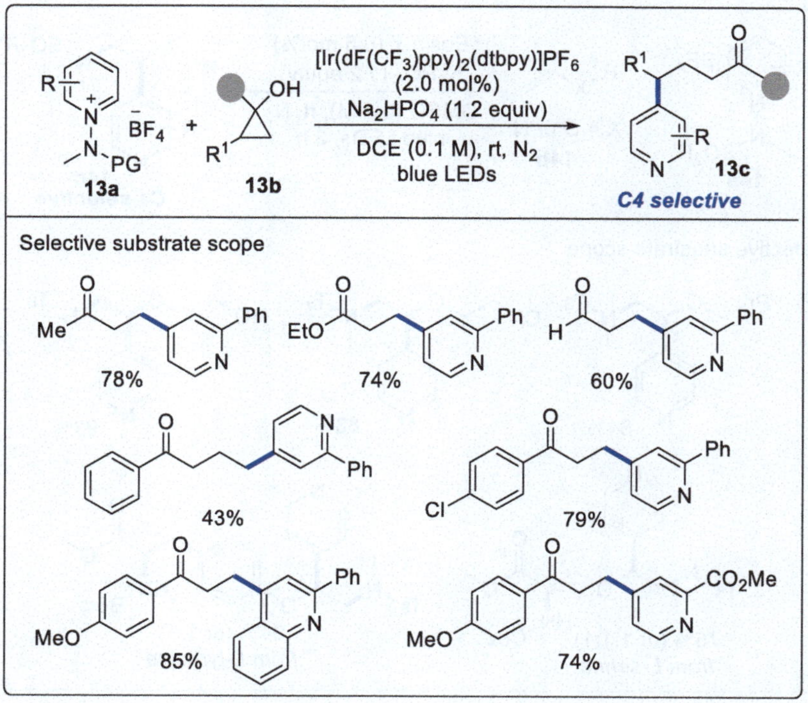

Scheme 13 β-Carbonyl alkylation using cyclopropanols

both aminating and pyridine sources (Scheme 14) [32]. Computational and frontier molecular orbital (FMO) analyses showed that amidyl radical preferentially adds to electron-rich olefins. The process initiates with the generation of an alkyl radical, which forms from the addition of an amidyl radical to the olefin. This alkyl radical then proceeds to add to the pyridinium substrate, resulting in the formation of a radical cation. Density functional theory (DFT) calculations revealed that the transition-state energy for alkyl radical addition leading to the C4 adduct is 1.3 kcal/mol lower than that for the C2 adduct. This energy difference explains the C4 site selectivity observed throughout the reaction scope. This approach demonstrates the potential of *N*-aminopyridinium derivatives in photocatalytic transformations, offering a powerful tool for the selective functionalization of heterocycles.

The use of *N*-centered radicals from *N*-aminopyridinium salts has become a powerful tool for C4 functionalization of pyridines. To address selective C2 functionalization in unblocked pyridine frameworks, we developed *N*-aminopyridinium ylide **15a** as an alternative approach. Our research introduced a photocatalytic method for *ortho*-selective aminopyridylation of olefins using ylides **15a** (Scheme 15a) [33]. The process begins with the direct oxidation of **15a** by a photocatalyst through single-electron transfer (SET). This leads to the formation of a nitrogen-centered radical, which then adds to the olefin. The resulting alkyl radical subsequently attaches to the C2 position of the pyridine core. This attachment

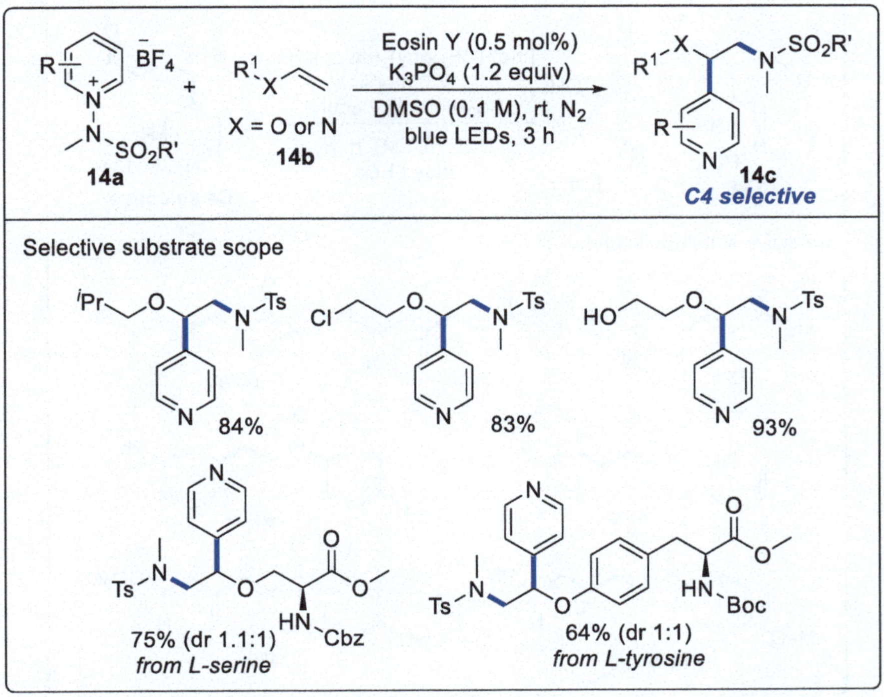

Scheme 14 Aminopyridylation using *N*-aminopyridinium salt as a bifunctional reagent

culminates in the formation of the cycloaddition adduct. Key features of this method include achievement of both *ortho*-selectivity and excellent diastereoselectivity, producing 1,2-*syn*-aminopyridylated compounds **15c**. The method demonstrates broad substrate scope, including various styrene derivatives, unactivated alkenes, challenging terminal aliphatic alkenes, internal alkenes, and substituted pyridinium ylides, along with broad functional group tolerance.

We extended this strategy to the aminopyridylation of ketones through an umpolung approach (Scheme 15b) [34]. The process begins with the formation of radical intermediate through oxidation of **15a** by a photocatalyst. This radical then adds to an enol silane generated in situ. The resulting α-silyloxy carbon radical undergoes intramolecular interception at the C2 position, forming a five-membered cycloadduct. Subsequently, the intermediate undergoes single-electron transfer (SET) reduction with the photocatalyst, followed by protonation, which generates a silyl ether. The mechanism concludes with the addition of tetrabutylammonium fluoride (TBAF), yielding the aminopyridylated product in a one-pot process. This method's scope was successfully expanded to include aryl and alkyl enamide derivatives, with good tolerance for both electron-withdrawing and -donating groups on the ylides. These approaches demonstrate the versatility of *N*-aminopyridinium ylides in achieving selective C2 functionalization of pyridines, opening new avenues for the synthesis of complex heterocyclic compounds.

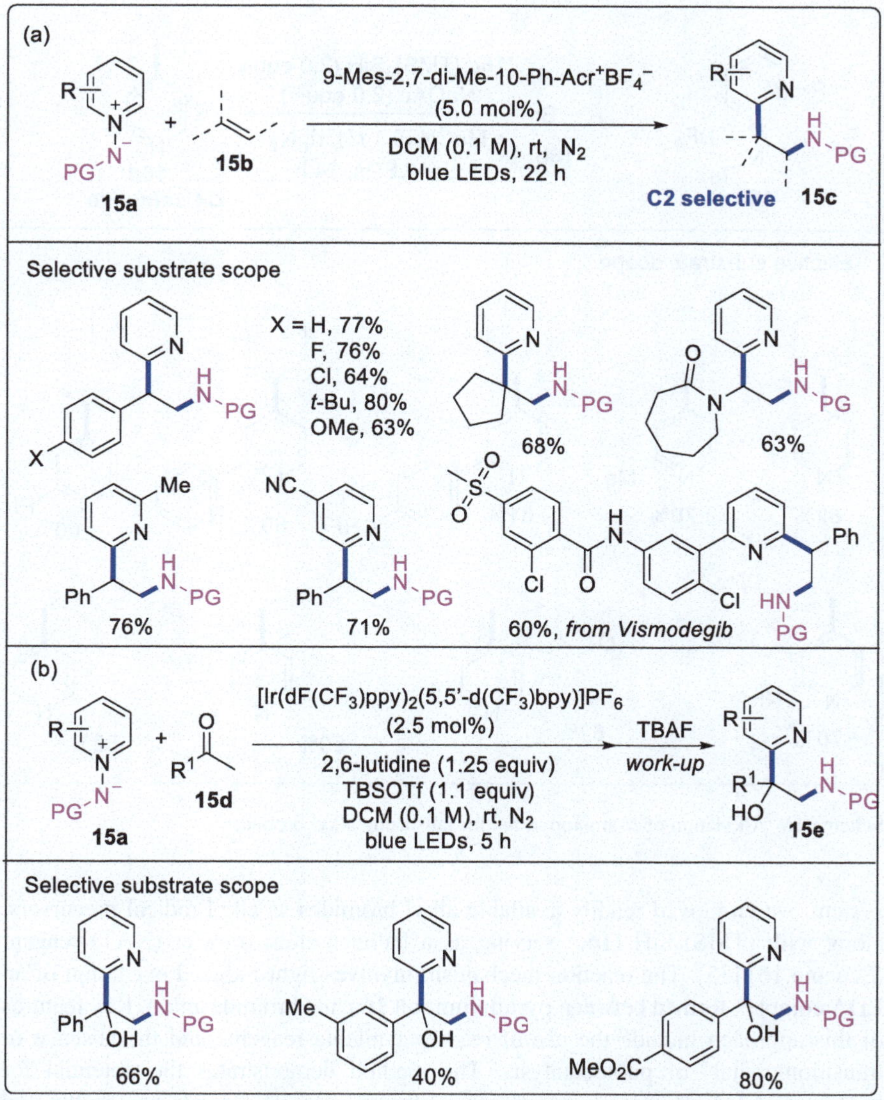

Scheme 15 *ortho*-Selective aminopyridylation of (**a**) alkenes and (**b**) ketones

3.2 *Reactions of N–N Pyridinium Salts Through EDA Complexes*

Our innovative method leverages the formation of a light-absorbing electron donor-acceptor (EDA) complex between the pyridinium salt and a suitable electron donor. This activation mode enables facile and environmentally friendly photochemical transformations without the need for transition metals or photocatalysts. In our

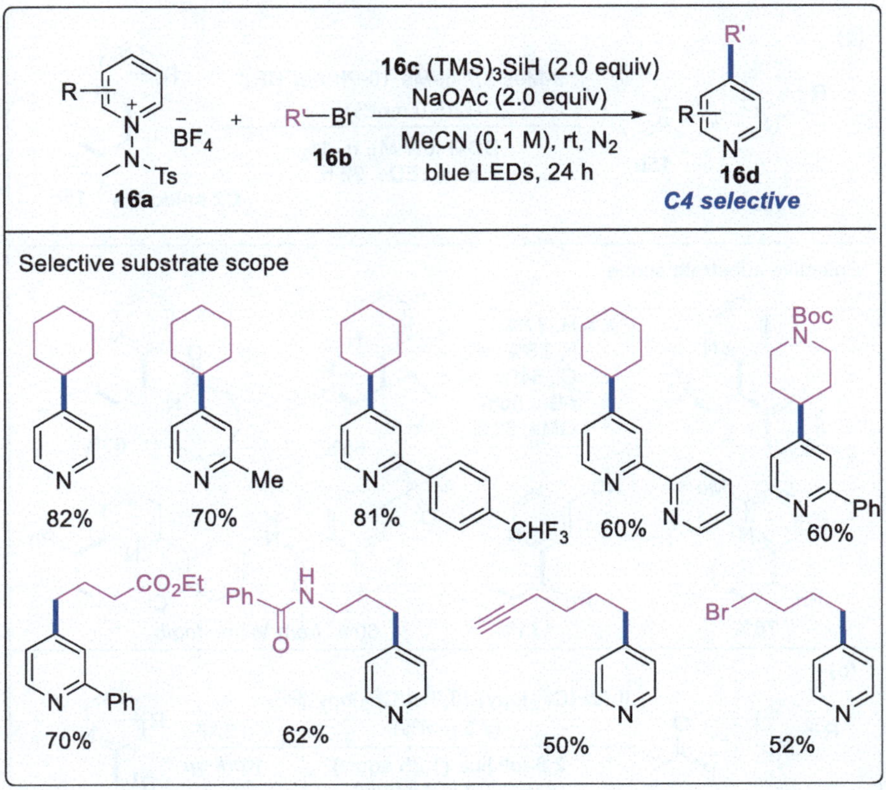

Scheme 16 Alkylation of *N*-aminopyridinium salts using alkyl bromides

system, we employed readily available alkyl bromides as alkyl radical precursors, along with $(TMS)_3SiH$ (**16c**) serving as a halogen-atom transfer (XAT) reagent (Scheme 16) [35]. The reaction mechanism involves light-induced excitation of an EDA complex formed between pyridinium salt **16a** and bromide anion. Key features of this approach include the use of readily available reagents and the absence of transition metals or photocatalysts. The method demonstrates the potential for harnessing EDA complexes in organic synthesis, offering a more sustainable and efficient alternative to traditional photoredox methods for pyridinium salt activation.

Our group further explored the capacity of *N*-aminopyridinium salts to form EDA complexes with 1,4-dihydropyridine (DHP) under light absorption (Scheme 17) [36]. This investigation led to an innovative strategy utilizing 1,4-DHP (**17b**) in a dual role: as a single-electron donor to activate the pyridinium salt and as a carbon-centered radical precursor, generating various alkyl, acyl, and carbamoyl radicals. The reaction mechanism involves light irradiation triggering SET within the EDA complex formed between the pyridinium salt and 1,4-DHP. Key features of this approach include a versatile synthetic platform for producing C4-alkylated pyridines with remarkable functional group tolerance, utilization of EDA complexes in

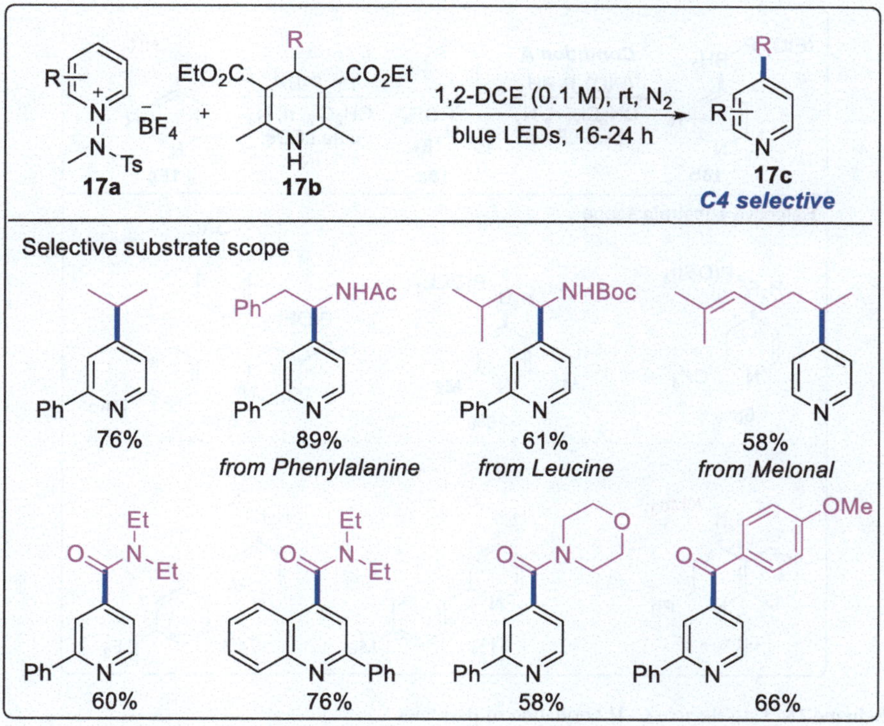

Scheme 17 Reaction of *N*-aminopyridinium salts with 1,4-DHPs

organic synthesis offering a more efficient and unified approach to C4-functionalized pyridines, and combination of the unique properties of *N*-aminopyridinium salts and 1,4-DHPs to achieve selective transformations under mild conditions. This method demonstrates the potential for harnessing EDA complexes in organic synthesis, showcasing how the synergistic properties of *N*-aminopyridinium salts and 1,4-DHPs can be leveraged to achieve selective transformations under mild conditions.

Borylated pyridines are crucial synthetic building blocks due to their ability to form new carbon–carbon and carbon–heteroatom bonds. Traditional methods for pyridine C–H borylation, such as metal-catalyzed processes, often face challenges related to regioselectivity and require expensive catalysts and stringent conditions. A photoinduced, C4-selective C–H borylation method for pyridines has been developed using EDA complexes derived from Lewis base and *N*-amidopyridinium salts, operating under conditions free of external oxidants and photocatalysts (Scheme 18) [37]. The nucleophilic nature of phosphite-ligated boryl radicals facilitates their addition to the C4 position of pyridinium salts, yielding C4-borylated heteroarenes, which are challenging to synthesize otherwise. This versatile approach enables the installation of both phosphite- and amine-coordinated boron groups on pyridines under mild conditions with excellent C4 positional selectivity. The optimized

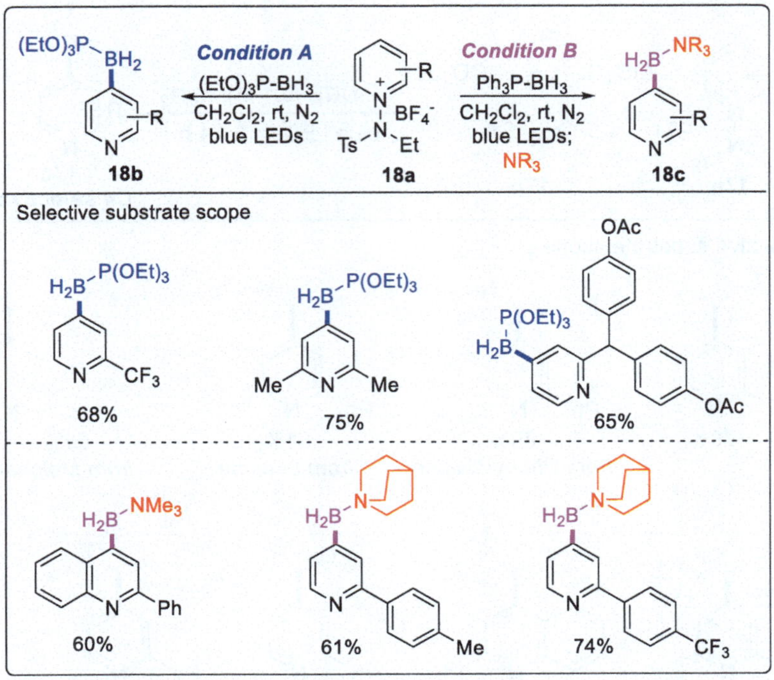

Scheme 18 C4-selective C–H borylation of pyridines

conditions involved blue-light irradiation and did not require an external base or photocatalyst. The reaction was efficient and regioselective, yielding stable borylated products. Mechanistically, the process involves the formation of EDA complexes between pyridinium salts and Lewis bases. Upon light irradiation, these complexes undergo SET to generate amidyl and boryl radicals, which then selectively add to the C4 position of pyridinium salts. The reaction proceeds through a radical chain pathway, ensuring high efficiency and selectivity. This method provides a straightforward approach to synthesizing C4-borylated pyridines with broad functional group tolerance under mild conditions.

Cyclic amines' deconstructive diversification involves breaking and forming new bonds to create acyclic structures. Recent advancements include using difluorocarbene reagents for C–N bond cleavage. However, these methods typically install only halogens or narrow nucleophiles. A one-pot umpolung method for ring-opening pyridylation of unstrained cyclic amines using *N*-amidopyridinium salts has been developed (Scheme 19) [38]. This process forms EDA complexes between bromide and *N*-amidopyridinium salts, functionalizing pyridines under visible-light conditions without an external photocatalyst. This new approach provides a valuable tool for synthesizing C4-alkylated pyridines from simple cyclic amines under mild conditions. It works with various 5- and 6-membered cyclic amines and pyridines, creating C4-functionalized pyridines. The method is notable for its mild conditions and broad compatibility with different cyclic amines, which are common in natural

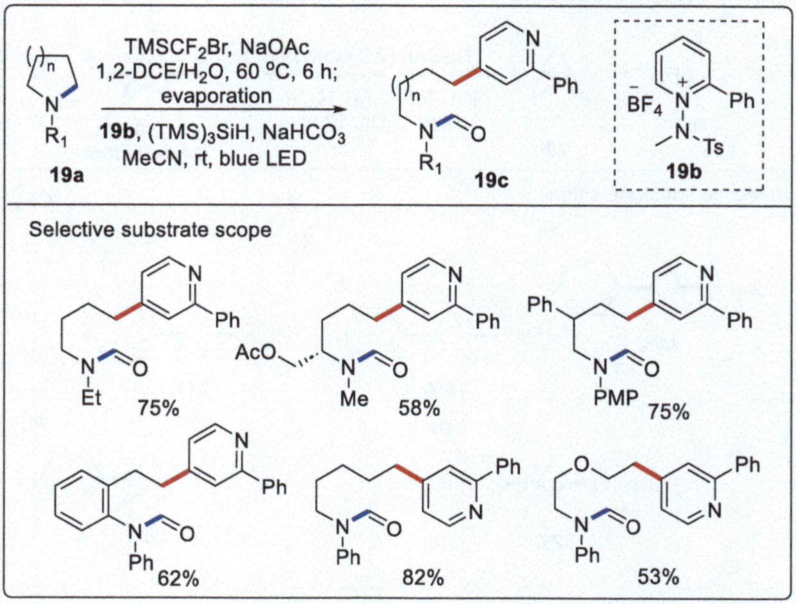

Scheme 19 C4-selective C–H alkylation of pyridines

products, pharmaceuticals, and bioactive compounds. This new strategy introduces heteroarenes, expanding chemical diversity and C–C bond formation. EDA complexes under visible light generate reactive species without a photocatalyst. Pyridinium salts and donors form radicals through SET. This protocol utilizes difluorocarbene reagents to form bromoalkylformamide intermediates, which generate bromine radicals, promoting a radical chain reaction. The method successfully converts cyclic amines into alkyl radicals, installing pyridyl groups at the α-position without external photocatalysts or oxidants. Optimization studies revealed the best conditions for this reaction, including using $PtBu_3$ for SET and visible light for acceleration. The method achieved high yields and regioselectivity across various substrates, including spirocyclic and fused cyclic amines, showcasing its generality and efficiency.

Our group further expanded the synthetic utility of *N*-aminopyridinium salts as bifunctional reagents for introducing both amino and pyridyl groups. This development led to a novel method for synthesizing 1,3-aminopyridylated bicyclo[1.1.1] pentane (BCP) through the ring-opening functionalization of propellane (Scheme 20) [39]. Our strategy revolves around the formation of an EDA complex between *N*-aminopyridinium salt **20a** and acetate. The reaction mechanism involves initiation of SET by the EDA complex and generation of amidyl radical through SET and N–N bond homolysis. To showcase the practical versatility of this protocol, we successfully applied it to three-component cascade reactions. By utilizing phosphinoyl and trifluoromethyl radicals, we were able to synthesize 1,3-phosphinoyl and trifluoromethyl-pyridylated BCP products, respectively (Scheme 20). Key features

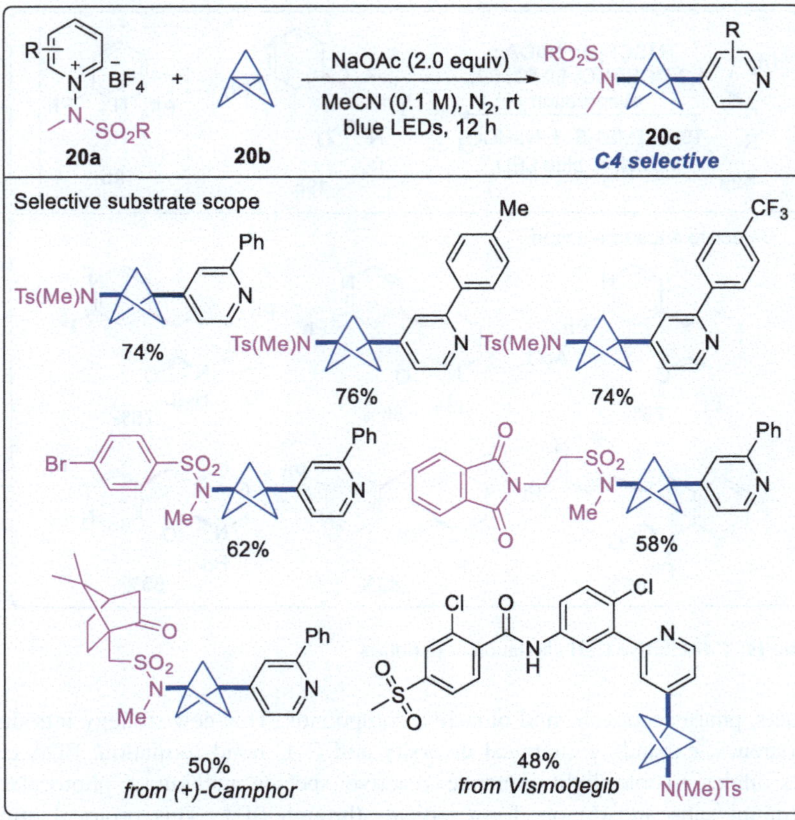

Scheme 20 Photochemical 1,3-aminopyridylation of propellanes

of this methodology include the use of *N*-aminopyridinium salts as bifunctional reagents, the formation of an EDA complex to initiate the reaction, and its applicability to three-component cascade reactions. This approach demonstrates the potential for *N*-aminopyridinium salts in creating complex molecular architectures through radical-mediated processes, opening new avenues for the synthesis of functionalized BCPs and related compounds.

We developed a visible light-mediated sulfonative pyridylation of olefins, leveraging the formation of an EDA complex between a sulfinate and *N*-aminopyridinium salt **21a** (Scheme 21) [40]. Remarkably, by carefully modulating reaction conditions in the absence of visible light, we could completely alter the reactivity pattern of these two compounds, switching between one- and two-electron pathways. The radical pathway mechanism involves SET within the EDA complex. Conversely, by adding a catalytic amount of base, we enabled C4 sulfonylation of the pyridinium salt via a two-electron polar pathway. Key features of this dual-reactivity approach include the versatility of *N*-aminopyridinium salts and sulfinates in organic synthesis and the ability to dramatically alter reaction pathways through

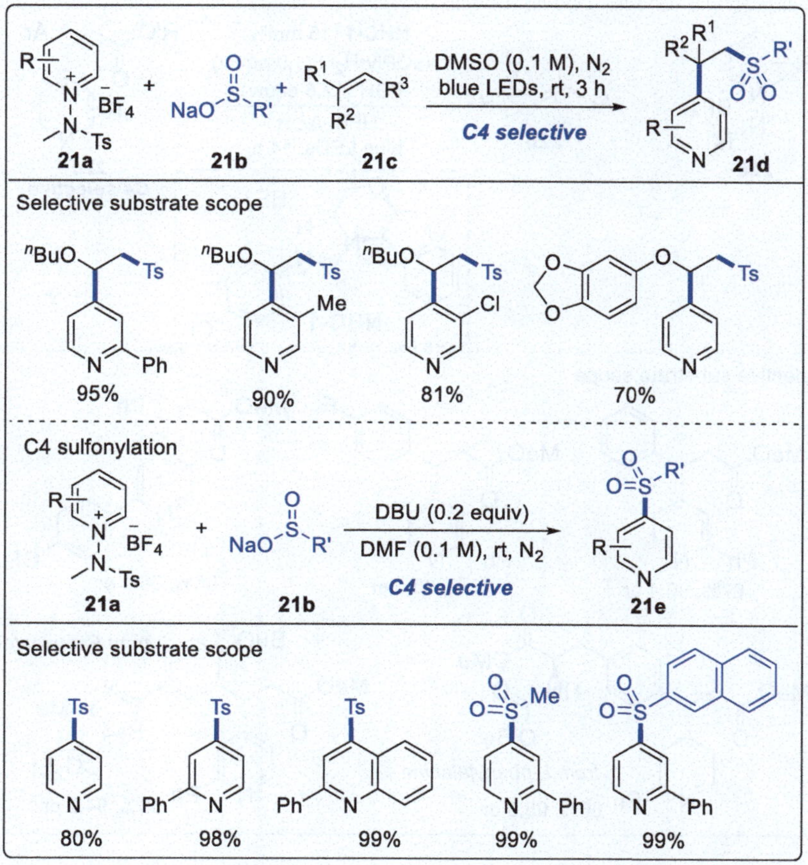

Scheme 21 Alkene sulfonative pyridylation using sulfinate anions

subtle changes in reaction conditions. This method allows for the selective synthesis of different products from the same starting materials, demonstrating the potential for controlled reactivity in complex organic transformations.

4 Enantioselective Variant and Pyridylic C–H Functionalization

In 2022, our group introduced an innovative methodology for the enantioselective functionalization of enals via C4 addition to pyridinium salts, utilizing single-electron *N*-heterocyclic carbene (NHC) catalysis (Scheme 22) [41]. This approach demonstrated remarkable site- and enantioselectivity under mild, metal-free conditions. The process initiates with a direct SET between the chiral enolate form and pyridinium salt **22a**, which generates an amidyl radical and a homoenolate radical.

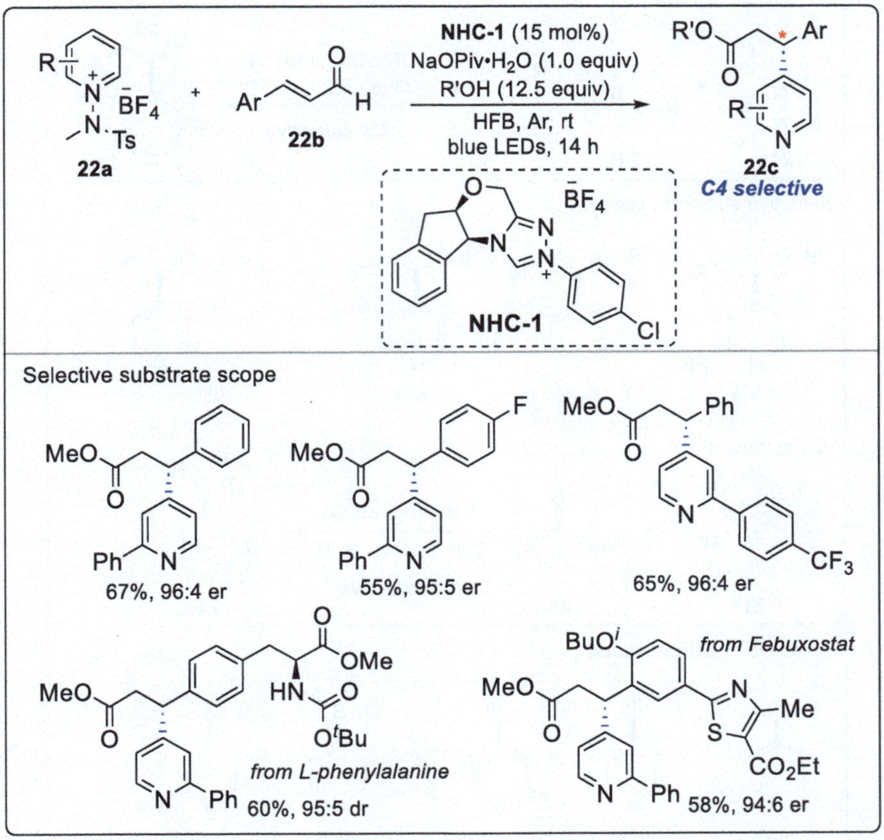

Scheme 22 Enantioselective NHC-catalyzed β-pyridylation of enals

The β-radical then adds to the C4 position of the pyridine core, which undergoes deprotonation and homolytic cleavage of the N–N bond. The mechanism concludes with the reaction of an alcohol, yielding the final β-pyridylated ester product **22c** while simultaneously regenerating the NHC catalyst. Key features of this methodology include its ability to achieve highly selective transformations under environmentally friendly conditions, combining the power of NHC catalysis with radical chemistry. It represents a significant advance in asymmetric synthesis, demonstrating the potential for creating complex, enantioenriched molecules.

The selective functionalization of the pyridylic position has significant implications for both synthetic and medicinal chemistry. We developed a versatile method for visible light-induced pyridylic C–H fluoroalkylation and cascade reactions (Scheme 23) [42]. This approach successfully yielded a wide array of pyridylic fluoroalkylated and β-fluoroalkylsulfonylated heteroarenes **23d**. The process begins with the reversible formation of an alkylidene dihydropyridine intermediate from the *N*-amidopyridinium salt **23a**. This intermediate then undergoes intermolecular trapping of external radicals, leading to the formation of new carbon-carbon or carbon-

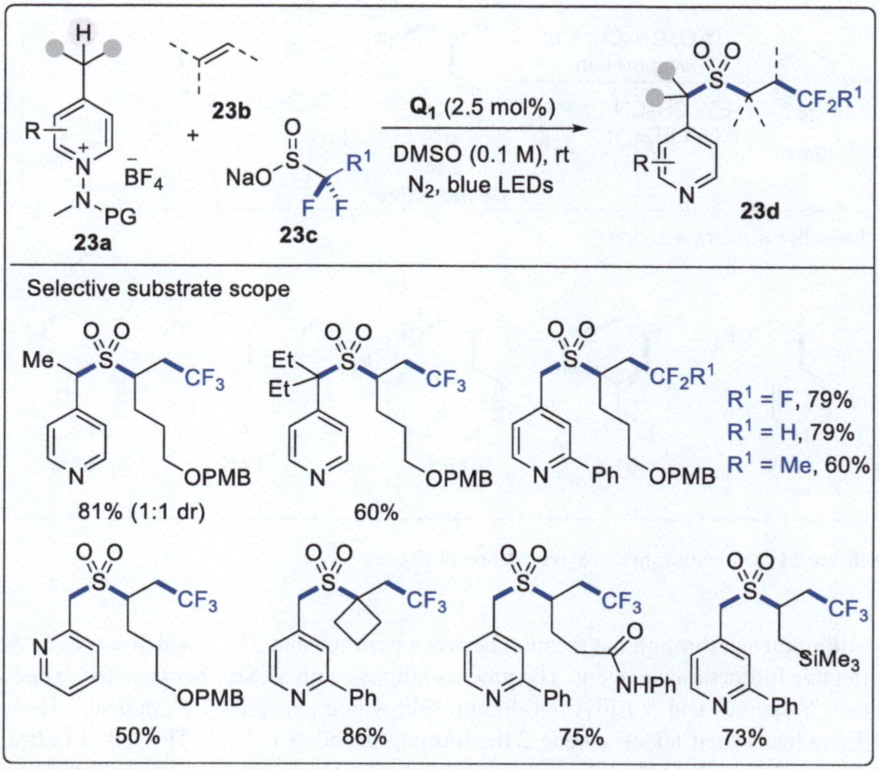

Scheme 23 C(sp^3)–H functionalization via radical cascades at the pyridylic position

sulfur bonds at the pyridylic position. The mechanism continues with the generation of a CF_3 radical through SET between the oxidized $\mathbf{Q_1}$ radical cation and the sulfinate salt substrate **23c**. Finally, in the presence of alkenes, the CF3 radical engages in cascade reactions, further expanding the scope of the transformation. Key features of this method include its ability to harness visible light for selective functionalization of pyridine derivatives, offering a powerful tool for introducing fluoroalkyl groups and creating complex heterocyclic structures under mild conditions. The method's versatility in enabling both direct functionalization and cascade reactions highlights its significance in expanding the toolkit for pyridine modification.

5 Reactions Involving In Situ-Generated Pyridinium Salts

Our group developed a broadly applicable strategy for the photocatalytic trifluoromethylative pyridylation of alkenes, utilizing pyridines and triflic anhydride (Scheme 24) [43]. This method is based on the in situ formation of *N*-triflyl

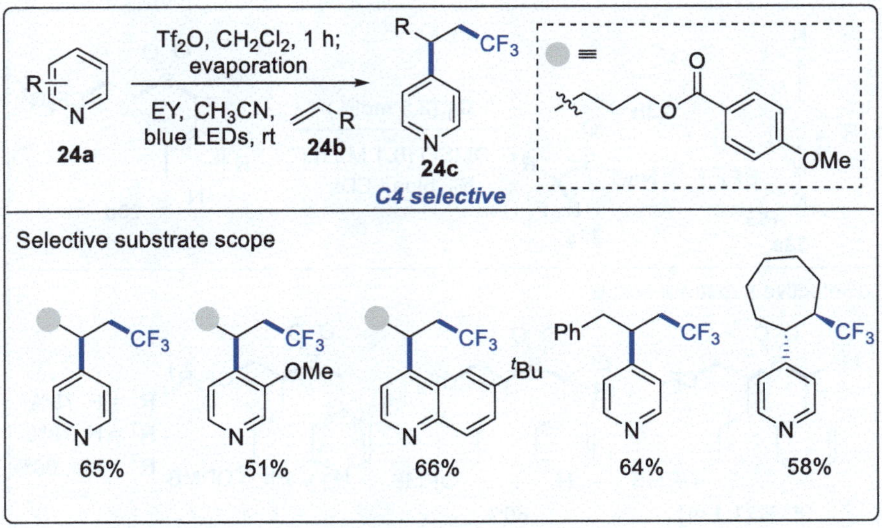

Scheme 24 Trifluoromethylative pyridylation of alkenes

pyridinium salt through the reaction between pyridine and Tf_2O, which serves as an effective bifunctional reagent. The process initiates with an SET between the excited eosin Y species and *N*-triflyl pyridinium salt, which generates CF_3 radicals. These CF_3 radicals then attack alkene **24b**, forming an alkyl radical. This alkyl radical subsequently adds to the C4 position of the pyridine core, resulting in the formation of a radical cation. The mechanism concludes with the deprotonation and C–S bond cleavage, yielding the final product **24c** while simultaneously regenerating the CF_3 radical. Key features of this methodology include the simultaneous introduction of both pyridyl and CF_3 groups into various unactivated alkenes with C4 selectivity, use of *N*-triflyl pyridinium salts as versatile reagents in organic synthesis, mild photocatalytic conditions for the transformation, and potential for a chain process due to CF_3 radical regeneration. This approach demonstrates the potential of *N*-triflyl pyridinium salts in creating complex fluorine-containing heterocycles under mild conditions.

We further expanded our research by developing a method for remote $C(sp^3)$–H pyridylation of hydroxamates, resulting in C2-cyanoalkylated pyridines (Scheme 25) [44]. This innovative approach relies on the generation of an iminyl radical through direct excitation of an in situ-formed oxime salt. The process begins with the electrophilic activation of amide substrate **25b**, resulting in the formation of an iminium triflate. This is followed by nucleophilic addition of the triflate. The mechanism then progresses with direct photoexcitation of intermediate, which leads to homolytic N–O bond cleavage and generates an iminyl radical. This iminyl radical undergoes a 1,5-hydrogen atom transfer (1,5-HAT), producing a nucleophilic alkyl radical. The mechanism concludes with an intramolecular endo addition at the C2 position of the pyridine ring, forming a radical cation. Key features of this

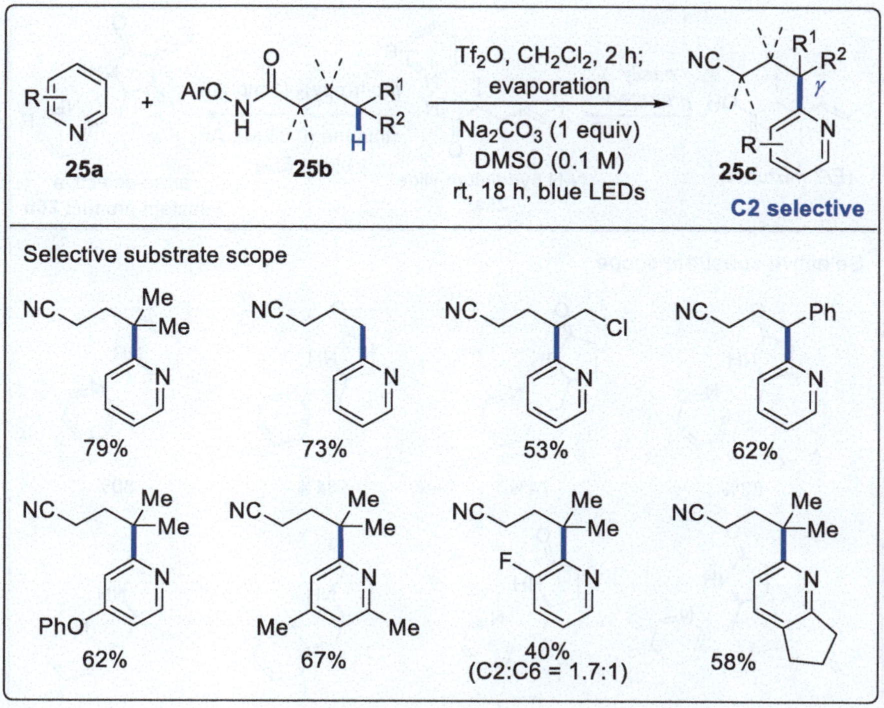

Scheme 25 Remote C–H pyridylation of hydroxamates

method include its ability to achieve remote $C(sp^3)$–H pyridylation, the generation of iminyl radicals through direct excitation, and its versatility in transforming alkyne-containing hydroxamates into structurally diverse pyridyl-allene adducts via C2 addition of allenyl radicals. This approach demonstrates the potential of hydroxamates as versatile precursors in radical chemistry, enabling selective functionalization of remote C–H bonds and offering a powerful tool for the synthesis of complex pyridine derivatives under mild photochemical conditions.

6 Energy Transfer-Induced [3+2] Cycloadditions of N–N Pyridinium Ylides

An innovative approach to diastereoselective pyridyl lactamization was developed through a photoinduced [3+2] cycloaddition, leveraging the unique triplet-state reactivity of N–N pyridinium ylides in the presence of a photosensitizer (Scheme 26) [45]. The method offers an efficient way to convert alkenes into high-value synthetic materials that are typically challenging to obtain under thermal conditions. This visible light-induced method allows for stepwise [3+2] cycloaddition with internal alkenes under mild conditions, efficiently forming unprotected γ- and

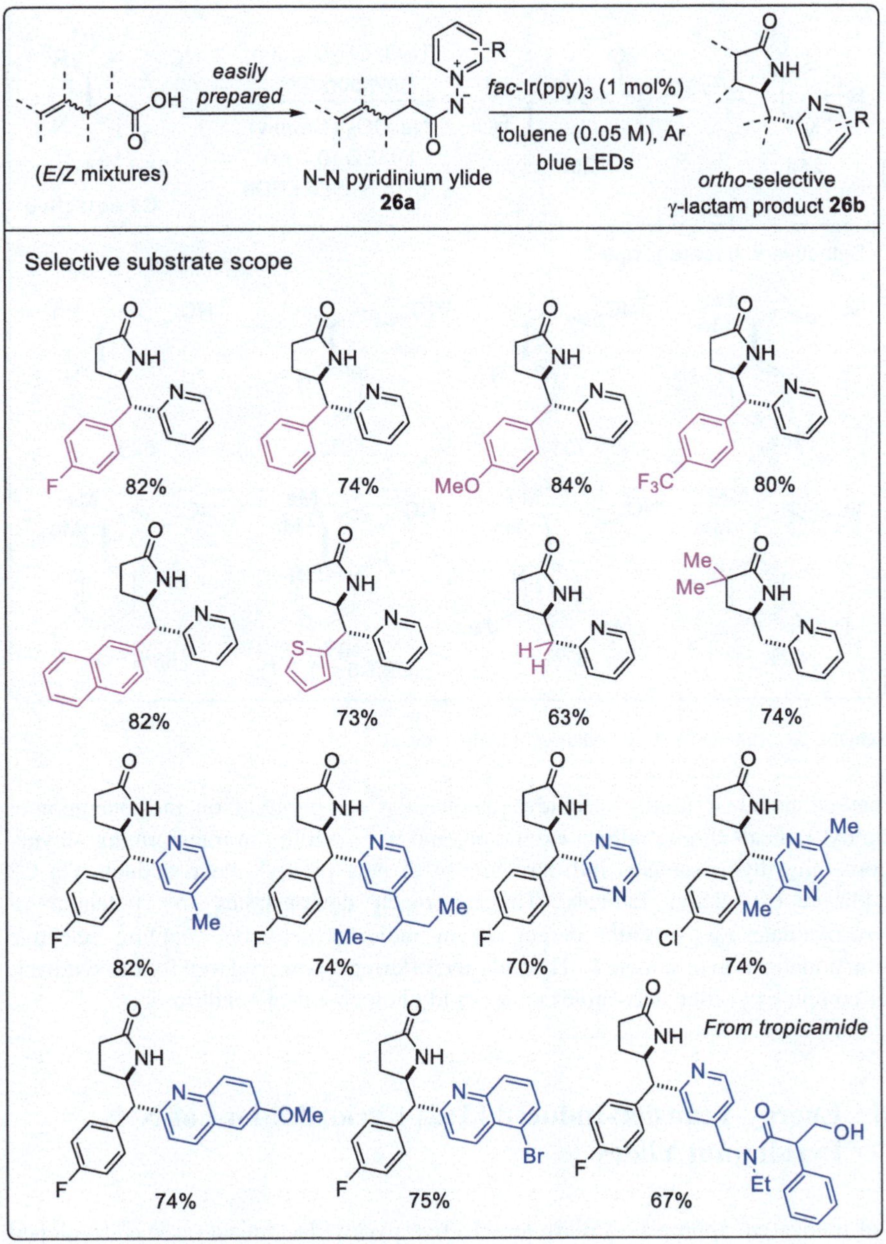

Scheme 26 [3+2] Cycloadditions of N–N pyridinium ylides

δ-lactam scaffolds with ortho-pyridyl groups. It achieves excellent 1,2-*syn* diastereoselectivity in a single step, influenced by the steric repulsion of olefin substituents. Key features of this approach include mild reaction conditions, high functional group tolerance, and broad scope for both activated and unactivated alkenes. The method demonstrates excellent efficiency and diastereoselectivity, providing a useful synthon for ortho-pyridyl γ- and δ-lactam scaffolds with *syn*-configuration. The process begins with an initiation step triggered by visible-light irradiation of the photocatalyst *fac*-[Ir(ppy)$_3$]. This leads to energy transfer to the N–N pyridinium ylide. The next crucial step involves C–N bond formation, which occurs via a low-energy barrier transition state. Concurrently, there exists a competing, less favored pathway that involves N–N bond cleavage. The mechanism then progresses to the formation and recombination of a diradical intermediate. The final stages of the reaction include a series of steps: proton transfer, N–N bond cleavage, and rearomatization of the pyridine ring. This innovative technique addresses previous synthetic challenges in accessing important lactam frameworks containing heteroarenes, offering a powerful tool for synthetic chemists in this area. The combination of experimental and computational studies reveals that the energy transfer process leads to a triplet-state diradical of N–N pyridinium ylides, which promotes the stepwise cycloaddition, thereby expanding our understanding of photocycloaddition reactions.

7 Conclusion and Future Perspectives

The radical-mediated functionalization of pyridine using *N*-functionalized pyridinium salts has emerged as a groundbreaking strategy in synthetic organic chemistry, transforming how we approach pyridine scaffold modifications. This innovative method offers enhanced reactivity and selectivity compared to conventional approaches, particularly when combined with visible-light activation, which enables selective functionalization under mild conditions. This methodology has demonstrated remarkable versatility in late-stage functionalization of biorelevant compounds, opening new possibilities in drug discovery and development. Additionally, the introduction of specially designed *N*-substituted pyridinium salts as bifunctional reagents has created a novel platform for complex molecule synthesis through alkene difunctionalization. A deeper understanding of how *N*-substituents influence reactivity and regioselectivity has led to more precise and efficient transformations of the pyridine core.

Looking ahead, the field holds immense promise for further advancement. There is significant potential for structural optimization and the design of new classes of *N*-functionalized pyridinium salts, which could dramatically expand synthetic diversity and utility. An exciting direction is the development of enantioselective variants to generate enantioenriched pyridine-containing compounds, addressing a crucial need in asymmetric synthesis. This methodology shows great potential in providing solutions to long-standing synthetic challenges, potentially offering more powerful disconnections and valuable transformations. The principles developed for pyridine functionalization could be extended to other important heterocyclic systems, broadening its impact across organic synthesis. Moreover, the mild conditions and visible-light activation align well with green chemistry principles, suggesting potential for more sustainable synthetic processes. Future research may focus on developing catalytic versions of these reactions to enhance efficiency and applicability. Deeper mechanistic studies could lead to more predictable and controllable transformations, further refining the methodology. As this field continues to evolve, it is expected to inspire new methodologies, address complex synthetic challenges, and contribute substantially to medicinal chemistry, materials science, and beyond. The radical-mediated functionalization of pyridine using *N*-functionalized pyridinium salts represents a significant leap forward in synthetic organic chemistry. Its ongoing innovation is likely to yield transformative tools for organic synthesis, potentially reshaping how chemists approach the construction of pyridine-containing molecules and related heterocyclic systems. This burgeoning area of research promises to provide powerful new strategies for tackling complex synthetic problems, ultimately advancing our ability to create diverse and functionally rich molecular architectures [46].

Acknowledgments This research was supported financially by the Institute for Basic Science (IBS-R010-A2).

References

1. Pozharskii AF, Soldatenkov AT, Katritzky AR (2011) Heterocycles in life and society: an introduction to heterocyclic chemistry, biochemistry, and applications, 2nd edn. John Wiley & Sons, Hoboken, pp 139–143
2. Vitaku E, Smith DT, Njardarson JT (2014) Analysis of the structural diversity, substitution patterns, and frequency of nitrogen heterocycles among U.S. FDA approved pharmaceuticals. J Med Chem 57:10257–10274
3. Zafar MN, Atif AH, Nazar MF, Sumrra SH, Gul-E-Saba, Paracha R (2016) Pyridine and related ligands in transition metal homogeneous catalysis. Russ J Coord Chem 42:1–18
4. Hoshikawa T, Inoue M (2013) Photoinduced direct 4-pyridination of C(sp^3)–H bonds. Chem Sci 4:3118–3123
5. O'Hara F, Blackmond DG, Baran PS (2013) Radical-based regioselective C–H functionalization of electron-deficient heteroarenes: scope, tunability, and predictability. J Am Chem Soc 135:12122–12134
6. Jin J, MacMillan DWC (2015) Alcohols as alkylating agents in heteroarene C–H functionalization. Nature 525:87–90
7. Li G-X, Morales-Rivera CA, Wang Y, Gao F, He G, Liu P, Chen G (2016) Photoredox-mediated Minisci C–H alkylation of *N*-heteroarenes using boronic acids and hypervalent iodine. Chem Sci 7:6407–6412
8. Garza-Sanchez RA, Tlahuext-Aca A, Tavakoli G, Glorius F (2017) Visible light-mediated direct decarboxylative C–H functionalization of heteroarenes. ACS Catal 7:4057–4061
9. Klauck FJR, James MJ, Glorius F (2017) Deaminative strategy for the visible-light-mediated generation of alkyl radicals. Angew Chem Int Ed 56:12336–12339
10. Matsui JK, Primer DN, Molander GA (2017) Metal-free C–H alkylation of heteroarenes with alkyltrifluoroborates: a general protocol for 1°, 2° and 3° alkylation. Chem Sci 8:3512–3522
11. Proctor RSJ, Phipps RJ (2019) Recent advances in Minisci-type reactions. Angew Chem Int Ed 58:13666–13699

12. Kim I, Kang G, Lee K, Park B, Kang D, Jung H, He Y-T, Baik M-H, Hong S (2019) Site-selective functionalization of pyridinium derivatives via visible-light-driven photocatalysis with quinolinone. J Am Chem Soc 141:9239–9248
13. Sowmiah S, Esperança JMSS, Rebelo LPN, Afonso CAM (2018) Pyridinium salts: from synthesis to reactivity and applications. Org Chem Front 5:453–493
14. He F-S, Ye S, Wu J (2019) Recent advances in pyridinium salts as radical reservoirs in organic synthesis. ACS Catal 9:8943–8960
15. Rössler SL, Jelier BJ, Magnier E, Dagousset G, Carreira EM, Togni A (2020) Pyridinium salts as redox-active functional group transfer reagents. Angew Chem Int Ed 59:9264–9280
16. Wang Y, Bao Y, Tang M, Ye Z, Yuan Z, Zhu G (2022) Recent advances in difunctionalization of alkenes using pyridinium salts as radical precursors. Chem Commun 58:3847–3864
17. Katz RB, Mistry J, Mitchell MB (1989) An improved method for the mono-hydroxymethylation of pyridines. A modification of the Minisci procedure. Synth Commun 19:317–325
18. Ma X, Herzon SB (2016) Intermolecular hydropyridylation of unactivated alkenes. J Am Chem Soc 138:8718–8721
19. Kim I, Park B, Kang G, Kim J, Jung H, Lee H, Baik M-H, Hong S (2018) Visible-light-induced pyridylation of remote C(sp^3)–H bonds by radical translocation of *N*-alkoxypyridinium salts. Angew Chem Int Ed 57:15517–15522
20. Kim Y, Lee K, Mathi GR, Kim I, Hong S (2019) Visible-light-induced cascade radical ring-closure and pyridylation for the synthesis of tetrahydrofurans. Green Chem 21:2082–2087
21. Rammal F, Gao D, Boujnah S, Hussein AA, Lalevée J, Gaumont A-C, Morlet-Savary F, Lakhdar S (2020) Photochemical C–H silylation and hydroxymethylation of pyridines and related structures: synthetic scope and mechanisms. ACS Catal 10:13710–13717
22. Rammal F, Gao D, Boujnah S, Gaumont AC, Hussein AA, Lakhdar S (2020) Visible-light-mediated C–H alkylation of pyridine derivatives. Org Lett 22:7671–7675
23. Xie D, Wang Y, Zhang X, Fu Z, Niu D (2022) Alkyl sulfoxides as radical precursors and their use in the synthesis of pyridine derivatives. Angew Chem Int Ed 61:e202204922
24. Huang H-M, Bellotti P, Ma J, Dalton T, Glorius F (2021) Bifunctional reagents in organic synthesis. Nat Rev Chem 5:301–321
25. Jeon J, He Y-T, Shin S, Hong S (2020) Visible-light-induced ortho-selective migration on pyridyl ring: trifluoromethylative pyridylation of unactivated alkenes. Angew Chem Int Ed 59: 281–285
26. Mathi GR, Jeong Y, Moon Y, Hong S (2020) Photochemical carbopyridylation of alkenes using *N*-alkenoxypyridinium salts as bifunctional reagents. Angew Chem Int Ed 59:2049–2054
27. Jung S, Lee H, Moon Y, Jung H-Y, Hong S (2019) Site-selective C–H acylation of pyridinium derivatives by photoredox catalysis. ACS Catal 9:9891–9896
28. Lee W, Jung S, Kim M, Hong S (2021) Site-selective direct C–H pyridylation of unactivated alkanes by triplet excited anthraquinone. J Am Chem Soc 143:3003–3012
29. Kim C, Jeong J, Vellakkaran M, Hong S (2022) Photocatalytic decarboxylative pyridylation of carboxylic acids using in situ-generated amidyl radicals as oxidants. ACS Catal 12:13225–13233
30. Vellakkaran M, Kim T, Hong S (2022) Visible-light-induced C4-selective functionalization of pyridinium salts with cyclopropanols. Angew Chem Int Ed 61:e202113658
31. Yu X-Y, Zhao Q-Q, Chen J, Xiao W-J, Chen J-R (2020) When light meets nitrogen-centered radicals: from reagents to catalysts. Acc Chem Res 53:1066–1083
32. Moon Y, Park B, Kim I, Kang G, Shin S, Kang D, Baik M-H, Hong S (2019) Visible light induced alkene aminopyridylation using *N*-aminopyridinium salts as bifunctional reagents. Nat Commun 10:4117
33. Moon Y, Lee W, Hong S (2020) Visible-light-enabled ortho-selective aminopyridylation of alkenes with *N*-aminopyridinium ylides. J Am Chem Soc 142:12420–12429
34. Im H, Choi W, Hong S (2020) Photocatalytic vicinal aminopyridylation of methyl ketones by a double umpolung strategy. Angew Chem Int Ed 59:17511–17516

35. Jung S, Shin S, Park S, Hong S (2020) Visible-light-driven C4-selective alkylation of pyridinium derivatives with alkyl bromides. J Am Chem Soc 142:11370–11375
36. Kim I, Park S, Hong S (2020) Functionalization of pyridinium derivatives with 1,4-dihydropyridines enabled by photoinduced charge transfer. Org Lett 22:8730–8734
37. Choi W, Kim M, Lee K, Park S, Hong S (2022) C4-selective C–H borylation of pyridinium derivatives driven by electron donor-acceptor complexes. Org Lett 24:9452–9457
38. Hong S, Kweon B, Lee W, Chang S, Hong S (2023) Deconstructive pyridylation of unstrained cyclic amines through a one-pot umpolung approach. Org Lett 25:2722–2727
39. Shin S, Lee S, Choi W, Kim N, Hong S (2021) Visible-light-induced 1,3-aminopyridylation of [1.1.1]propellane with *N*-aminopyridinium salts. Angew Chem Int Ed 60:7873–7879
40. Kim M, You E, Park S, Hong S (2021) Divergent reactivity of sulfinates with pyridinium salts based on one- versus two-electron pathways. Chem Sci 12:6629–6637
41. Choi H, Mathi GR, Hong S, Hong S (2022) Enantioselective functionalization at the C4 position of pyridinium salts through NHC catalysis. Nat Commun 13:1776
42. Kim M, You E, Kim J, Hong S (2022) Site-selective pyridylic C–H functionalization by photocatalytic radical cascades. Angew Chem Int Ed 61:e202204217
43. Lee K, Lee S, Kim N, Kim S, Hong S (2020) Visible-light-enabled trifluoromethylative pyridylation of alkenes from pyridines and triflic anhydride. Angew Chem Int Ed 59:13379–13384
44. Kweon B, Kim C, Kim S, Hong S (2021) Remote C–H pyridylation of hydroxamates through direct photoexcitation of O-aryl oxime pyridinium intermediates. Angew Chem Int Ed 60: 26813–26821
45. Lee W, Koo Y, Jung H, Chang S, Hong S (2023) Energy-transfer-induced [3+2] cycloadditions of N-N pyridinium ylides. Nat Chem 15:1091–1099
46. Kim M, Koo Y, Hong S (2022) N-functionalized pyridinium salts: a new chapter for site-selective pyridine C–H functionalization via radical-based processes under visible light irradiation. Acc Chem Res 55(20):3043–3056

Top Heterocycl Chem (2025) 62: 101–120
https://doi.org/10.1007/7081_2025_83

Published online: 13 April 2025

Photoinduced Synthesis of Imidazopyridines and Imidazothiazoles

Sumit Ghosh, Madhusudan Mondal, Sudip Laru, and Alakananda Hajra

Contents

Abstract In the realm of synthetic organic chemistry, photoinduced synthesis has been skyrocketing in the last two decades due to harnessing of eco-friendly potential light energy to perform complex chemical reactions. At the same time, imidazopyridine and imidazothiazole are two eminent heterocycles, which are widely seen in medicinal as well as industrial chemistry. Therefore, the utilization of photo energy for the synthesis of these two heterocycles is highly commendable. This book chapter summarizes all the light-induced synthesis of imidazopyridines and imidazothiazoles published so far.

Keywords Heterocycle synthesis · Imidazopyridine · Imidazothiazole · Photoinduced synthesis · Visible-light-induced synthesis

Sumit Ghosh and Madhusudan Mondal have first authorship and made equal contributions.

S. Ghosh, M. Mondal, S. Laru, and A. Hajra (✉)
Department of Chemistry, Visva-Bharati (A Central University), Santiniketan, India
e-mail: alakananda.hajra@visva-bharati.ac.in

1 Introduction

Nitrogen-containing heterocyclic drugs stand out as the most popular and commercially successful therapeutic agents [1, 2]. Among the various *N*-containing heterocycles, imidazopyridines [3–6] and imidazothiazoles [7–10] are two important classes of heterocyclic compounds (Fig. 1), which play crucial roles in the development of pharmaceuticals, agrochemicals, and materials with customized properties due to their diverse biological activities. Therefore, a bunch of reports are available on synthesis as well as functionalization of these two prominent heterocycles [11–16]. However, traditional methods for synthesizing these compounds typically require multistep processes and harsh reaction conditions and result in significant waste generation. Therefore, it is crucial to urgently develop more effective and environmentally friendly synthetic methods for these distinguished heterocycles.

On the other hand, the adoption of photo-mediated functionalization has surged in popularity because of its sophistication, proficiency, and energy-efficient and eco-friendly approach as a pathway for generating new chemical compounds [17–24]. Furthermore, the revolutionary contributions of pioneering scientists, such as MacMillan [25–27], Nicewicz [28, 29], Stephenson [30, 31], and Yoon, in the field of visible light-induced synthetic methodologies have not only showcased an impressive level of superiority but also inspired the exploration of this topic as a research probe. Therefore, use of light as the energy source in the synthesis of imidazopyridines and imidazothiazoles could be obviously a decent choice in this domain. In addition to that, there is no review still reported on this topic, despite the presence of a bunch of research articles, highlighting the need for a comprehensive overview of the topic.

Our group contributed significantly to both photoinduced synthetic chemistry [32–41] and imidazopyridine chemistry [42–48] for more than a decade. In continuous interest on visible light-induced synthetic chemistry, herein, we report the first review on photoinduced synthesis of imidazopyridines and imidazothiazoles with literature coverage up to October 2023. For better understanding of the readers, we have divided the book chapter based on the type of heterocycle synthesis, and each division is further subdivided based on the type of catalyst used for the synthesis.

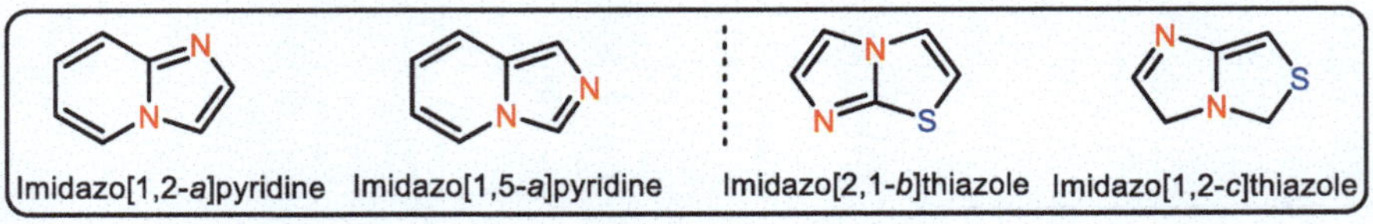

Fig. 1 Structures of imidazopyridines and imidazothiazoles

2 Photoinduced Synthesis of Imidazopyridines

2.1 Catalyst-Free Synthesis of Imidazopyridines

In 2018, Siddiqui et al. [49] enlisted a visible light-mediated multicomponent one-pot reaction strategy for the synthesis of 3-aminoimidazo[1,2-*a*]pyridines under catalyst- and solvent-free conditions. These imidazo-fused scaffolds are very crucial in medicinal chemistry due to their broad range of pharmacological and biological activities, such as antibacterial, antiviral, anti-inflammatory, and fungicidal activities. According to this methodology, all three reactants, such as 2-aminopyridine (**1a**), aldehyde (**1b**), and isocyanide (**1c**), were taken together in a closed round-bottom flask under the irradiation of 24 W CFL for 3 h at room temperature. The desired product (**1d**) was achieved with up to 98% yield (Scheme 1). It was observed that in the absence of CFL at room temperature, i.e., in dark conditions, a trace amount of product was obtained, which suggested the essential role of visible light to activate the reaction. With the optimal condition in hand, the authors looked over the generality and versatility for this method with various substituted 2-aminopyridines, aldehydes, and isocyanides and got the corresponding product with excellent yields irrespective of the electronic nature on substrates. They applied this protocol for the synthesis of imidazothiazoles also. Mild conditions, appreciable yields, green method, operational accessibility, atom efficiency, and wide substrate scopes were vital advantages for this process. The mechanistic studies suggested a light-mediated free radical pathway for this transformation, as confirmed by the radical trapping experiment.

In 2019, Komogortsev group [50] devised a visible light-induced reagent- and solvent-free method for the synthesis of imidazo[1,5-*a*]pyridine-5,8-dione

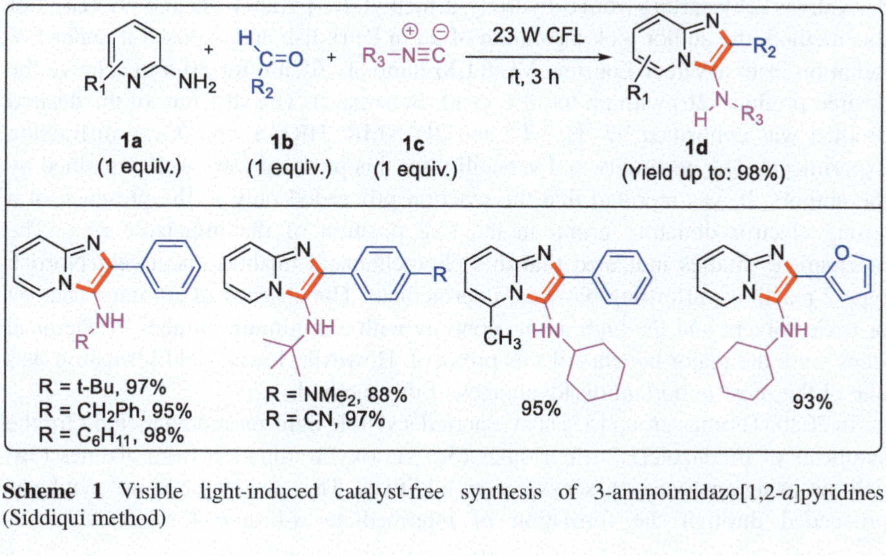

Scheme 1 Visible light-induced catalyst-free synthesis of 3-aminoimidazo[1,2-*a*]pyridines (Siddiqui method)

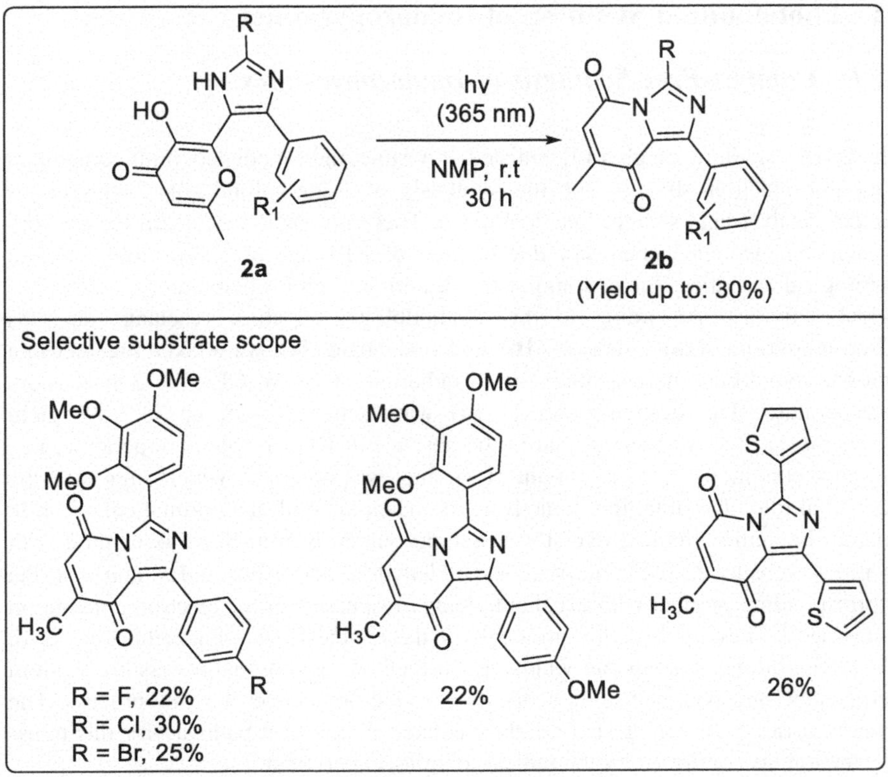

Scheme 2 Visible light-induced synthesis of imidazo[1,5-*a*]pyridine-5,8-dione derivatives (Komogortsev method)

derivatives (**2b**) starting from 5-hydroxy-2-methyl-4*H*-pyran-4-one moiety (**2a**). For this method, the author took a solution of **2a** in Petri dish and exposed it under UV radiation from a Vilber Lourmat VL-6.LM lamp at 365 nm for 30 h to achieve the desired product (**2b**) with up to 30% yield (Scheme 2). The structure of the desired product was confirmed by ^{1}H, ^{13}C, and 2D NMR, HRMS, and X-ray diffraction experiments. The generality and versatility for this protocol were well explained by the authors. It was reported that the reaction proceeded only in the presence of a strong electron-donating group in the C-2 position of the imidazole ring. The mechanistic studies indicated that this photochemical method underwent Norrish type 2 reactions, affording biradical intermediate. The absence of complex catalyst or toxic solvent and the high atom economy with a minimum number of chemical steps were the major benefits of this protocol. However, lower yield formation was one of the most important disadvantages of this method.

In 2020, Thomas group [51] also reported a visible light-promoted method for the synthesis of imidazole[1,2-*a*]pyridines (**3c**) via oxybromination from alkenes (**3a**) without using any catalyst, sensitizer, or additives. This two-step one-pot synthesis proceeded through the formation of intermediate α-bromo ketones from *N*-

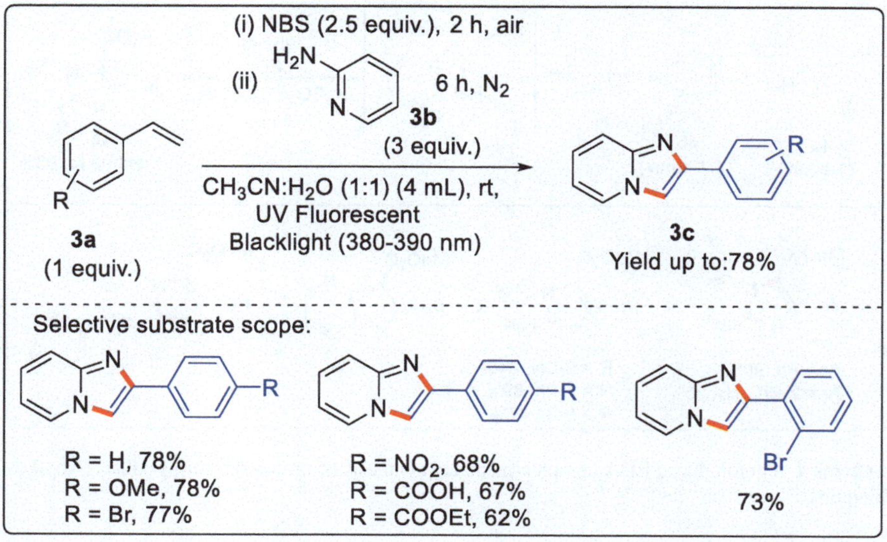

Scheme 3 Visible light-promoted synthesis of imidazole[1,2-*a*]pyridines (Thomas method)

bromosuccinimide (NBS) and alkene under the irradiation of 380–390 nm of UV LED fluorescent black light in acetonitrile/water mixture, which then reacted with 2-amino pyridine (**3b**) to form the desired product (**3c**) with up to 78% yield under N_2 atmosphere for 6 h (Scheme 3). The author found that dark conditions decreased the reaction yield significantly, which suggested that the reactions required light activation. They observed that a variety of aryl alkenes containing electron-donating or electron-withdrawing groups were well compatible for this protocol and afforded the anticipated product with moderate to high yields. The gram-scale synthesis signified the practical applicability of this methodology and got up to 74% yield under the same reaction conditions. This transformation underwent the radical pathway, as confirmed by the radical trapping experiments. Mild and catalyst-free conditions, gram-scale applicability, and wide substrate scopes were vital profits for the industrial process.

Recently, electron donor-acceptor (EDA) complex-mediated synthetic chemistry has been getting tremendous importance [52]. In 2022, Novikov et al. [53] developed an EDA complex-mediated one-pot synthesis of imidazo[1,2-*a*]pyridines (**4c**) from 2-bromoazirines (**4a**) and pyridines (**4b**) without using any catalyst. The desired product was formed with up to 93% yield within 14 h via the formation of intermediate (2*H*-azirin-2-yl)pyridinium bromides followed by dehydrobrominative UV light-induced cyclization (Scheme 4). The author studied this reaction using different types of substituted pyridines under optimized reaction conditions. They reported that isoquinoline and pyridazine also participated in this protocol and furnished the target product with moderate to excellent yields. This reaction proceeded through the formation of the EDA complex followed by radical generation, as confirmed by radical trapping experiments and DFT calculations.

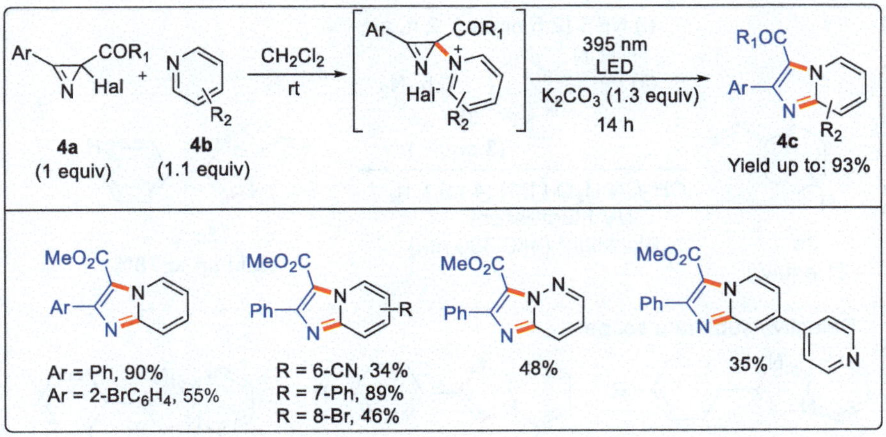

Scheme 4 Photoinduced EDA complex-mediated synthesis of imidazo[1,2-*a*]pyridines (Novikov method)

2.2 *Metal-Based Photocatalyzed Synthesis of Imidazopyridines*

In 2017, Maurya and Kamal groups [54] reported a metal-catalyzed, greener and sustainable protocol for the synthesis of imidazo[1,2-*a*]pyridine (**5c**) from the coupling of 2-aminopyridine (**5a**) and α-keto vinyl azide (5**b**) under visible-light condition. For this method, the authors used 2 mol% of $Ru(bpy)_3Cl_2{\cdot}6H_2O$ as photocatalyst in 10 mL of ethyl acetate solvent under the irradiation of 15 W white LED bulb, which was kept at a distance of approx. 10 cm from the reaction vessel to obtain the desired product with up to 96% yield (Scheme 5). It has been observed that no conversion took place in the absence of light or photocatalyst. They investigated a wide range of substrate scope for this transformation under the optimized reaction condition. A variety of α-keto vinyl azides containing various substituents (aryl, alkyl, and heterocyclic moieties) as well as functional group like –MeO, $–NO_2$, –Cl, –Br, and –CN were well tolerated without affecting the reaction yield. In addition, several functional groups such as halogen, alkyl, and esters in 2-aminopyridine performed very smoothly in this reaction under the same reaction conditions. These wide substrate scopes clearly indicated the potential for industrial applications of this synthetic strategy. These synthesized imidazopyridine compounds showed cytotoxic activity in the selected human cancer cell lines, i.e., A549 (lung cancer), DU-145 (prostate cancer), MCF-7 (breast cancer), and Hela (cervical cancer). The controlled experiment suggested that the reaction proceeded through the photosensitized decomposition of vinyl azides.

In 2021, Liu and Sun groups [55] demonstrated a method for the synthesis of C-3 alkylated imidazopyridines (**6c**) through one-pot condensation and alkylation of α-bromocarbonyl compounds (**6b**) with 2-aminopyridines (**6a**). About 2 mol% of fac-$Ir(ppy)_3$ and 2 equiv. of Et_3N were used as photocatalyst and base, respectively,

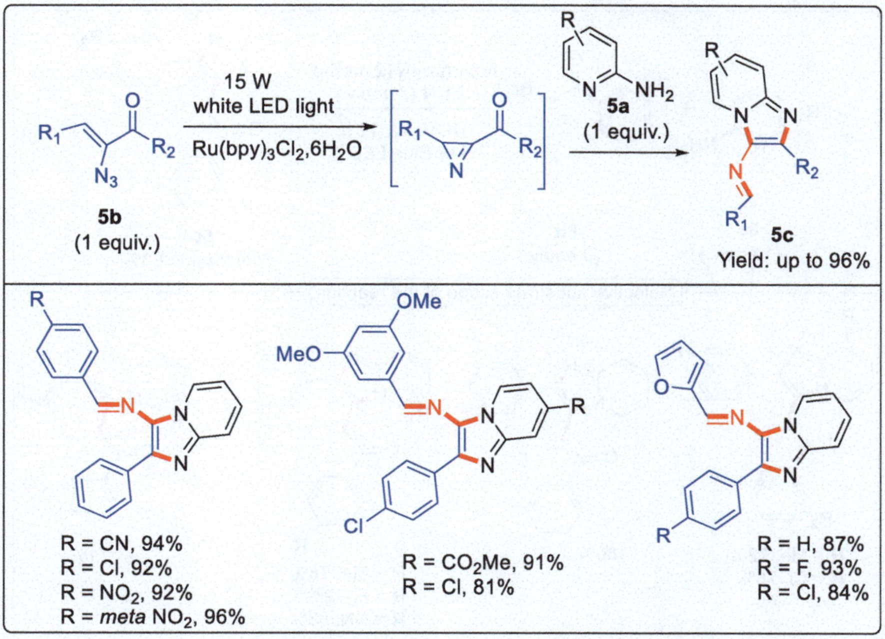

Scheme 5 Visible light-assisted synthesis of imidazopyridines (Maurya and Kamal method)

for this reaction. The reaction was carried out in DCM solvent under the irradiation of 5 W blue LED at rt in the Ar atmosphere. The desired alkylated products were achieved with up to 80% yield within 12 h (Scheme 6). It was reported that in the absence of visible-light irradiation or photocatalyst, no desired product was obtained. The author observed that organic photocatalysts like eosin Y and 4CzIPN also have comparable catalytic activity in this transformation. With optimized conditions in hand, the substrate scope of this protocol was examined. They observed that various substituents at the C-3, C-4, and C-5 positions of 2-aminopyridine underwent the reaction very smoothly and afforded the corresponding products in moderate to high yields, whereas substituents at C-6 position of the pyridine ring failed to proceed the reaction due to steric hindrance. In addition, α-bromo ketone containing functional groups without considering the electronic nature of electron donating or withdrawing also performed well in this transformation, leading the desired product in good to excellent yields. Interestingly, α-bromocarbonyl compounds and alkyl bromide were added successively to achieve the C-3 alkylated imidazopyridines. This method was applicable for the gram-scale synthesis of C-3 alkylated imidazopyridine products. This method provided an attractive synthetic route for the rapid synthesis of zolpidem. This reaction proceeded via a radical pathway, as confirmed by the Stern-Volmer and radical trapping experiments.

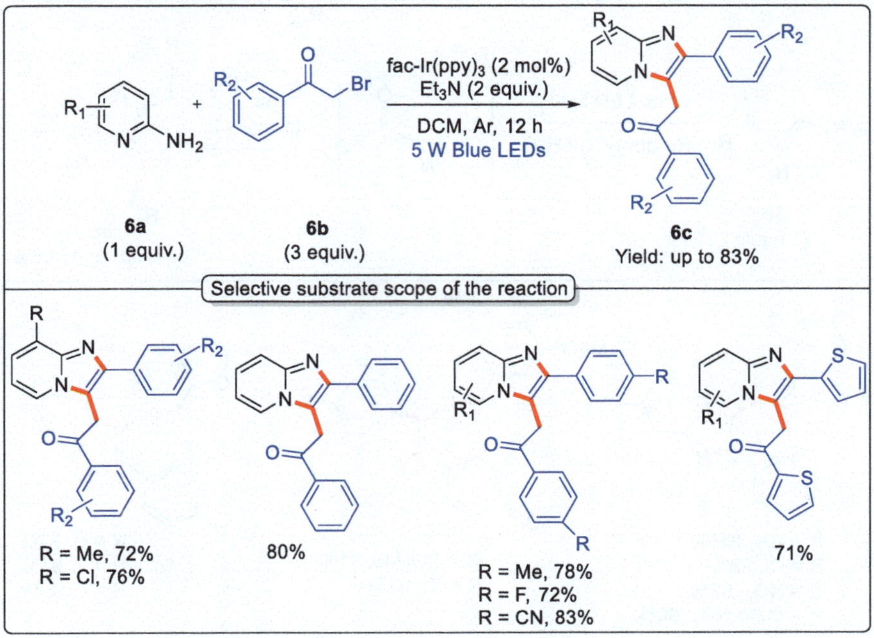

Scheme 6 Visible light-induced synthesis of C-3 alkylated imidazopyridines (Liu and Sun method)

2.3 *Organophotocatalyzed Synthesis of Imidazopyridines*

In 2020, Singh group [56] devised a visible light-assisted mild and environment-friendly strategy for the synthesis of 3-aminoimidazo[1,2-*a*]pyridine (**7d**) via hydrogen atom transfer (HAT) process through the one-pot multicomponent reaction of benzylamine (**7a**), 2-aminopyridine (**7b**), and t-butyl isocyanide (**7c**) without using any additive or metal catalyst. According to this reaction methodology, the author used 2 mol% of eosin Y as HAT photocatalysis in EtOH/H_2O (2:1) solvent under the irradiation of 22 W white LED. The desired amino-imidazopyridine products were obtained with up to 97% yield at rt in open air (Scheme 7). The reaction failed to proceed under dark conditions at rt, which indicated the essentiality of visible light for this transformation. The author observed that other organic photocatalysts, such as rhodamine B, acridine red, and rose bengal, were inefficient for this transformation. They further viewed that under the Ar-atmosphere, a negligible amount of reaction yield was observed. With the optimized conditions in hand, they examined the generality and versatility of this protocol. It has been observed that benzylamine containing an electron-withdrawing group led to a higher yield as well as reacted at a faster rate, whereas the electron-donating group on benzylamine reduced the reaction yield with a slower reaction rate. The gram synthesis signified the practical applicability of this methodology scale. Low cost, nontoxicity, effortless accessibility of

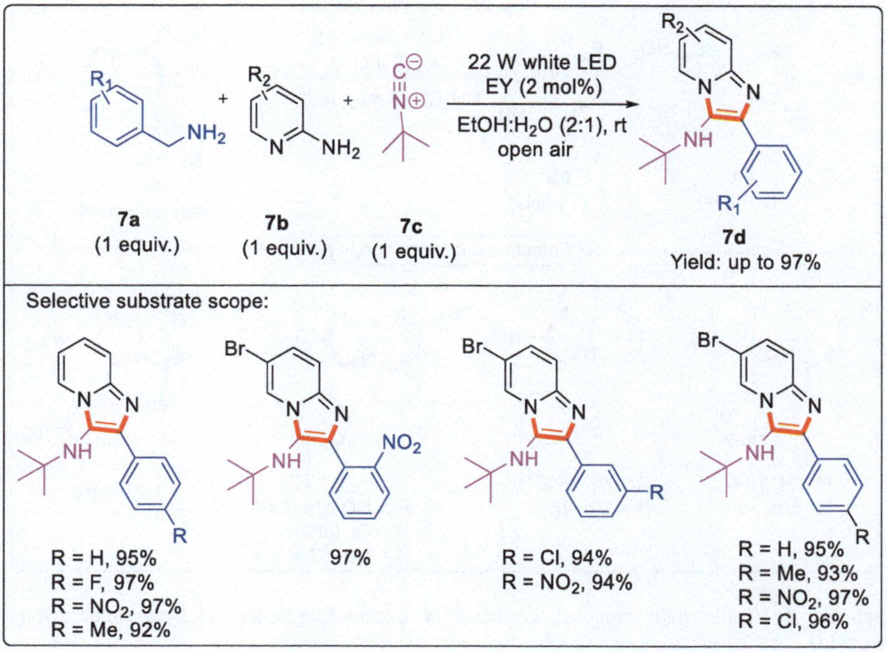

Scheme 7 Visible light-driven eosin Y-catalyzed synthesis of 3-aminoimidazo[1,2-*a*]pyridine (Singh method)

starting materials, mild conditions, environment friendliness, and wide substrate scope were important advantages for this methodology. The author suggested a photoredox pathway in which radical initiation through HAT process occurred.

In 2016, Singh and coworkers [57] enlisted a visible light-induced eco-friendly one-pot synthesis of 2-nitro-3-arylimidazo[1,2-*a*]pyridine (**8c**) starting from nitroalkene (**8a**) and 2-aminopyridine (**8b**) at room temperature. No metal was used in this method. According to this methodology, this method required 2 mol% of eosin Y as a photoredox catalyst and acetonitrile as solvent under the irradiation of green LED and an open atmosphere to achieve the desired product with up to 79% yield within 5 h (Scheme 8). They observed that only a trace amount of target product was formed under the N_2 atmosphere, indicating that oxygen had a critical role for this reaction. The substrate scope study suggested that nitroalkene containing electron-donating and -withdrawing groups underwent this transformation very smoothly, affording the desired product with good yields, while aliphatic nitroalkene did not respond in this reaction. Moreover, 2-aminopyridines with methyl substitution in different positions were well tolerated and gave good yields under the same optimized reaction conditions. They have also applied this strategy for Friedel-Crafts alkylation of the nitroalkene derivatives with indole to synthesize 3-substituted indoles. The controlled experiment suggested a photoredox pathway in which radical initiation through the SET process occurred.

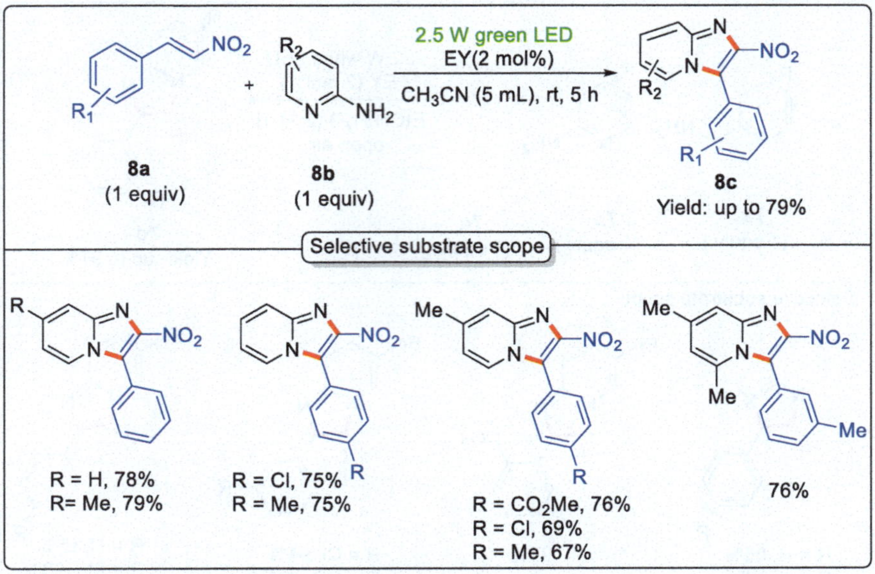

Scheme 8 Visible light-promoted synthesis of 2-nitro-3-arylimidazo[1,2-*a*]pyridine (Singh method)

In 2019, Chuah et al. [58] reported a green strategy for the synthesis of imidazo [1,2-*a*]pyridines from the coupling of 1,3-dicarbonyls (**9a**) with 2-aminopyridines (**9b**) under visible light-mediated photoredox catalysis. This method required 5 mol% Erythrosine B as photocatalyst, 20 mol% KBr as halogenating agent for in situ bromination, and acetonitrile as solvent. The process was completed at room temperature within 24 h in open air under the irradiation of 15 W CFL. The desired product (**9c**) achieved up to 89% yield (Scheme 9). The reaction failed to proceed in the absence of photocatalyst or visible light, confirming the photocatalytic nature of the reaction. The author reported that no desired product was obtained in the absence of KBr, indicating that the reaction proceeded via in situ bromination of the 1,3-dicarbonyl moiety. A bunch of β-ketoesters containing various alkyl or aryl substituents at the keto moiety were well tolerated in this transformation and furnished the desired product with moderate to excellent yields. A wide array of 2-aminopyridines substituted with methyl, chloro, bromo, etc. as well as unsubstituted 2-aminopyridine performed very smoothly in this protocol. However, in the case of electron-withdrawing group NO_2 at 2-aminopyridine, only a trace amount of product was observed. This protocol was applied for large-scale synthesis also. The reaction proceeded through the formation of free radicals by SET followed by a reductive quenching cycle as confirmed by radical trapping experiment and cyclic voltammetry studies, respectively. This strategy was the first example involving the use of a catalytic amount of halogenating agent, regenerated by the photocatalyst, which led to a smaller E-factor.

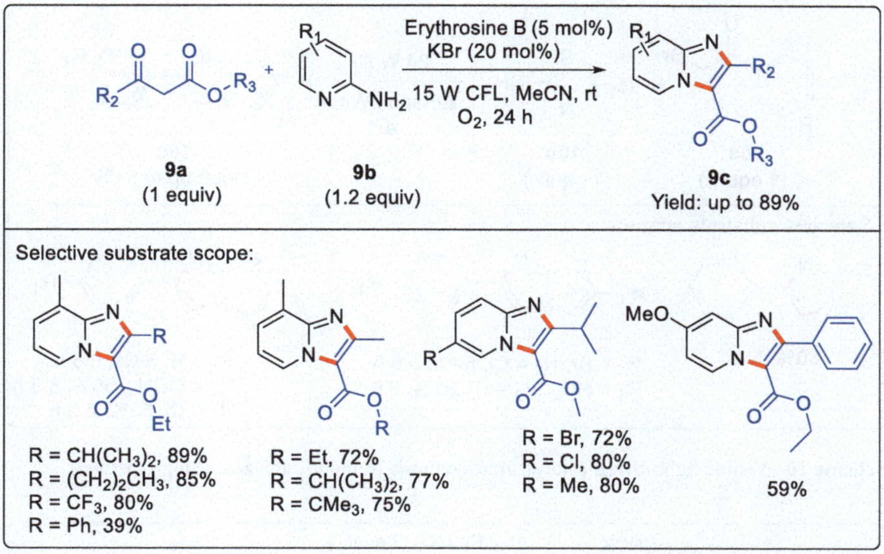

Scheme 9 Visible light-assisted green synthesis of imidazopyridines (Chuah method)

3 Photoinduced Synthesis of Imidazothiazole

The imidazothiazole scaffold widely exhibits biological activities like antitumor, antifungal, antimicrobial, anti-inflammatory, anticonvulsant, diuretic, and antioxidant activities; moreover, they have huge applications in medicinal chemistry such as the treatment of allergies, inflammation, HIV infections, bacterial infection, and treatment of pain. Therefore, interest for the synthesis of imidazothiazole is getting intense nowadays. In this part, all the photoinduced methodologies for the synthesis of this eminent heterocycle are discussed.

3.1 Catalyst-Free Synthesis of Imidazothiazole

In 2016, Singh group [59] first reported the visible light-assisted synthetic strategy for the synthesis of imidazothiazoles without using any base, ligand, heat, or an additional oxidant. This reaction proceeded via oxidative coupling of phenacyl bromide (**10a**) with 2-amino thiazole (**10b**) under open air at room temperature. They used 23 W CFL as a visible light source and EtOH:H_2O as a green solvent and obtained the target imidazothiazole product **10c** with up to 61% yield (Scheme 10). Under the optimal reaction conditions, the authors examined the substrate scope with a series of different derivatives of phenacyl bromides and thiazoles and got satisfactory yields in each case. The main attractive features of this protocol were eco-efficiency, cost-effectiveness, use of greener reagent, short reaction time, easy

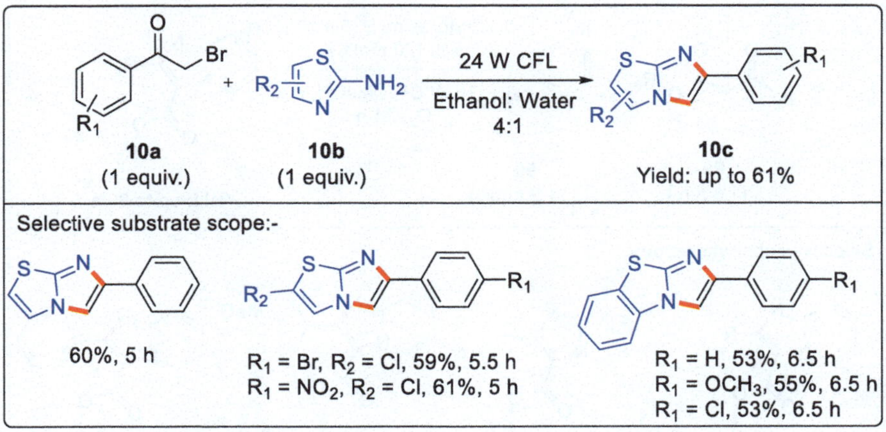

Scheme 10 Visible light-driven catalyst-free synthesis of imidazothiazole (Singh method)

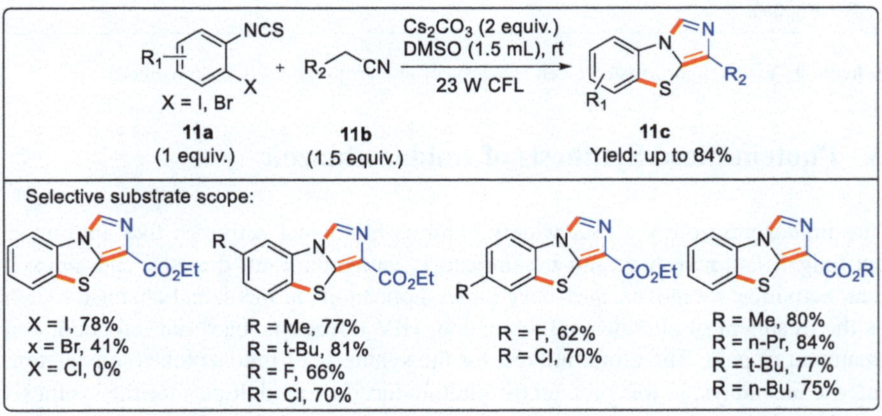

Scheme 11 Visible light-induced synthesis of benzoimidazothiazole (Wang and Yan method)

workup, and high yield formation. The reaction went through the formation of radical intermediate, as confirmed by the radical trapping experiment.

In 2021, Wang and Yan groups [60] enlisted a mild and efficient visible light-induced synthetic protocol for the preparation of benzo[*d*]imidazo[5,1-*b*]thiazoles. This protocol proceeded under metal- and photocatalyst-free conditions. According to this report, the cascade cyclization of 2-haloaryl isothiocyanates (**11a**) and isocyanides (**11b**) required 2 equiv. of Cs_2CO_3 as a base in DMSO solvent. The desired product (**11c**) achieved up to 84% yield within 15 h at room temperature (Scheme 11). The generality and versatility were well studied for this transformation by the authors. A wide range of 2-haloaryl isothiocyanates containing electron-withdrawing as well as electron-donating groups performed well and furnished the target product in good to excellent yields. Moreover, the different substituents on isothiocyanates also participated in this reaction. However, phenyl- and nitro-

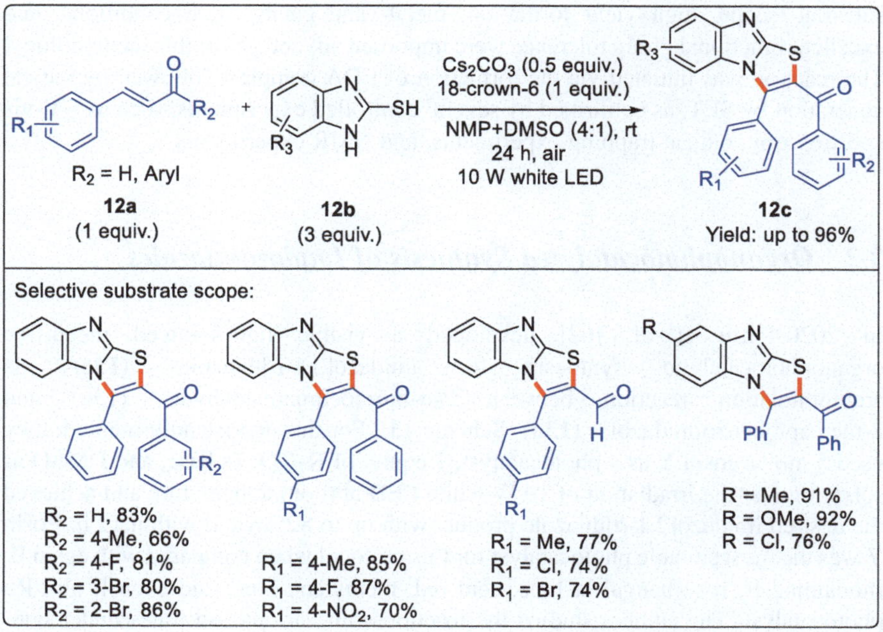

Scheme 12 Visible light-induced EDA complex-mediated synthesis of imidazo[2,1-b]thiazoles (Liu method)

substituted isonitrile substrates failed to proceed in the reaction. The gram-scale synthesis signified the practical applicability of this methodology. In addition, they calculated E-factor to study the environmental impact of the formation of benzo[*d*]imidazo[5,1-*b*]thiazoles in this transformation. Moreover, they conducted several controlled experiments, such as UV-vis spectroscopic measurements, light ON/OFF experiment, HRMS, and radical trapping experiment to gain insight into the mechanism and reported that this protocol proceeded through the formation of the electron-donor acceptor (EDA) complex, followed by radical generation.

In 2022, Liu and coworkers [61] developed an EDA complex-promoted external photocatalyst-, transition metal-, and oxidant-free strategy for the synthesis of imidazo[2,1-*b*]thiazoles via [3+2] cyclization of chalcones (**12a**) with 2-mercaptobenzimidazoles (**12b**) under visible-light condition. According to this methodology, it required 0.5 equiv. of Cs_2CO_3, and 1 equiv. of 18-crown-6 in 1 mL of DMSO/NMP solvent under the irradiation of 10 W white LED. This method worked well at room temperature under an aerial atmosphere and afforded the desired product **12c** with up to 96% yield within 24 h (Scheme 12). A wide range of electron-withdrawing and electron-donating groups on the benzene ring of chalcone **12b** responded well in this protocol and provided the functionalized products in moderate to good yields. Furthermore, the authors investigated the substrate scope of 2-mercaptobenzimidazoles, and it tolerated a bunch of functional groups on imidazole moiety. Simple and mild reaction conditions, extensive substrate scope, high yield formation, metal- and catalyst-free conditions, and excellent functional group tolerance were important advantages of this methodology. The reaction was initiated via the formation of EDA complex, followed by radical generation by SET, as confirmed by several controlled experiments, such as UV-vis spectroscopy, radical trapping experiments, and NMR experiments.

3.2 *Organophotocatalyzed Synthesis of Imidazothiazoles*

In 2020, Liu et al. [62] developed a visible light-induced metal-free organophotocatalytic synthesis for imidazo[2,1-*b*]thiazoles (**13c**) via aminothiolation reaction between *Z*-α-bromocinnamaldehydes (**13a**) and 2-mercaptobenzoimidazoles (**13b**) (Scheme 13). For this eco-friendly method, they used 5 mol% eosin Y as a photocatalyst, 1 equiv. of K_2CO_3 as base, and DMSO as solvent under the irradiation of 10 W white LED at room temperature and achieved the desired imidazo[2,1-*b*]thiazole product with up to 82% yield within 12 h. Eosin Y was the most suitable photocatalyst for this protocol when compared with eosin B, rhodamine B, rose bengal lactone, acid red, methylene blue, and even Ir and Ru photocatalyst. The authors studied the reaction with various substrates under optimized reaction conditions. It was found that *Z*-α-bromocinnamaldehydes containing both electron-donating and electron-withdrawing groups at different positions of the

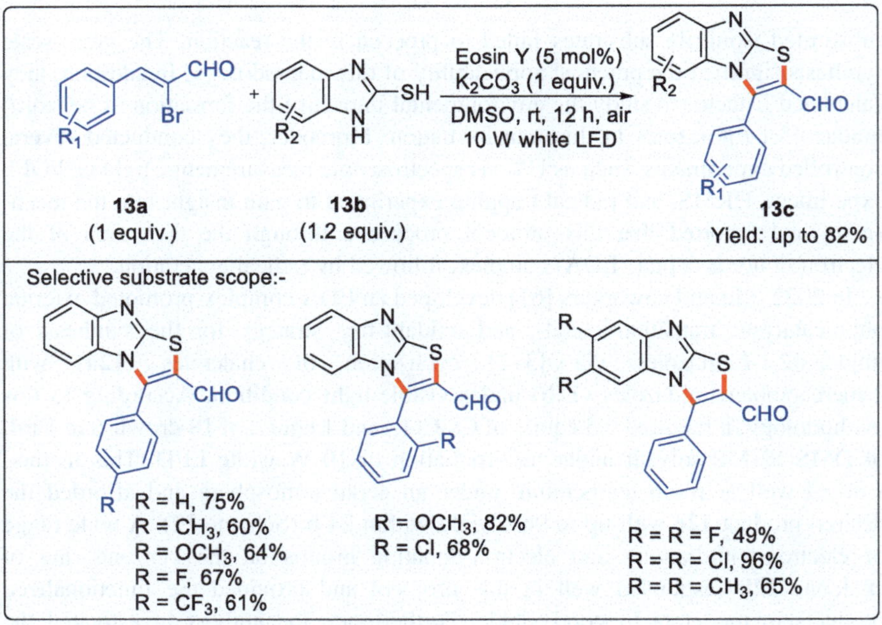

Scheme 13 Visible light-driven eosin Y-catalyzed synthesis of imidazo[2,1-*b*]thiazoles (Liu method)

benzene ring smoothly participated in this protocol and provided the desired product with moderate to good yields. Furthermore, various 2-mercaptobenzoimidazole derivatives were studied under the same reaction condition, and the corresponding product was obtained with good yields. This method was applied for large-scale synthesis also. To predict the reaction pathway, they conducted several controlled experiments, such as radical trapping experiments, light ON/OFF experiments, and UV-vis spectroscopic measurements. On the basis of these experiments and earlier reports, they suggested that the reaction proceeded through the formation of the EDA complex, followed by radical intermediate generation by SET.

4 Conclusion

In conclusion, we have summarized the recent advances in light-induced synthesis of two very important heterocycles known as imidazopyridine and imidazothiazole. It is obvious that although an immense number of reports are available on various metal-catalyzed syntheses of these heterocycles, the photoinduced synthesis of imidazopyridine and imidazothiazole is in infancy state. Only a few reports are available so far. In this scenario, chances and challenges are very high. Therefore, more research is needed urgently. Moreover, overcoming hurdles, such as insufficient mechanistic insights, use of costly catalysts, and extended reaction durations, is crucial for achieving optimal efficiency in these processes. The use of modern technologies, such as flow process and batch reactor technology, is still not reported, but these are expected to come very soon. Only a few precious metal (Ir and Ru) photocatalysts are known so far on these synthetic methodologies, and hence cheap first-line 3d-transition metal photocatalysts should also be considered. Mechanistic investigation of these reactions should be carefully done as complete understanding of the mechanism leads to innovative methodology. Finally, we hope that this book chapter will pull a good interest among synthetic organic chemists to carry out more research on photoinduced synthesis of imidazopyridine and imidazothiazole.

Acknowledgements A.H. acknowledges the financial support from CSIR and SERB, New Delhi.

References

1. Mills AD, Nazer MZ, Haddadin MJ, Kurth MJ (2006) N,N-bond-forming heterocyclization: synthesis of 3-alkoxy-2H-indazoles. J Org Chem 71(7):2687–2689. https://doi.org/10.1021/jo0524831
2. Ghosh S, Mondal S, Hajra A (2020) Direct catalytic functionalization of indazole derivatives. Adv Synth Catal 362(18):3768–3794. https://doi.org/10.1002/adsc.202000423
3. Cecile E-G, Alain G (2007) Recent progress in the pharmacology of imidazo[1,2-a]pyridines. Mini Rev Med Chem 7(9):888–899. https://doi.org/10.2174/138955707781662645

4. Nisha D, Dharmender S, Ravindra KR, Jitender B, Virender S (2016) Medicinal attributes of imidazo[1,2-a]pyridine derivatives: an update. Curr Top Med Chem 16(26):2963–2994. https://doi.org/10.2174/1568026616666160506145539
5. Aakash D, Richa Kaur B, Ramanjot K, Sanjiv K, Upendra Kumar J, Harinder S et al (2017) Imidazo[1,2-a]pyridine scaffold as prospective therapeutic agents. Curr Top Med Chem 17(2): 238–250. https://doi.org/10.2174/1568026616666160530153233
6. Samanta S, Ghosh AK, Ghosh S, Ilina AA, Volkova YA, Zavarzin IV et al (2020) Fe(III)-catalyzed synthesis of steroidal imidazoheterocycles as potent antiproliferative agents. Org Biomol Chem 18(29):5571–5576. https://doi.org/10.1039/D0OB01241F
7. Fascio ML, Errea MI, D'Accorso NB (2015) Imidazothiazole and related heterocyclic systems. Synthesis, chemical and biological properties. Eur J Med Chem 90:666–683. https://doi.org/10.1016/j.ejmech.2014.12.012
8. Afsina Abdulla CM, Neetha M, Aneeja T, Anilkumar G (2020) Synthesis and applications of imidazothiazoles: an overview. ChemistrySelect 5(33):10374–10386. https://doi.org/10.1002/slct.202002842
9. Shareef MA, Khan I, Babu BN, Kamal A (2020) A comprehensive review on the therapeutic versatility of imidazo [2,1-b]thiazoles. Curr Med Chem 27(40):6864–6887. https://doi.org/10.2174/0929867326666190729152440
10. Guo M, Yu X, Zhu YZ, Yu Y (2023) From bench to bedside: what do we know about imidazothiazole derivatives so far? Molecules 28(13):5052
11. Andreani A, Granaiola M, Leoni A, Locatelli A, Morigi R, Rambaldi M (2001) Synthesis and antitubercular activity of imidazo[2,1-b]thiazoles. Eur J Med Chem 36(9):743–746. https://doi.org/10.1016/S0223-5234(01)01266-1
12. Park J-H, El-Gamal MI, Lee YS, Oh C-H (2011) New imidazo[2,1-b]thiazole derivatives: synthesis, in vitro anticancer evaluation, and in silico studies. Eur J Med Chem 46(12): 5769–5777. https://doi.org/10.1016/j.ejmech.2011.08.024
13. Ghosh S, Laru S, Hajra A (2022) Ortho C−H functionalization of 2-arylimidazo[1,2-a]pyridines. Chem Rec 22:e202100240. https://doi.org/10.1002/tcr.202100240
14. Bagdi AK, Santra S, Monir K, Hajra A (2015) Synthesis of imidazo[1,2-a]pyridines: a decade update. Chem Commun 51(9):1555–1575. https://doi.org/10.1039/C4CC08495K
15. Bagdi AK, Hajra A (2020) Visible light promoted C–H functionalization of imidazoheterocycles. Org Biomol Chem 18(14):2611–2631. https://doi.org/10.1039/D0OB00246A
16. Ghosh D, Ghosh S, Hajra A (2021) Electrochemical functionalization of imidazopyridine and indazole: an overview. Adv Synth Catal 363(22):5047–5071. https://doi.org/10.1002/adsc.202100981
17. Cheung KPS, Sarkar S, Gevorgyan V (2022) Visible light-induced transition metal catalysis. Chem Rev 122(2):1543–1625. https://doi.org/10.1021/acs.chemrev.1c00403
18. Shaw MH, Twilton J, MacMillan DWC (2016) Photoredox catalysis in organic chemistry. J Org Chem 81(16):6898–6926. https://doi.org/10.1021/acs.joc.6b01449
19. Chan AY, Perry IB, Bissonnette NB, Buksh BF, Edwards GA, Frye LI et al (2022) Metallaphotoredox: the merger of photoredox and transition metal catalysis. Chem Rev 122(2):1485–1542. https://doi.org/10.1021/acs.chemrev.1c00383
20. Bagdi AK, Rahman M, Bhattacherjee D, Zyryanov GV, Ghosh S, Chupakhin ON et al (2020) Visible light promoted cross-dehydrogenative coupling: a decade update. Green Chem 22: 6632–6681. https://doi.org/10.1039/D0GC02437F
21. Ghosh A, Pyne P, Ghosh S, Ghosh D, Majumder S, Hajra A (2022) Visible-light-induced metal-free coupling of C(sp3)–H sources with heteroarenes. Green Chem 24(8):3056–3080. https://doi.org/10.1039/D1GC04384F
22. Ghosh S, Lai D, Hajra A (2021) Visible-light-induced silylation: an update. Org Biomol Chem 19(11):2399–2415. https://doi.org/10.1039/D1OB00082A

23. Staveness D, Bosque I, Stephenson CRJ (2016) Free radical chemistry enabled by visible light-induced electron transfer. Acc Chem Res 49(10):2295–2306. https://doi.org/10.1021/acs.accounts.6b00270
24. Zhou Q-Q, Zou Y-Q, Lu L-Q, Xiao W-J (2019) Visible-light-induced organic photochemical reactions through energy-transfer pathways. Angew Chem Int Ed 58(6):1586–1604. https://doi.org/10.1002/anie.201803102
25. Prier CK, Rankic DA, MacMillan DWC (2013) Visible light photoredox catalysis with transition metal complexes: applications in organic synthesis. Chem Rev 113(7):5322–5363. https://doi.org/10.1021/cr300503r
26. Nicewicz DA, MacMillan DWC (2008) Merging photoredox catalysis with organocatalysis: the direct asymmetric alkylation of aldehydes. Science 322(5898):77–80. https://doi.org/10.1126/science.1161976
27. Twilton J, Le C, Zhang P, Shaw MH, Evans RW, MacMillan DWC (2017) The merger of transition metal and photocatalysis. Nat Rev Chem 1(7):0052. https://doi.org/10.1038/s41570-017-0052
28. Romero NA, Nicewicz DA (2016) Organic photoredox catalysis. Chem Rev 116(17): 10075–10166. https://doi.org/10.1021/acs.chemrev.6b00057
29. Holmberg-Douglas N, Nicewicz DA (2022) Photoredox-catalyzed C–H functionalization reactions. Chem Rev 122:1925–2016. https://doi.org/10.1021/acs.chemrev.1c00311
30. Narayanam JMR, Stephenson CRJ (2011) Visible light photoredox catalysis: applications in organic synthesis. Chem Soc Rev 40(1):102–113. https://doi.org/10.1039/B913880N
31. Tucker JW, Stephenson CRJ (2012) Shining light on photoredox catalysis: theory and synthetic applications. J Org Chem 77(4):1617–1622. https://doi.org/10.1021/jo202538x
32. Mondal S, Ghosh S, Hajra A (2024) Visible-light-induced redox-neutral difunctionalization of alkenes and alkynes. Chem Commun 60:9659. https://doi.org/10.1039/D4CC03552F
33. Hajra A, Mondal M, Ghosh S, Lai D (2024) C–H functionalization of heteroarenes via electron donor-acceptor complex photoactivation. ChemSusChem 17:e202401114. https://doi.org/10.1002/cssc.202401114
34. Gupta S, Kundu A, Ghosh S, Chakraborty A, Hajra A (2023) Visible light-induced organophotoredox-catalyzed difunctionalization of alkenes and alkynes. Green Chem 25(21): 8459–8493. https://doi.org/10.1039/D3GC03291D
35. Ghosh S, Majumder S, Ghosh D, Hajra A (2023) Redox-neutral carbon–heteroatom bond formation under photoredox catalysis. Chem Commun 59:7004–7027. https://doi.org/10.1039/D3CC01873C
36. Ghosh S, Pyne P, Ghosh A, Choudhury S, Hajra A (2023) Visible-light-induced cascade reaction: a sustainable approach towards molecular complexity. Org Biomol Chem 21(8): 1591–1628. https://doi.org/10.1039/D2OB02062A
37. Mohar M, Ghosh S, Hajra A (2023) Visible light induced three-component 1,2-dicarbofunctionalization of alkenes and alkynes. Chem Rec 23:e202300121
38. Neogi S, Ghosh AK, Mandal S, Ghosh D, Ghosh S, Hajra A (2021) Three-component carbosilylation of alkenes by merging iron and visible-light photocatalysis. Org Lett 23(16): 6510–6514. https://doi.org/10.1021/acs.orglett.1c02322
39. Neogi S, Ghosh AK, Majhi K, Samanta S, Kibriya G, Hajra A (2020) Organophotoredox-catalyzed direct C–H amination of 2H-indazoles with amines. Org Lett 22(14):5605–5609. https://doi.org/10.1021/acs.orglett.0c01973
40. Kibriya G, Ghosh D, Hajra A (2020) Visible-light-promoted oxidative coupling of styrene with cyclic ethers. Sci China Chem 63(1):42–46. https://doi.org/10.1007/s11426-019-9609-9
41. Mitra S, Ghosh M, Mishra S, Hajra A (2015) Metal-free thiocyanation of imidazoheterocycles through visible light photoredox catalysis. J Org Chem 80(16):8275–8281. https://doi.org/10.1021/acs.joc.5b01369
42. Singsardar M, Mondal S, Laru S, Hajra A (2019) Organophotoredox-catalyzed C(sp2)–H difluoromethylenephosphonation of imidazoheterocycles. Org Lett 21(14):5606–5610. https://doi.org/10.1021/acs.orglett.9b01954

43. Samanta S, Mondal S, Ghosh D, Hajra A (2019) Rhodium-catalyzed directed C–H amidation of imidazoheterocycles with dioxazolones. Org Lett 21(12):4905–4909. https://doi.org/10.1021/acs.orglett.9b01832
44. Samanta S, Hajra A (2019) Mn(II)-catalyzed C–H alkylation of imidazopyridines and N-heteroarenes via decarbonylative and cross-dehydrogenative coupling. J Org Chem 84(7):4363–4371. https://doi.org/10.1021/acs.joc.9b00366
45. Ghosh P, Samanta S, Ghosh S, Jana S, Hajra A (2020) Aminomethylation of imidazopyridines using N,N-dimethylformamide as an aminomethylating reagent under Cu(II)-catalysis. Tetrahedron Lett 61(49):152581. https://doi.org/10.1016/j.tetlet.2020.152581
46. Ghosh S, Kundu D, Dey A, Majee A, Hajra A (2020) Functionalized ionic liquid tagged Cu (II) catalyst: design, characterization, and application in synthesis of imidazo[1,2-a]pyridine. J Indian Chem Soc 97(12a):2533–2539
47. Kibriya G, Bagdi AK, Hajra A (2018) Visible light induced tetramethylethylenediamine assisted formylation of imidazopyridines. Org Biomol Chem 16(18):3473–3478. https://doi.org/10.1039/C8OB00532J
48. Ghosh D, Ghosh S, Ghosh A, Pyne P, Majumder S, Hajra A (2022) Visible light-induced functionalization of indazole and pyrazole: a recent update. Chem Commun 58(28):4435–4455. https://doi.org/10.1039/D2CC00002D
49. Shivhare KN, Jaiswal MK, Srivastava A, Tiwari SK, Siddiqui IR (2018) Visible-light-activated C–C and C–N bond formation in the synthesis of imidazo[1,2-a]pyridines and imidazo[2,1-b] thiazoles under catalyst and solvent-free conditions. New J Chem 42(20):16591–16601. https://doi.org/10.1039/c8nj03339k
50. Melekhina VG, Komogortsev AN, Lichitsky BV, Mityanov VS, Fakhrutdinov AN, Dudinov AA et al (2019) Unexpected photochemical transformation of imidazole derivatives containing the 5-hydroxy-2-methyl-4H-pyran-4-one moiety. Environmentally friendly method for the synthesis of substituted imidazo[1,5-a]pyridine-5,8-diones. Tetrahedron Lett 60(39):151080. https://doi.org/10.1016/j.tetlet.2019.151080
51. Das A, Thomas KRJ (2020) Light promoted synthesis of quinoxalines and imidazo[1,2-a] pyridines via oxybromination from alkynes and alkenes. Asian J Org Chem 9(11):1820–1825. https://doi.org/10.1002/ajoc.202000388
52. Lima CGS, de M. Lima T, Duarte M, Jurberg ID, Paixão MW (2016) Organic synthesis enabled by light-irradiation of EDA complexes: theoretical background and synthetic applications. ACS Catal 6(3):1389–1407. https://doi.org/10.1021/acscatal.5b02386
53. Filippov IP, Agafonova AV, Titov GD, Smetanin IA, Rostovskii NV, Khlebnikov AF et al (2022) Synthesis of imidazo[1,2-a]pyridines via near UV light-induced cyclization of azirinylpyridinium salts. J Org Chem 87(9):6514–6519. https://doi.org/10.1021/acs.joc.2c00514
54. Adiyala PR, Sastry KNV, Kovvuri J, Nagarajan A, Reddy VG, Sayeed IB et al (2017) Visible light driven coupling of 2-aminopyridines and α-keto vinyl azides for the synthesis of imidazo [1, 2-a]pyridines and their cytotoxicity. ChemistrySelect 2(26):8158–8161. https://doi.org/10.1002/slct.201700952
55. Tong J, Zhan Y, Li J, Liu P, Sun P (2021) One-pot synthesis of C3-alkylated imidazopyridines from α-bromocarbonyls under photoredox conditions. Eur J Org Chem 2021(32):4541–4545. https://doi.org/10.1002/ejoc.202100922
56. Singh HK, Kamal A, Kumari S, Kumar D, Maury SK, Srivastava V et al (2020) Eosin Y-catalyzed synthesis of 3-aminoimidazo[1,2-a]pyridines via the HAT process under visible light through formation of the C–N bond. ACS Omega 5(46):29854–29863. https://doi.org/10.1021/acsomega.0c03941
57. Yadav S, Srivastava M, Rai P, Tripathi BP, Mishra A, Singh J et al (2016) Oxidative organophotoredox catalysis: a regioselective synthesis of 2-nitro substituted imidazopyridines and 3-substituted indoles, initiated by visible light. New J Chem 40(11):9694–9701. https://doi.org/10.1039/c6nj02365g

58. Roslan II, Ng K-H, Jaenicke S, Chuah G-K (2019) Photocatalytic regeneration of brominating agent in the visible light-mediated synthesis of imidazo[1,2-a]pyridines. Catal Sci Technol 9(6): 1528–1534. https://doi.org/10.1039/c9cy00141g
59. Mishra A, Srivastava M, Rai P, Yadav S, Tripathi BP, Singh J et al (2016) Visible light triggered, catalyst free approach for the synthesis of thiazoles and imidazo[2,1-b]thiazoles in EtOH:H2O green medium. RSC Adv 6(54):49164–49172. https://doi.org/10.1039/c6ra05385h
60. Yan K, Liu M, Wen J, Liu X, Wang X, Chen X et al (2021) Visible-light-promoted cascade cyclization towards benzo[d]imidazo[5,1-b]thiazoles under metal- and photocatalyst-free conditions. Green Chem 23(3):1286–1291. https://doi.org/10.1039/d0gc04135a
61. Chen Z, Xue F, Zhang Y, Jin W, Wang B, Xia Y et al (2022) Visible-light-promoted [3 + 2] cyclization of chalcones with 2-mercaptobenzimidazoles: a protocol for the synthesis of imidazo[2,1-b]thiazoles. Org Lett 24(17):3149–3154. https://doi.org/10.1021/acs.orglett.2c00867
62. Chen Z, Jin W, Xia Y, Zhang Y, Xie M, Ma S et al (2020) Aminothiolation of α-bromocinnamaldehydes to access imidazo[2,1-b]thiazoles by incorporation of two distinct photoinduced processes. Org Lett 22(21):8261–8266. https://doi.org/10.1021/acs.orglett.0c02907

Top Heterocycl Chem (2025) 62: 121–138
https://doi.org/10.1007/7081_2025_93

Published online: 3 June 2025

Photoinduced Wolff Rearrangement for Synthesizing Chiral Heterocycles

Yu-Qing Xiao, Dan Liu, Jun Chen, Wen-Jing Xiao, and Liang-Qiu Lu

Contents

Abstract Ketenes serve as versatile synthons in organic chemistry due to their distinctive physical properties and exceptional chemical reactivity. The asymmetric cyclization of ketenes has emerged as a powerful strategy for the efficient construction of chiral heterocyclic scaffolds. In contrast to conventional synthetic approaches, the photoinduced Wolff rearrangement of α-diazoketones offers a milder and more sustainable alternative for the traceless generation of ketenes. This chapter summarizes recent advances in the asymmetric cyclization of photogenerated ketenes via transition metal catalysis and organocatalysis, with particular emphasis on the underlying mechanisms and stereochemical control. In particular, the mild photochemical conditions that allow for the traceless and transient formation of ketenes via Wolff rearrangement are critical to the success of these cyclization reactions.

Keywords Asymmetric catalysis · Heterocyclic synthesis · Ketenes · Photoinduced Wolff rearrangement

Y.-Q. Xiao, D. Liu, J. Chen, W.-J. Xiao, and L.-Q. Lu (✉)
Engineering Research Center of Photoenergy Utilization for Pollution Control and Carbon Reduction, Ministry of Education, College of Chemistry, Central China Normal University, Wuhan, Hubei, People's Republic of China
e-mail: luliangqiu@ccnu.edu.cn

1 Introduction

Ketenes, as highly electron-deficient species, are widely used in synthetic chemistry [1]. More than ten named reactions are related to ketene intermediates, such as Staudinger cycloaddition, Arndt-Eistert homologation, and Danheiser benzannulation [2]. In 1905, Staudinger reported the first generation of diphenylketene **2**, a stable and isolable example, from the dehalogenation of 2-chloro-benzyl acetyl chloride **1** by metal zinc. He then found that ketene can react with (*E*)-*N*,1-diphenylmethanimine **3** by stepwise [2+2] cycloaddition to form β-lactams **4** via zwitterionic intermediate **A** (Scheme 1a) [3]. With the rapid development of ketene chemistry, many efficient methods [1, 4] including the thermal decomposition of ketene dimers **5** [4a], the decarboxylation of malonic acid compounds **7** under the action of trifluoroacetic anhydride (TFAA) [4b], and the dehydrohalogenation of acyl chlorides **9** under basic condition have been commonly used to generate relatively stable ketenes up to now [4d]. In particular, the Wolff rearrangement of α-diazoketones **10** is also an efficient transformation that allows the synthesis of stable and relatively unstable ketenes in situ (Scheme 1b) [5].

Ketene has various different reactivities due to its special molecular structure [6]. Most of the reactivity can be explained by the polarization properties of this molecule [7]. The lowest unoccupied molecular orbital (LUMO) of the molecule is mainly on the α-carbon atom, and the highest occupied molecular orbital (HOMO) is mainly on the β-carbon atom and the oxygen atom. Therefore, nucleophiles are more likely to react with the α-carbon atom, while electrophiles are more likely to react with the terminal atoms (Scheme 2).

Over the past few decades, organic chemists have conducted in-depth research on the synthesis and reactions of ketenes [1]. Among them, the asymmetric cyclization of ketenes is considered to be one of the important strategies to efficiently construct

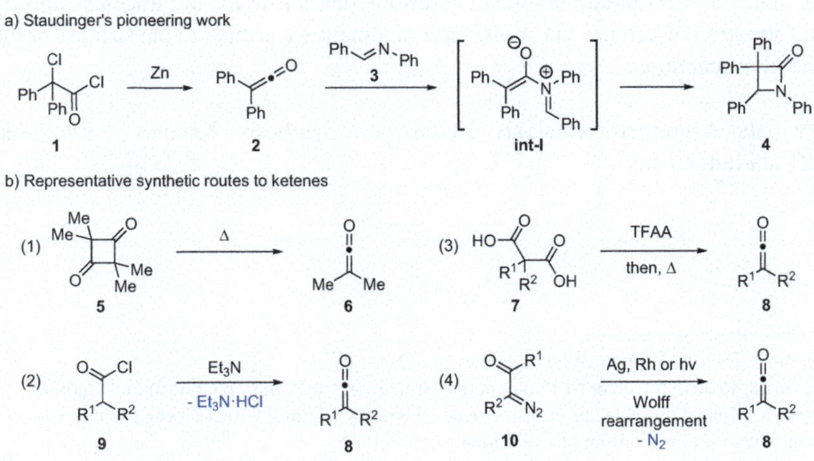

Scheme 1 Representative synthetic routes to ketenes

Scheme 2 Frontier molecular orbital analysis and resonance structures of ketenes

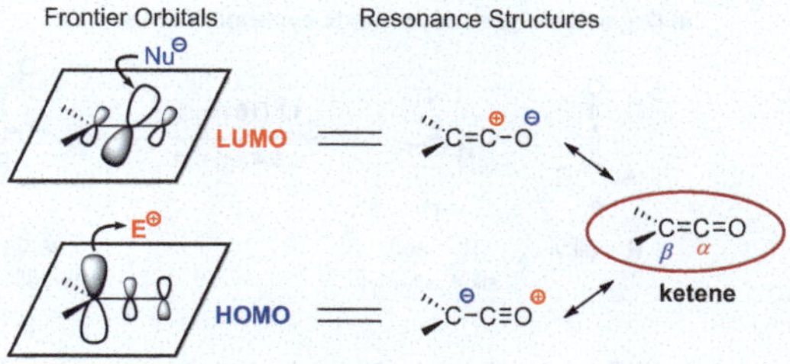

chiral heterocyclic compounds and has attracted widespread attention from chemists worldwide [8]. In the early 2000s, most of these reactions involved ketenes obtained by dehydrohalogenation of acyl chloride [9–11]. For example, in 2002, Fu reported an asymmetric cyclization of disubstituted ketene **8** and imines **11** by using the planar chiral ferrocene-derived tertiary amine catalyst **12** to afford chiral β-lactams **13** with good yields and high enantioselectivities and diastereoselectivities (Scheme 3a) [9b]. In 2008, the Ye group disclosed an asymmetric dimerization of disubstituted ketenes **8** by using chiral *N*-heterocyclic carbene (NHC) catalyst **14** without the addition of extra electrophiles (Scheme 3b) [9c]. In addition to organocatalyzed cyclizations, chiral Lewis acids can also be used as efficient catalysts to achieve the asymmetric cyclization of ketenes. For example, in 2014, the Feng group reported asymmetric [2+2] cycloadditions of disubstituted ketenes **8** in the presence of the self-developed chiral *N*,*N*′-dioxide ligand **17** and $Sc(OTf)_3$ as a Lewis acid to produce a series of highly functionalized β-lactones **18** bearing vicinal chiral centers in excellent yields and high diastereo- and enantioselectivities (Scheme 3c) [10c]. However, the asymmetric cyclization of ketenes via transition metal catalysis is relatively difficult to realize due to the instability of ketenes towards transition metals [12–14]. In 2011, the Louie group realized a Ni-catalyzed [2+2+2] annulation of diyne **19** and ketene **8a** with the assistance of (*R*)-BINAP as a chiral ligand, which afforded cyclohexadienone **20** in moderate yield and excellent enantioselectivity (Scheme 3d) [12b].

Despite these apparent advances, the exploration of reaction with ketenes still faces some challenges: the dehydrohalogenation of acyl chloride under basic condition to obtain ketenes is usually accompanied by the production of equivalent amount of ammonium salts, which often has an unfavorable effect on subsequent reactions [13], and it is therefore necessary to separate the corresponding by-products, which increases the difficulty of experimental operation [4d]. However, photoinduced Wolff rearrangement can provide milder and more sustainable reaction conditions for the efficient and trace-free production of ketenes, yet the potential of this process has not been fully exploited in ketene chemistry [15]. In this chapter, we will briefly review the recent advances in the asymmetric cyclization with photogenerated ketenes, which generally include transition metal-catalyzed and organocatalyzed asymmetric cyclizations.

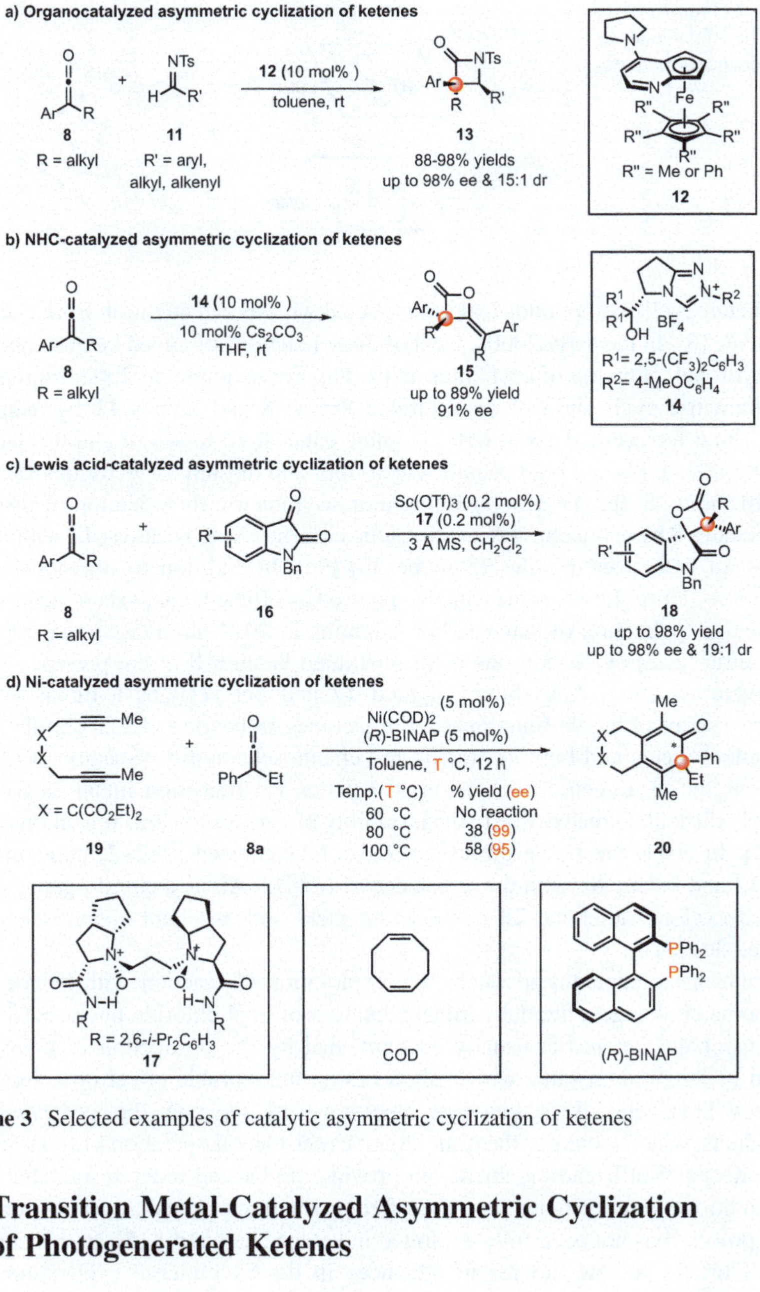

Scheme 3 Selected examples of catalytic asymmetric cyclization of ketenes

2 Transition Metal-Catalyzed Asymmetric Cyclization of Photogenerated Ketenes

Transition metal-catalyzed cyclizations, which are capable of architecturing complex heterocyclic compounds from readily available starting materials, have emerged as a cornerstone of modern synthetic organic chemistry [16]. However, the transition metal-catalyzed asymmetric cyclization of ketenes still remains underexploited due to some of the following reasons. First, ketene, as a kind of highly reactive intermediate, is very susceptible to decarbonylation or self-dimerization in the presence of transition metals [14]. In addition, ketenes are generally obtained by in situ dehydrohalogenation of acyl chlorides in the presence of base, leading to the formation of salts, which has a harmful effect on subsequent transition metal-catalyzed transformations [13]. Therefore, in the past few decades, there have been few reports on the transition metal-catalyzed cyclization of ketenes [12]. As a comparison, photoinduced Wolff rearrangement enables the efficient, traceless synthesis of ketenes under mild conditions [5]. This method provides a new possibility to realize transition metal-catalyzed ketene cyclization, and a series of sequential visible light-promoted Wolff rearrangement/transition metal-catalyzed asymmetric cyclizations have been reported in recent years.

Palladium-catalyzed cyclization has achieved great progresses since 1987 [17]. In 2017, Lu et al. proposed a visible-light photoactivation and palladium catalysis sequence strategy to realize an asymmetric [4+2] cyclization reaction of photogenerated ketenes. This protocol enables the enantioselective synthesis of chiral quinolinones **23** containing all carbon quaternary stereocenter from vinyl benzoxazinanones **21** and α-diazoketones **10** (Scheme 4) [18]. In particular, they found that when using a chiral hybrid P,S ligand **22** developed by themselves, excellent yields and stereoselectivities could be accessed. To demonstrate the advantages of the photogenerated ketenes, two control experiments were conducted by using prepared ketene **8** with $Et_3N{\cdot}HCl$ or using acyl chloride and Et_3N to form ketene **8** in situ. However, quinolinone **23** was formed in very low yield. These results suggested that $Et_3N{\cdot}HCl$, the by-product of the in situ generation of ketene

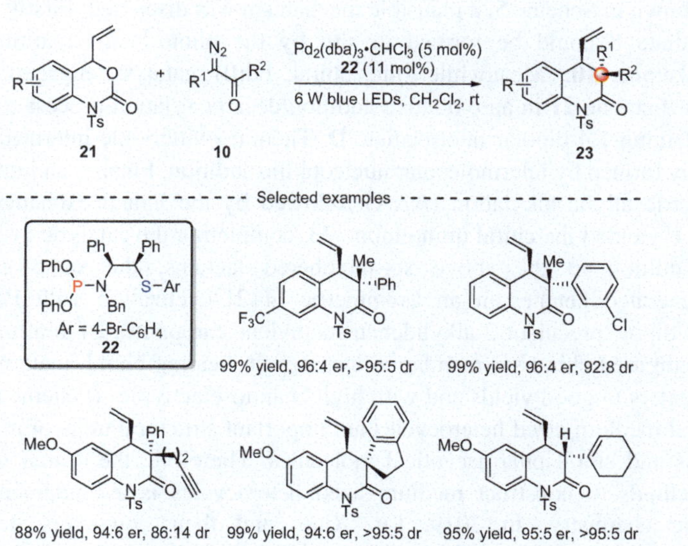

Scheme 4 Pd-catalyzed asymmetric [4+2] cyclization of vinyl benzoxazinanone with photogenerated ketenes

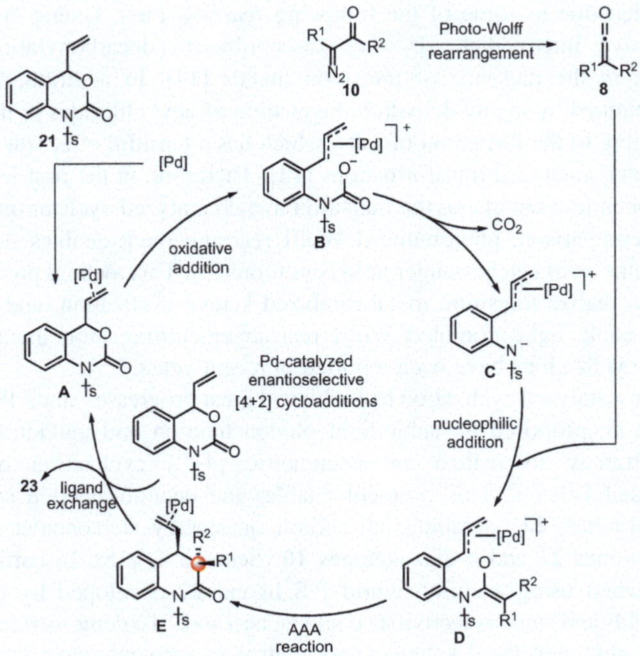

Scheme 5 Mechanism of Pd-catalyzed [4+2] cyclization

from acyl chloride, actually has a detrimental effect on this transformation, clearly demonstrating the advantage of the traceless in situ photogeneration of ketenes.

As shown in Scheme 5, a plausible mechanism was discussed. Firstly, the ketene intermediate **8** could be formed in situ by the photo-Wolff rearrangement of α-diazoketone **10**. Meanwhile, the chiral Pd(0) catalyst engaged the vinyl benzoxazinanone **21** in an oxidative addition/decarboxylation process to afford the Pd-containing 1,4-dipolar intermediate **D**. Then, a zwitterionic intermediate **E** was smoothly formed by intermolecular nucleophilic addition. Finally, an intramolecular asymmetric allylic alkylation (AAA) followed by the ligand exchange from the species **F** yielded the chiral quinolinone **23**, completing the catalytic cycle.

In addition to the above six-membered lactams, the same group used photogenerated ketenes in an asymmetric [4+2] cyclization with Pd-stabilized oxo-1,4-dipole precursor 2-alkylidenetrimethylene carbonates **24** to afford a variety of *Z*-configured chiral six-membered lactones **26** bearing chiral quaternary carbon stereocenters in good yields and with high enantioselectivities (Scheme 6) [19].

Chiral medium-sized heterocycles are important structural units of many natural products and chiral pharmaceutical molecules. Therefore, the search for efficient new methods to construct medium-sized heterocycles is a fundamental task in synthetic chemistry. In 2019, Lu, Xiao, and their coworkers developed a Pd-catalyzed asymmetric [5+2] cyclization of vinylethylene carbonates **27** with photogenerated ketenes. However, unlike the aforementioned works, the

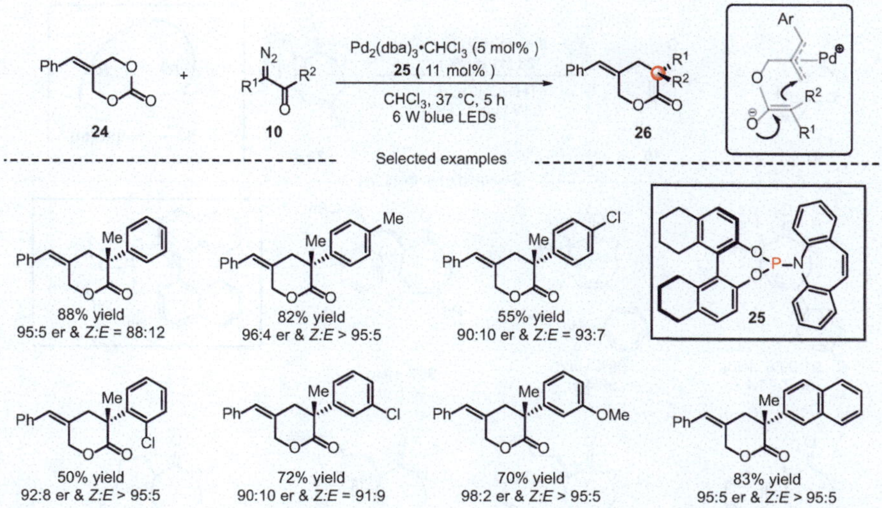

Scheme 6 Pd-catalyzed asymmetric [4+2] cyclization of 2-alkylidenetrimethylene carbonates with photogenerated ketenes

regioselectivity of this strategy in the final ring closure step is terminal attack to form the 7-membered lactones **29** due to the steric effect. The desired products could be formed with good efficiency and high enantioselectivity by using the phosphoramidite ligand **28**. Nonlinear experiments revealed that a palladium atom coordinated with two chiral ligands forms an effective catalyst involved in this catalytic asymmetric cycloaddition. Furthermore, characteristic mass signals of the key π-allyl Pd intermediates were detected, elucidating the proposed reaction mechanism (Scheme 7) [20].

In 2020, the same group successfully achieved another Pd-catalyzed, visible light-driven, asymmetric [5+2] cyclization of vinylcyclopropane **30** with α-diazoketones **10**. Enantioenriched seven-membered lactone compounds **32** bearing chiral quaternary carbon centers were synthesized with high efficiency and selectivity (Scheme 8) [21]. The key feature of this designed reaction is the generation of Pd-containing oxa-1,5-dipoles by using the second-generation chiral P,S ligands **31** developed by their group; this intermediate could be detected by high-resolution mass spectrometry. Computational investigations were performed to explain the high chemo- and periselectivity, with the results suggesting that the intermolecular [5+2] cyclization would be the most kinetically favorable process.

In the same year, Lu and Xiao introduced an enantioselective high-order [8+2] dipolar cyclization via the merger of visible-light photoactivation and asymmetric palladium catalysis. By using vinyl carbamates **33** as the dipole precursors and hybrid P,S ligand **34** as the optimal ligand, a series of enantioenriched ten-membered monocyclic compounds **35** with chiral quaternary stereocenters were achieved with high efficiency and excellent enantioselectivity (Scheme 9) [22]. In this reaction, the CO_2 unit of vinyl carbamates **33** could be retained, in

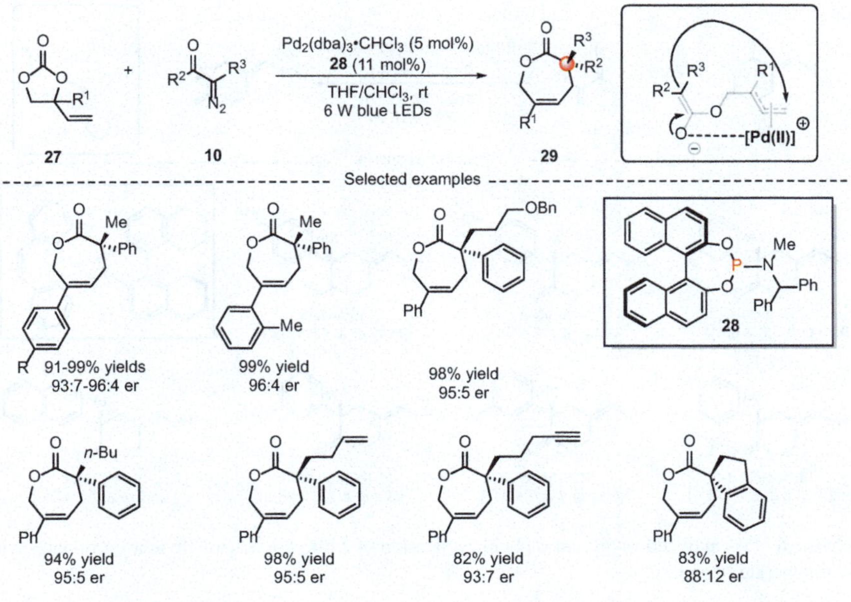

Scheme 7 Pd-catalyzed asymmetric [5+2] cyclization of vinylethylene carbonates with ketenes

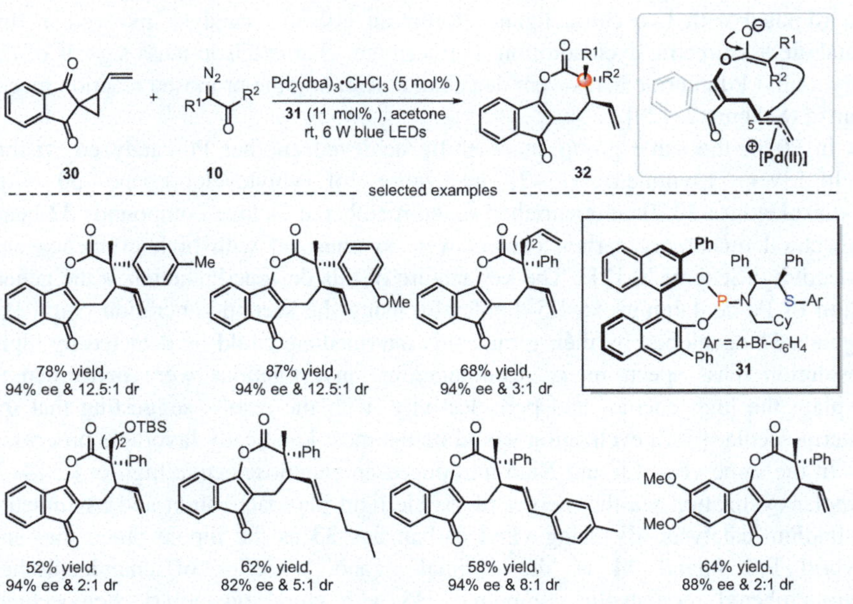

Scheme 8 Pd-catalyzed asymmetric [5+2] cyclization of vinylcyclopropane with photogenerated ketenes

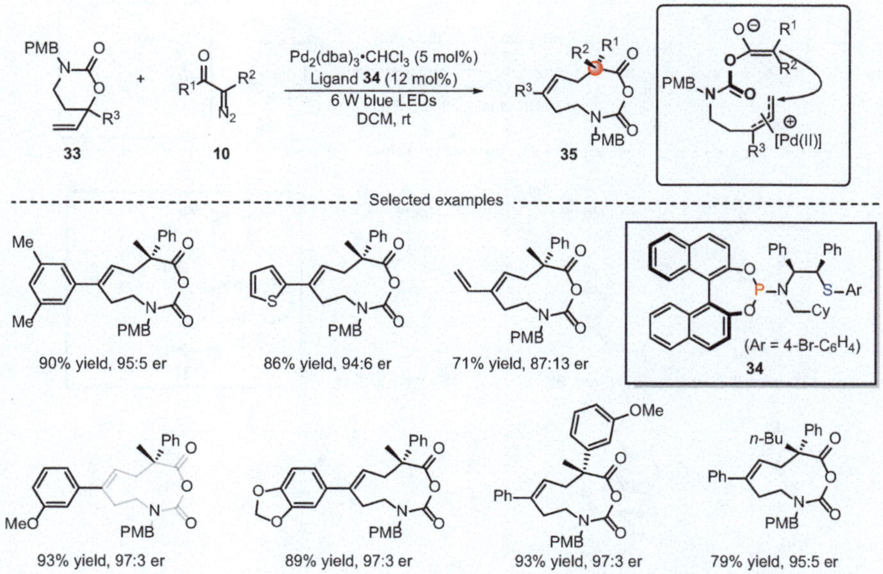

Scheme 9 Pd-catalyzed asymmetric [8+2] cyclization of vinyl carbamates with photogenerated ketenes

sharp contrast to the previously developed dipolar cyclization. The control experiment indicated that the protected group on the nitrogen atom of vinyl carbamates is crucial for the retention of the CO_2 unit. Furthermore, DFT calculations were performed to illustrate the high chemo- and peri-selectivities, which were attributed to the stronger Lewis basicity of aliphatic amines than tosyl amides and the less steric repulsion of the terminal carbon of the π-allyl-Pd unit.

More recently, Lu et al. reported a Pd-catalyzed enantioselective [6+2] dipolar cyclization between vinyl oxetanes **36** and photogenerated ketenes. This protocol allows the efficient synthesis of eight-membered lactones **38** bearing chiral all-carbon quaternary stereocenters. A possible stereoinduction mode was proposed, and the newly developed chiral hybrid P,S ligand **37** was believed to be the key to realize the linear selectivity and high enantioselectivity (Scheme 10) [23].

In the previous work, ketenes were identified as efficient C2 synthons for [*n*+2] cyclization by addition via their C=C bonds. In 2019, Lu and co-workers used vinyl cyclopropanes as the Pd-containing dipole precursors to realize a novel cascade photoinduced Wolff rearrangement/Pd-catalyzed [3+2] cyclization. Selective *O*-allylic alkylation was observed over usual *C*-allylic alkylation, surprisingly generating a series of highly functionalized tetrahydrofurans rather than the expected cyclopentanones [24]. Four years later, the same group chose cyano- and ester-substituted vinyl cyclopropane **39** as the Pd-containing 1,3-dipole precursor to realize the asymmetric variant. After screening a variety of ligands, they found that when the chiral H-bonding ligand **40** [25] was used, the desired tetrahydrofuran

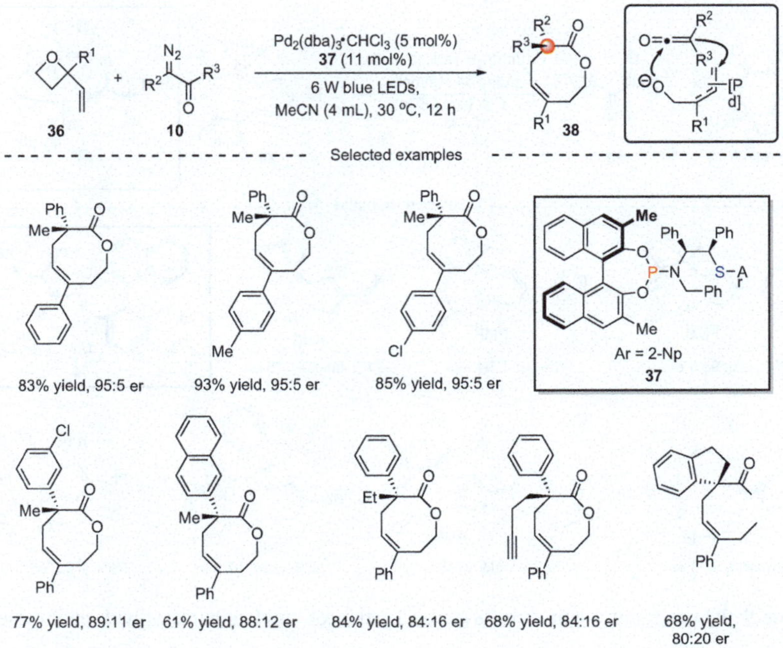

Scheme 10 Pd-catalyzed [6+2] asymmetric cyclization of vinyl oxetanes with photogenerated ketenes

products **41** could be formed in good yields with high diastereo- and enantioselectivities. They proposed a stereocontrol mode which revealed that the key to the success of this stereoselective transformation is the hydrogen bond from the amide moiety of the chiral ligand **40**, through remote H-bonding interactions between the chiral ligands and substrates to govern the high stereoselectivity (Scheme 11) [26].

Aforementioned impressive advances have proven that Pd-catalyzed asymmetric cyclization with photogenerated ketenes is an efficient strategy to form complexity medium-sized chiral heterocycles. In 2021, Lu and co-workers further developed a three-component reaction sequence merging the photoinduced Wolff rearrangement with dipolar cyclization and fragmentation through multiple Ir/Lewis acid/base catalysis, affording a series of chiral eight-membered lactams **47** with high reaction efficiency and selectivity in one pot. Based on these results, they proposed a possible reaction pathway as shown in Scheme 12. First, chiral Ir catalyst coordinated with allyl alcohol **46** to form intermediate **A** and then underwent oxidative addition to form intermediate **B** under the action of Lewis acid. Meanwhile, the ketene intermediate **48** could be formed in situ by the photo-Wolff rearrangement of diazoketone **42** and then trapped by alcohols or thiols **43** to form 2-acyl cyclobutanone **44**. After that, an asymmetric allylic alkylation of the enol intermediate **C**, which was generated from **44** and Lewis acid, afforded intermediate **D**. Subsequently, ligand

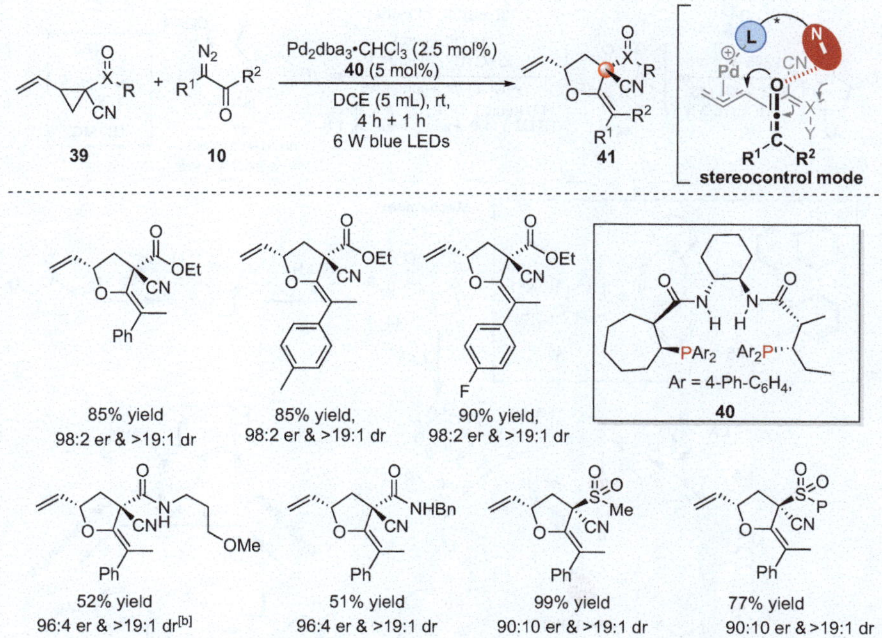

Scheme 11 Pd-catalyzed asymmetric [3+2] cyclization of vinyl cyclopropanes with photogenerated ketenes

exchange formed intermediate **E** and furnished the Ir catalysis cycle. Next, treatment of **E** with base afforded intermediate **F**, which underwent intramolecular cyclization to afford the intermediate **H**. Finally, the fragmentation of **H** driven by strain release followed by protonation produced the desired eight-membered heterocycle **47**. Clearly, this catalytic asymmetric dipolar cyclization/fragmentation strategy provides a new avenue for the synthesis of medium-sized chiral heterocycles by using photogenerated ketenes (Scheme 12) [27].

3 Organocatalyzed Asymmetric Cyclizations of Photogenerated Ketenes

In addition to transition metal catalysis, the asymmetric cyclization of photogenerated ketenes can also be achieved through organocatalysis. Due to the highly electron-deficient character of ketenes, the α-carbon atom of ketenes is electrophilic, which can react with chiral nucleophiles to produce chiral enolates, further realizing following asymmetric cyclization [28]. Therefore, the selection of chiral nucleophilic catalysts is very critical. Recently, the most commonly used

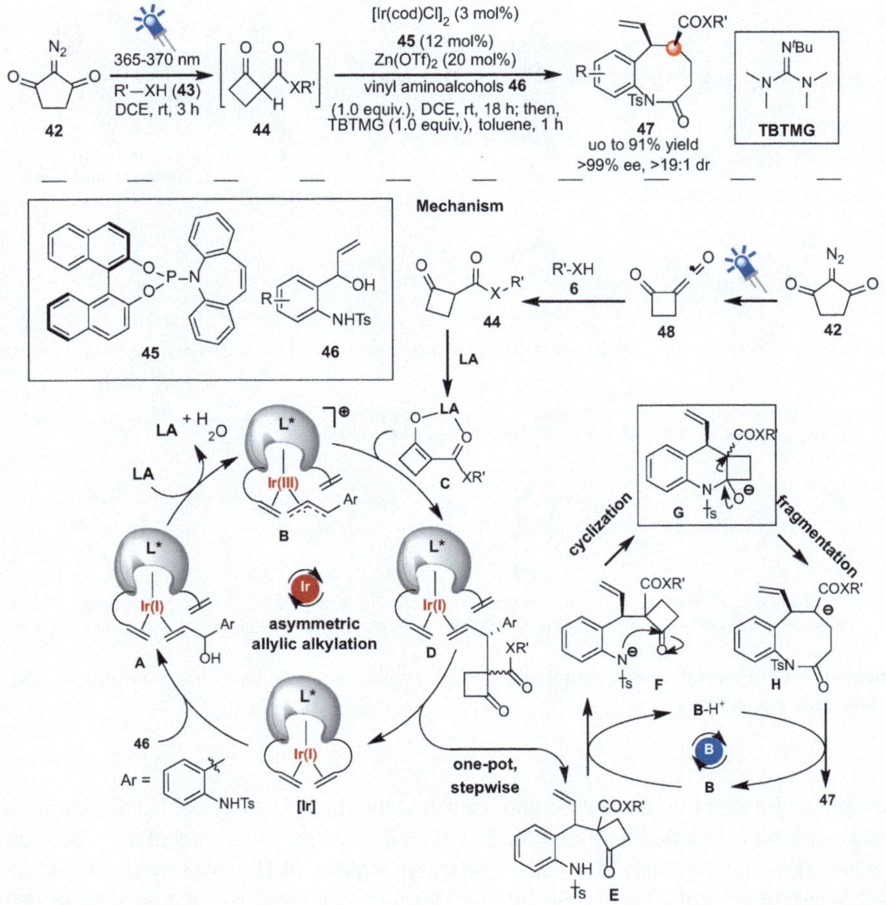

Scheme 12 Enantioselective dipolar cyclization/fragmentation sequence with photogenerated ketenes via multiple Ir/Lewis acid/base catalysis

nucleophilic catalysts include chiral carbene catalysts, chiral isothiourea catalysts, and so on [9, 29, 30]. The first research on asymmetric organocatalyzed cyclization of photogenerated ketenes dates back to 2002. Lectka used commercially available cinchona alkaloid **BQ** as nucleophilic catalyst to achieve the [2+2] asymmetric cyclization of photogenerated ketene **50** with imines **51** under mild conditions; chiral β-lactam product **52** was synthesized in moderate yield and with high stereoselectivity. This is believed to be the first time that photogenerated ketene has been used to achieve asymmetric catalytic cyclizations, but no further research was conducted in this area (Scheme 13) [4d].

Until 2019, Song achieved a visible light-promoted Wolff-rearrangement/organocatalytic asymmetric [4+2] cyclization cascade of azadienes **53** and α-diazoketone **10** by using chiral isothioure (**ITU**) catalyst **54**. This protocol

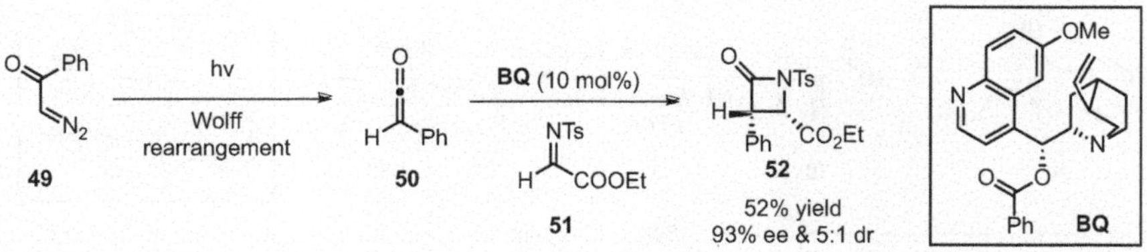

Scheme 13 Chiral amine-catalyzed asymmetric [4+2] cyclization of imines with photogenerated ketenes

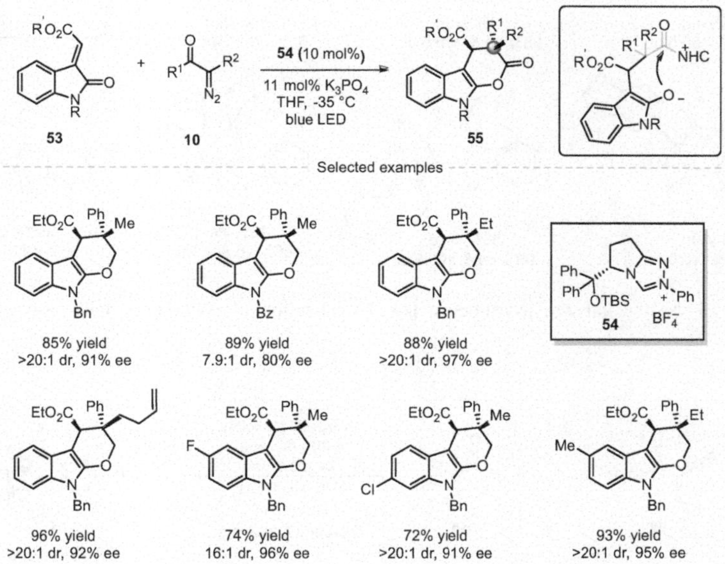

Scheme 14 ITU-catalyzed asymmetric [4+2] cyclization of azadienes with photogenerated ketenes

successfully produced a series of structurally diverse benzofuran derivatives **55** in good yields with high enantio- and diastereoselectivities (Scheme 14) [30]. Independently, Hui reported an enantioselective [4+2] cyclization of enamides **56** and α-diazoketones **10** under visible-light irradiation by using chiral *N*-heterocyclic carbene catalyst (**NHC**). A series of tetrahydropyranoindoles **58** with an all-carbon quaternary stereocenter could be achieved successfully in good yields with excellent stereoselectivities (Scheme 15) [29]. Obviously, the sustained release of ketenes by photoinduced Wolff rearrangement and rearomatization-driven cyclization of zwitterionic intermediates are important to the success of these two cyclization reactions.

In 2024, Xie, Lu, and Xiao developed an asymmetric cyclization of enynamides **59** and α-diazoketones **10** through the combination of visible-light photoactivation and the relay of gold and *N*-oxide catalysis with up to 95% yield, 98% ee, and >19:1 dr (Scheme 16) [31]. The chiral *N*-oxides were identified for the first time as efficient

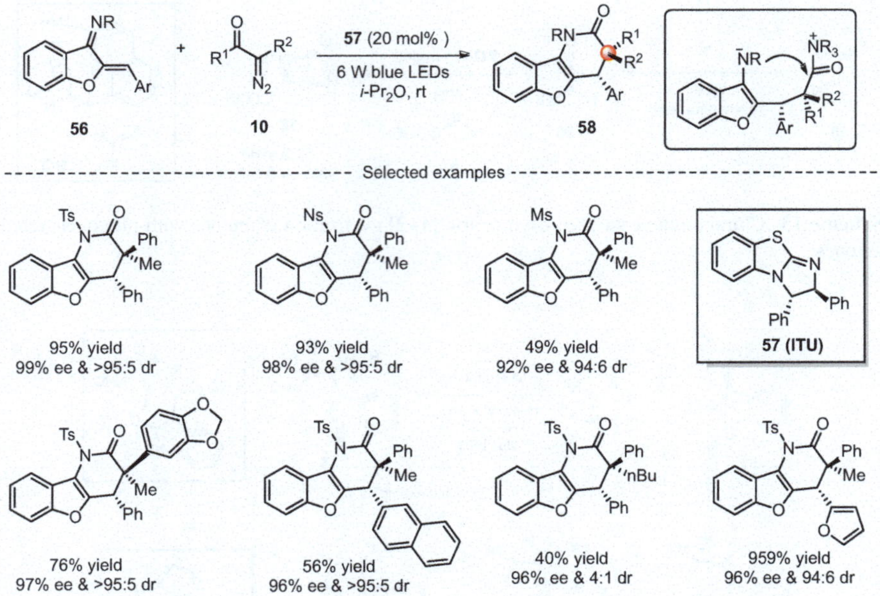

Scheme 15 NHC-catalyzed asymmetric [4+2] cyclization of oxo-dienes with photogenerated ketenes

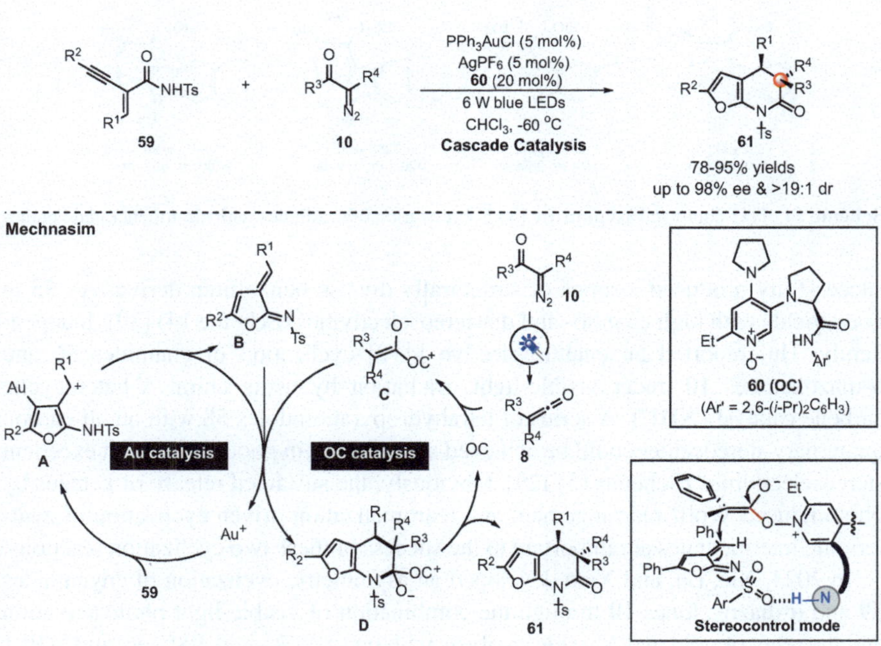

Scheme 16 Asymmetric cascade cyclization of enynamides with photogenerated ketenes via relay gold and *N*-oxide catalysis

catalysts for asymmetric ketene cyclization. Theoretical calculations revealed a stepwise [4+2] cyclization mechanism. First, α-diazoketone **10** undergoes photoinduced Wolff rearrangement to generate the ketene **8** in situ. Meanwhile, enynamide **59** undergoes intramolecular cyclization to form *aza-o*-quinone **B** in situ via cation intermediate **A** under the action of gold catalyst. After that, these two reactive species undergo asymmetric stepwise [4+2] cyclization via intermediates **C** and **D** by chiral *N*-oxide catalyst **60**. The stereocontrol mode proposed by the author suggested that the stereochemistry was controlled by the amide group of the organocatalyst. This cascade cyclization provides an efficient approach to construct molecular complexity and underscores its future potential in chemical synthesis.

4 Conclusion

In this chapter, we have summarized the recent advances in the asymmetric cyclization of photogenerated ketenes by employing transition metal catalysis and organocatalysis. It is well established that the cascade-photoinduced Wolff rearrangement/catalytic asymmetric cyclization is an efficient strategy to construct many chiral heterocyclic compounds that are difficult to form by classical methods. Notably, the traceless and transient generation of ketenes from α-diazoketones under mild conditions by photoinduced Wolff rearrangement is crucial for the success of the aforementioned cyclizations.

Despite these great achievements, future efforts are still expected in the following three areas. First, the exploitation of new ketene analogues, in which oxygen is replaced by sulfur or nitrogen, is necessary; they could enable the construction of heterocyclic molecules that are difficult to obtain by traditional methods. Second, most asymmetric cyclizations involving ketenes undergo *C*-allylic alkylation in the ring-closing step. *O*-allylic alkylation has only one example of reaction with vinyl cyclopropanes by Lu and Xiao. For the selectivity between *O*-allylic alkylation and *C*-allylic alkylation in the ring-closing step, more researches are still required to discover the rules behind this selectivity in order to realize more achievements through *O*-allylic alkylation. Third, at present, ketenes are usually involved in the reaction as C2 synthons, so achieving the cyclization of α,β-unsaturated ketenes as C4 synthons is the focus of the next research. We have enough reasons to believe that the development of catalytic asymmetric cyclization of ketenes will not only contribute to the research areas of heterocyclic synthesis and synthetic methodology, but also benefit the discovery of therapeutic candidates.

References

1. For selected reviews and books, see: (a) Tidwell TT (2006) Ketenes II, 2nd edn. Wiley. (b) Tidwell TT (2005) Angew Chem Int Ed 44:5778. (c) Allen AD, Tidwell TT (2012) Eur J Org Chem 2012:1081. (d) Allen AD, Tidwell TT (2013) Chem Rev 113:7287

2. Hassner A, Namboothiri I (2012) Organic syntheses based on name reactions, 3rd edn. Elsevier
3. (a) Staudinger H (1905) Chem Ber 38:1735. (b) Staudinger H (1907) Chem Ber 40:1145. (c) Staudinger H (1907) Justus Liebigs Ann Chem 356:51
4. (a) Moore HW, Wilbur DS (1980) J Org Chem 45:4483. (b) Kresze G, Runge W, Ruch E (1972) Justus Liebigs Ann Chem 756:112. (c) Taggi AE, Hafez AM, Wack H, Young B, Ferraris D, Lectka T (2002) J Am Chem Soc 124:6626. (d) Chiang Y, Kresge AJ, Popik VV (1999) J Am Chem Soc 121:5930. (e) Wagner BD, Arnold BR, Brown GS, Lusztyk J (1998) J Am Chem Soc 120:1827. (f) Zhang Z, Liu Y, Ling L, Li Y, Dong Y, Gong M, Zhao X, Zhang Y, Wang J (2011) J Am Chem Soc 133:4330
5. For selected reviews, see: (a) Kirmse W (2002) Eur J Org Chem 14:2193. (b) Ford A, Miel H, Ring A, Slattery CN, Maguire AR, McKervey MA (2015) Chem Rev 115:9981. For selected works, see: (c) Yang H, Foster K, Stephenson CRJ, Brown W, Roberts E (2000) Org Lett 2: 2177. (d) Bernardim B, Hardman-Baldwin AM, Burtoloso ACB (2015) RSC Adv 5:13311. (e) Chapman LM, Beck JC, Wu L, Reisman SE (2016) J Am Chem Soc 138:9803. (f) Liu D, Ding W, Zhou Q-Q, Wei Y, Lu L-Q, Xiao W-J (2018) Org Lett 20:7278. (g) Meng J, Ding WW, Han Z-Y (2019) Org Lett 21:9801
6. (a) Snider BB (1988) Chem Rev 88:793. (b) France S, Guerin DJ, Miller SJ, Lectka T (2003) Chem Rev 103:2985. (c) Tidwell TT (2005) Angew Chem Int Ed 44:5778. (d) Paull DH, Weatherwax A, Lectka T (2009) Tetrahedron 65:6771
7. Geoffroy GL, Bassner SL (1998) Adv Organometal Chem 28:1
8. For selected reviews, see: (a) Fu GC (2004) Acc Chem Res 37:542. (b) Gaunt MJ, Johansson CCC (2007) Chem Rev 107:5596. (c) Paull DH, Weatherwax A, Lectka T (2009) Tetrahedron 65:6771
9. For selected work, see: (a) Taggi AE, Hafez AM, Wack H, Young B, Drury WJ III, Lectka T (2000) J Am Chem Soc 122:7831. (b) Hodous BL, Fu GC (2002) J Am Chem Soc 124:1578. (c) Lv H, Zhang Y-R, Huang X-L, Ye S (2008) Adv Synth Catal 350:2715. (d) Huang X-L, He L, Shao P-L, Ye S (2009) Angew Chem Int Ed 48:192. (e) Ibrahim AA, Wei P-H, Harzmann GD, Kerrigan NJ (2010) J Org Chem 75:7901
10. For selected works, see: (a) Evans DA, Janey JM (2001) Org Lett 3:2125. (b) Gnanadesikan V, Corey E (2006) J Org Lett 8:4943. (c) Hao X-Y, Liu X-H, Li W, Tan F, Chu Y-Y, Zhao X-H, Lin L-L, Feng X-M (2014) Org Lett 16:134
11. For a review, see: (a) Paull DH, Abraham CJ, Scerba MT, Alden-Danforth E, Lectka T (2008) Acc Chem Res 41:655. For selected examples, see: (b) France S, Shah MH, Weatherwax A, Wack H, Roth JP, Lectka T (2005) J Am Chem Soc 127:1206. (c) Abraham CJ, Paull DH, Bekele T, Scerba MT, Dudding T, Lectka T (2008) J Am Chem Soc 130:17085. (d) Paull DH, Scerba MT, Alden-Danforth E, Widger LR, Lectka T (2008) J Am Chem Soc 130:17260. (e) Erb J, Paull DH, Dudding T, Belding L, Lectka T (2011) J Am Chem Soc 133:7536.
12. (a) Staudaher ND, Arif AM, Louie J (2016) J Am Chem Soc 138:14083. (b) Kumar P, Troast DM, Cella R, Louie J (2011) J Am Chem Soc 133:7719
13. Ouyang K, Xi Z (2013) Huaxue Xuebao 71:13
14. For metal-ketene complexes, see: (a) Miyashita A, Grubbs RH (1981) Tetrahedron Lett 22: 1255. (b) Miyashita A, Shitara H, Nohira H (1985) Organometallics 4:1463. (c) Strau DA, Grubbs RH (1982) J Am Chem Soc 104:5499. (d) Grotjahn DB, Hoerter JM, Hubbard JL (2004) J Am Chem Soc 126:8866. (e) Grotjahn DB, Hoerter JM, Hubbard JL (1999) Organometallics 18:5614
15. Kirmse W (2002) Eur J Org Chem 14:2193
16. (a) Zhang M-M, Qu B-L, Shi B, Xiao W-J, Lu L-Q (2022) Chem Soc Rev 51:4146. (b) Trowbridge A, Walton SM, Gaunt M (2020) J Chem Rev 120:2613
17. Trost BM, Mata G (2020) Acc Chem Res 53:1293
18. Li M-M, Wei Y, Liu J, Chen H-W, Lu L-Q, Xiao W-J (2017) J Am Chem Soc 139:14707
19. Li M-M, Qu B-L, Xiao Y-Q, Xiao W-J, Lu L-Q (2021) Sci Bull 66:1719
20. Wei Y, Liu S, Li M-M, Li Y, Lan Y, Lu L-Q, Xiao W-J (2019) J Am Chem Soc 141:133

21. Li M-M, Xiong Q, Qu B-L, Xiao Y-Q, Lan Y, Lu L-Q, Xiao W-J (2020) Angew Chem Int Ed 59:17429
22. Zhang Q-L, Xiong Q, Li M-M, Xiong W, Shi B, Lan Y, Lu L-Q, Xiao W-J (2020) Angew Chem Int Ed 59:14096
23. Wang B-C, Wei Y, Lu L-Q, Xiao W-J (2022) Sci China Chem 65:2437
24. Liu J, Li M-M, Qu B-L, Lu L-Q, Xiao W-J (2019) Chem Commun 55:2031
25. Cao M-Y, Ma B-J, Lao Z-Q, Wang H-L, Wang J, Liu J, Xing K, Huang Y-H, Gan K-J, Gao W, Wang H-M, Hong X, Lu H-H (2020) J Am Chem Soc 142:12039
26. Xiao Y-Q, Li M-M, Zhou Z-X, Li Y-J, Cao M-Y, Liu X-P, Lu H-H, Rao L, Lu L-Q, Beauchemin AM, Xiao W-J (2023) Angew Chem Int Ed 62:e202212444
27. Zhang M-M, Chen P, Xiong W, Hui X-S, Lu L-Q, Xiao W-J (2022) CCS Chem 4:2620
28. Gevorgyan V (2024) Chem Rev. https://doi.org/10.1021/acs.chemrev.3c00869
29. Wang C, Wang Z, Yang J, Shi S-H, Hui X-P (2020) Org Lett 22:4440
30. Fan T, Zhang Z-J, Zhang Y-C, Song J (2019) Org Lett 21:7897
31. Liu D, Shu Z-C, Zhang Z, Wang Z-T, Wang L, Xie M-S, Guo H-M, Lu L-Q, Xiao W-J (2024) ACS Catal 14:1741

Top Heterocycl Chem (2025) 62: 139–160
https://doi.org/10.1007/7081_2025_86

Published online: 13 April 2025

Heterocycle Synthesis Through Photochemical Transformation of 2*H*-Azirines

Bao-Gui Cai and Jun Xuan

Contents

Abstract Heterocyclic compounds play a crucial role in various industries such as pharmaceuticals, agrochemicals, and dyes. In recent years, photocatalytic technology has gained popularity among synthetic chemists due to its environmentally friendly and sustainable characteristics. At the same time, numerous studies have focused on the photochemical transformation of 2*H*-azirines, leading to the synthesis

---

B.-G. Cai
Anhui Province Key Laboratory of Chemistry for Inorganic/Organic Hybrid Functionalized Materials and Key Laboratory of Functional Inorganic Materials of Anhui Province, College of Chemistry and Chemical Engineering, Anhui University, Hefei, Anhui, People's Republic of China

J. Xuan (✉)
Anhui Province Key Laboratory of Chemistry for Inorganic/Organic Hybrid Functionalized Materials and Key Laboratory of Functional Inorganic Materials of Anhui Province, College of Chemistry and Chemical Engineering, Anhui University, Hefei, Anhui, People's Republic of China

Key Laboratory of Structure and Functional Regulation of Hybrid Materials, Anhui University, Ministry of Education, Hefei, Anhui, People's Republic of China
e-mail: xuanjun@ahu.edu.cn

of diverse heterocyclic compounds. Therefore, this book chapter aims to provide a summary of heterocycle synthesis through the photochemical transformation of 2*H*-azirines published so far.

Keywords 2*H*-azirines · Heterocycle · Organic synthesis · Photochemistry

1 Introduction

Heterocyclic compounds, characterized by their cyclic structure consisting of atoms from at least two different elements, represent a significant class of organic molecules with wide-ranging applications in pharmaceuticals, agrochemicals, dyes etc. [1–6]. As such, the methodologies for synthesizing heterocyclic compounds are of great interest in the field of organic synthesis.

2*H*-azirines, although being a strained three-membered ring, are stable on the bench and readily available reagents that were first reported by Neber and Burgard in 1932 [7, 8]. Due to their reactivity, 2*H*-azirines have been extensively used as versatile precursors to access various aza-heterocycles, such as pyrroles, oxazoles, and 1,2,4-triazoles through ring-opening reactions (Fig. 1) [9–12]. The traditional methods for ring opening of 2*H*-azirines involved thermal conditions, UV irradiation, or use of acid catalysts or transition metals. However, visible-light photocatalysis has emerged as a cutting-edge platform for developing innovative synthetic transformations and is considered a key tool in green chemistry [13–20]. Over the past decade, significant progress has been made in the ring-opening functionalization of 2*H*-azirines under visible-light irradiation, with numerous research groups contributing to this area. It is anticipated that more novel ring-opening functionalization reactions of 2*H*-azirines will be discovered, prompting the need for a comprehensive review to further advance this vibrant research field.

In recent years, our research group has made significant contributions to the fields of photochemical synthesis and heterocycle synthesis. To further focus on visible light-induced synthetic chemistry, this book chapter provides a comprehensive review of the synthesis of heterocycles under photochemical conditions from 2*H*-azirines. The literature reviewed in this chapter is up to date as of May 2024. To

Fig. 1 Heterocycle synthesis from 2*H*-azirines

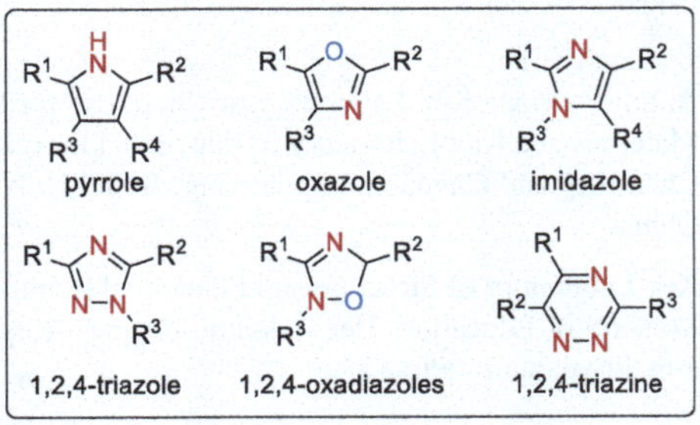

enhance readers' understanding of this area, the chapters of this book are organized based on different light sources, including ultraviolet light and visible light.

2 Photochemical Transformation of 2*H*-Azirines by UV Light

2.1 Direct Photoexcitation of 2H-Azirines by UV Light

In 1971, Padwa and Smolanoff discovered that exposing 2*H*-azirines **1a** under UV light irradiation in methyl acrylate or acrylonitrile resulted in a series of pyrroline derivatives (Scheme 1) [21]. Interestingly, when methyl methacrylate was used as the solvent, two isomeric products were obtained in a ratio of approximately 2:3. The authors suggested that this could be attributed to the ring-opening reaction of 2*H*-azirines **1a** under UV light irradiation, leading to the formation of nitrile ylide **1e** intermediates that then participate in a [3+2] cycloaddition reaction with electron-deficient alkenes. Subsequently, they expanded the substrate scope of the reaction and delved deeper into the reaction mechanism [22]. In 2010, Lin's research group demonstrated that 2*H*-azirines, along with equivalent amounts of methyl acrylate, could also undergo a [3+2] cycloaddition reaction in ethanol/water mixed solvents [23]. The authors propose that this technology holds significant promise for spatially controlled protein functionalization.

In 1973, Padwa et al. conducted an experiment in which 2*H*-azirine **2a** was dissolved in an inert solvent and exposed to UV light irradiation. They observed the dimerization of 2*H*-azirines, leading to the formation of tetraphenylpyrazine **2b** (Scheme 2a) [24]. The author proposes that the process initiates with the ring

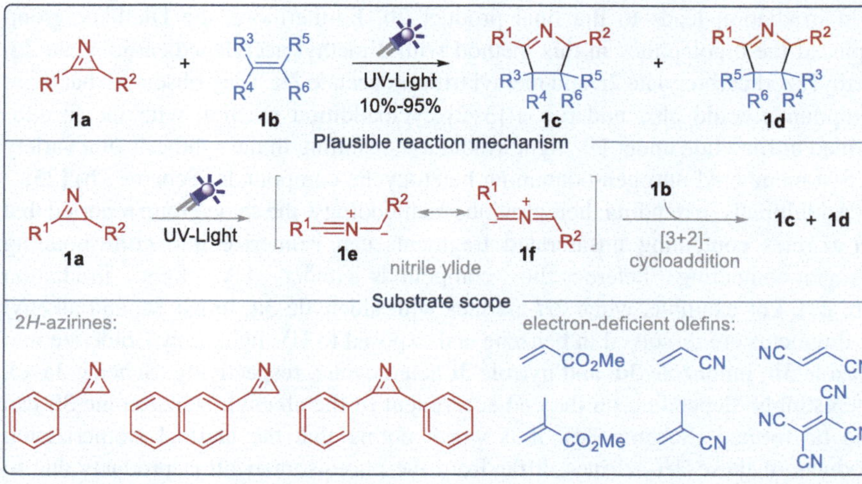

Scheme 1 UV light-induced [3+2] cycloaddition of 2*H*-azirines with electron-deficient olefins

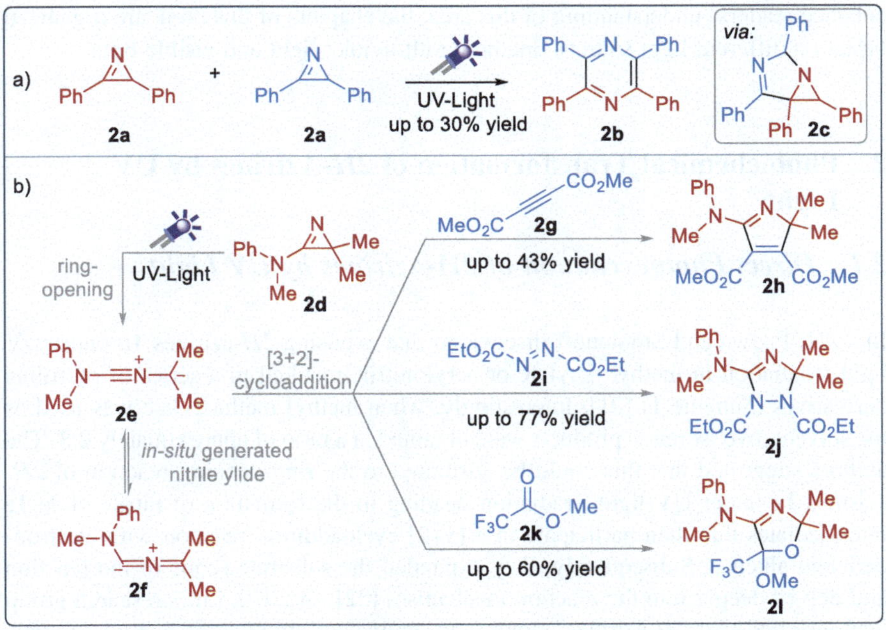

Scheme 2 UV light-induced [3+2] cycloaddition of 2*H*-azirines with different dipolephiles. a) UV-induced dimerization of 2*H*-azirine; b) [3+2] Cycloaddition of nitrile ylide with various dipolephiles

opening of 2*H*-azirines under UV light irradiation, resulting in the formation of a nitrile ylide intermediate. This intermediate then undergoes a [3+2] cycloaddition reaction with another molecule of 2*H*-azirines to generate 1,3-diazabicyclo[3.1,0] hex-3-ene intermediate **2c**. Subsequent rearrangement of **2c** under continuous UV light irradiation leads to the final product **2b**. Furthermore, the Dietliker group replaced the dipolephiles in this method with dimethyl acetylenedicarboxylate **2g**, diethyl azodicarboxylate **2i**, and methyl trifluoroacetate **2k**. They observed that these compounds could also undergo a [3+2] cycloaddition reaction with the in situ-formed nitrile ylide under UV light irradiation, resulting in the synthesis of a variety of five-membered nitrogen-containing heterocyclic compounds (Scheme 2b) [25].

Additionally, extending their previous methodology, the same group reported that 2*H*-azirines containing unsaturated fragments can isomerize into corresponding nitrogen-containing heterocyclic compounds under UV light irradiation [26, 27]. For example, when 2*H*-azirines with aldehyde **3a**, imine **3c**, and alkenyl **3e** fragments are dissolved in benzene and exposed to UV light, they isomerize into oxazole **3b**, imidazole **3d**, and pyrrole **3f** heterocyclic, respectively (Scheme 3a–c). Interestingly, depending on the end substituent of the alkenyl, benzazepine **3h** can also be formed (Scheme 3d). It is worth noting that the thermal isomerization products of these 2*H*-azirines differ from the photoisomerization products due to different ring-opening and bond-breaking mechanisms. Furthermore, Weragoda

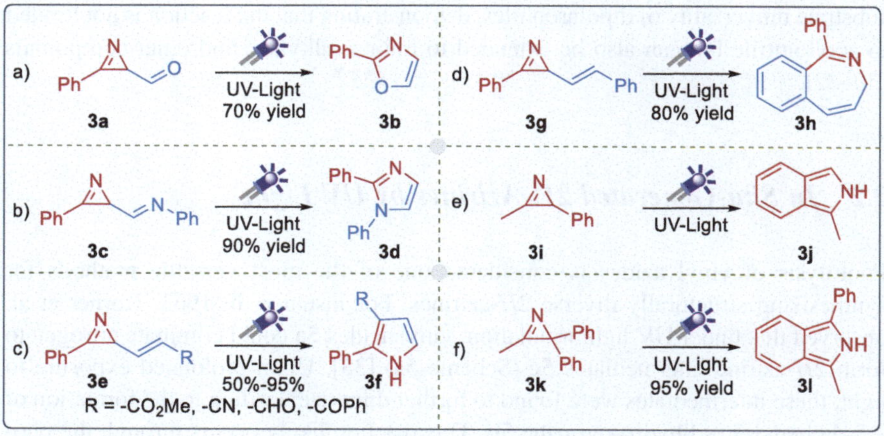

Scheme 3 Intramolecular [3+2] cycloaddition of 2*H*-azirines. a) Intramolecular isomerization of 2*H*-azirine with aldehyde; b) Intramolecular isomerization of 2*H*-azirine with aldehyde imine; c) Intramolecular isomerization of 2*H*-azirine with alkenyl; d) Formation of benzazepine from alkenyl-substituted 2*H*-azirine; e) C–C Bond cleavage of 3-methyl-2-phenyl-2*H*-azirine leading to isoindole derivative; f) C–C Bond cleavage and isomerization into isoindole derivative

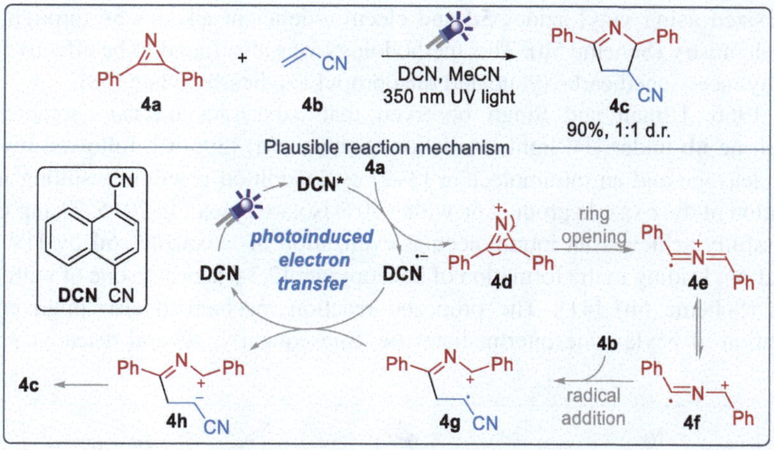

Scheme 4 Ring opening of 2*H*-azirines promoted by SET

group observed that 3-methyl-2-phenyl-2*H*-azirines can also undergo C–C bond cleavage under UV light and isomerize into corresponding isoindole derivatives (Scheme 3e, f) [28].

In 1991, Mattay discovered that 2*H*-azirines can undergo ring opening in the presence of DCN via the SET pathway to form the radical cationic intermediate **4e** (Scheme 4) [29]. Subsequently, a [3+2] cycloaddition reaction was carried out with acrylonitrile, resulting in the desired pyrroline derivative **4c** with a 90% isolated yield and a 1:1 diastereomeric ratio. Additionally, the author investigated the substrate universality of dipolarophiles, demonstrating that the reaction is not limited to acrylonitrile but can also be extended to imines, alkynes, and other compounds [30–32].

2.2 In Situ-Generated 2H-Azirines by UV Light

Photolysis of vinyl azides is considered one of the most versatile methods for synthesizing structurally diverse 2*H*-azirines. For instance, in 1963, Horner et al. observed that under UV light irradiation, vinyl azides **5a** could eliminate nitrogen to form 2*H*-azirine intermediates **5c** (Scheme 5a) [33]. Upon prolonged exposure to light, these intermediates were found to further dimerize, leading to the formation of 2,5-diphenyl-3,6-dihydro-pyrazine **5b**. This reaction likely occurs through the photoinduced in situ generation of the 2*H*-azirine intermediates. Nitrile ylides, which are produced from the UV-induced ring opening of 2*H*-azirines, were directly captured by another 2*H*-azirine to yield [3+2] cycloaddition products. Ultimately, the desired product **5b** can be achieved through intramolecular rearrangement. Subsequently, similar findings were reported by different researchers [34–37]. Furthermore, Kirschning discovered that a series of dihydropyrroles **5f** could be efficiently synthesized using vinyl azides **5d** and electron-deficient alkenes **5e** through flow photochemistry (Scheme 5b). This methodology was also found to be effective with dimethyl acetylenedicarboxylate and diisopropyl azodicarboxylate [38].

In 1966, Ulman and Singh observed that isoxazoles **6a** can isomerize to acylazirine **6b** under UV light irradiation (Scheme 6a) [39, 40], followed by C–C bond cleavage and an intramolecular [3+2] cycloaddition reaction, resulting in the formation of the oxazole product **6c** with a 50% isolated yield. In 2015, Zhang group successfully achieved an intramolecular cyclization of isoxazoles **6d** by UV light irradiation, leading to the formation of benzopyrano[2,3-*b*]azirin-7-one **6f** with good yields (Scheme 6b) [41]. The proposed reaction mechanism also involved the generation of acylazirine intermediates **6e**. Subsequently, several research groups

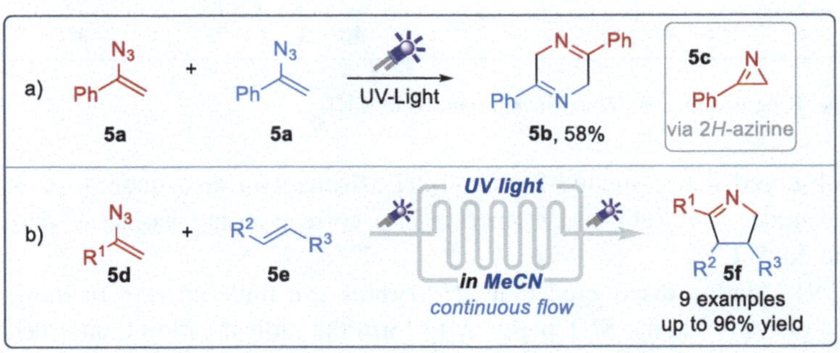

Scheme 5 [3+2] Cycloaddition of in situ-generated 2*H*-azirines from vinyl azides. a) Vinyl azide dimerization under UV-light-irradiation; b) [3+2] Cycloaddition of vinyl azides and alkenes through flow photochemistry

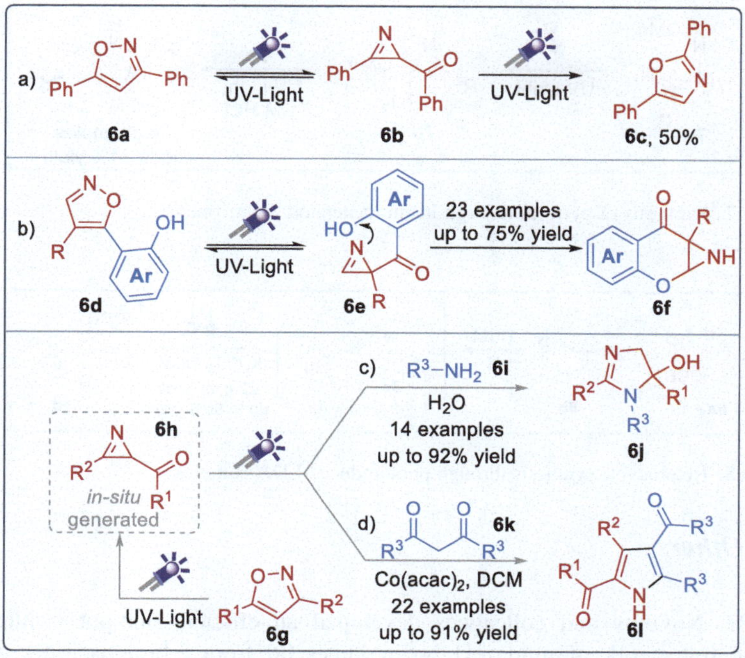

Scheme 6 Heterocycle synthesis through in situ-generated 2*H*-azirines. a) Isomerization of isoxazoles to acylazirine and oxazole; b) UV-induced intramolecular cyclization of isoxazoles to benzopyrano[2,3-*b*]azirin-7-one; c) Amine trapping of in situ-generated acylazirine to form 5-hydroxyimidazoline derivatives; d) Cobalt-catalyzed [3+2] cycloaddition of in situ-generated acylazirine with 1,3-diketones to form pyrrole derivatives

have explored the potential application of isoxazoles in the production of acylazirine under photochemical conditions, including dimerization of acylazirine and cycloaddition reactions of in situ-generated nitrile ylides with dipolephiles [42, 43]. Additionally, Nakamura synthesized a series of 5-hydroxyimidazoline derivatives **6j** by utilizing in situ-generated acylazirine captured by amine **6i** (Scheme 6c) [44]. In 2016, Opatz et al. utilized cobalt catalysis to synthesize a series of pyrrole derivatives **6l** through the [3+2] cycloaddition reaction of in situ-generated acylazirine with 1,3-diketones **6k** (Scheme 6d) [45]. Furthermore, they successfully synthesized the cycloheptadienone core by optimizing the structure of isoxazoles [46].

Diazo compounds can undergo nitrogen removal under light irradiation to form free carbenes, which then participate in carbene transfer reactions. In 2012, Park and colleagues utilized α-diazo oxime ethers **7a** to generate free carbenes under UV light irradiation, resulting in the formation of a series of 2*H*-azirines **7b** through intramolecular cyclization (Scheme 7) [47]. Notably, by introducing electron-deficient alkenes **7c** into the reaction mixture and extending the reaction time, they were able to synthesize pyrrole derivatives **7d** in a single step.

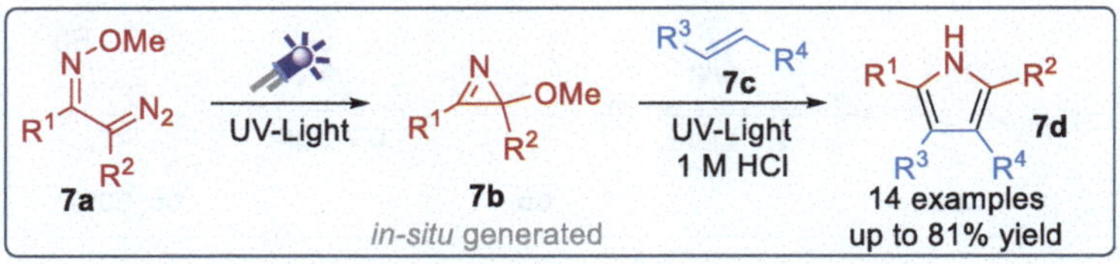

Scheme 7 Synthesis of pyrroles through in situ-generated 2*H*-azirines

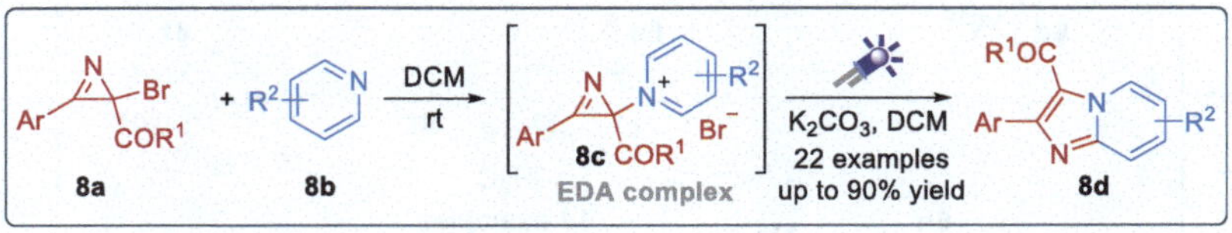

Scheme 8 Heterocycle synthesis through photoinduced EDA complex

2.3 Others

In 2022, Novikov and colleagues developed an efficient one-pot synthesis of bicyclic framework of imidazo[1,2-*a*]pyridines **8d** from 2-bromoazirines **8a** and pyridines **8b** without the use of a photocatalyst (Scheme 8) [48]. Mechanism verification experiments revealed that the formation of the bicyclic skeleton **8d** occurs in two steps. Initially, 2-bromoazirines **8a** and pyridines **8b** combine to form EDA complex (2*H*-azirin-2-yl)pyridinium bromides **8c**. Upon UV light irradiation, **8c** undergoes dehalogenation, ring opening, and intramolecular cyclization reactions.

3 Photochemical Transformation of 2*H*-Azirines by Visible Light

3.1 Direct Photoexcitation of 2H-Azirines by Visible Light

In 2014, Xiao et al. employed visible-light catalysis to achieve the formal [3+2] cycloaddition reaction of 2*H*-azirines **9a** with electron-deficient alkynes **9b** (Scheme 9) [49]. This strategy offers an effective approach for the synthesis of multi-substituted pyrrole **9c** and achieves the formal synthesis of HMG-CoA reductase inhibitor. The success of this approach is attributed to the reasonable choice of 9-mesityl-10-methyl-acridinium perchlorate **PC** as a photocatalyst, which possesses a high oxidation potential (+2.06 V). This potential enables the oxidation of 2*H*-azirines (R^1, R^2 = Ph, +1.56 V) under visible-light irradiation, leading to ring opening and formation of free radical cations **9e**. These radical cations then participate in the [3+2] cycloaddition with electron-deficient alkynes **9b**, ultimately yielding the desired pyrrole derivatives **9c**.

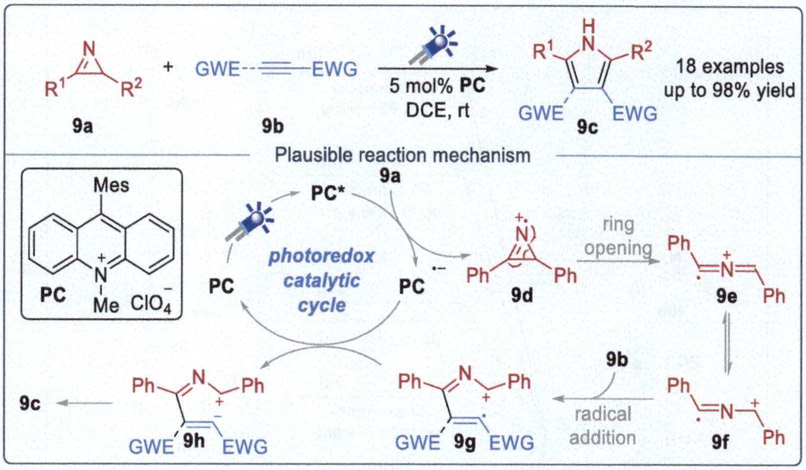

Scheme 9 Visible light-induced [3+2] cycloaddition of 2*H*-azirines with alkynes

Following the publication of Xiao's article, there was a notable increase in the use of 2*H*-azirines for visible light-induced reactions. For instance, Rastogi et al. effectively employed nitroalkenes **10c** as dipolarophiles to synthesize trisubstituted pyrroles **10d** (Scheme 10a) [50]. Wang and Li et al. utilized quinones **10e** as dipolarophiles to successfully generate a variety of benzo[*f*]isoindole-4,9-dione derivatives **10f** (Scheme 10b) [51]. Zhang employed maleimides **10g** as dipolarophiles for the synthesis of 4,6*a*-dihydropyrrolo[3,4-*c*]pyrrole-1,3(2*H*,3*aH*)-dione derivatives **10h** (Scheme 10c) [52]. In 2015, Xiao proposed a successful approach for oxazole **10j** synthesis by utilizing aldehydes **10i** as dipolarophiles (Scheme 10d) [53]. In 2017, Tang's group achieved the synthesis of 1,2,4-triazolines **10l** by employing azodicarboxylates **10k** as dipolarophiles, which could further be converted into 1,2,4-triazoles (Scheme 10e) [54]. Rastogi also managed to synthesize 1,2,4-triazoles **10n** using aryldiazonium salts **10m** as dipolarophiles (Scheme 10f) [55]. Additionally, in 2019, our research team utilized nitrosoarenes **10o** as free radical acceptors, presenting an effective strategy for synthesizing a variety of 2,5-dihydro-1,2,4-oxadiazole derivatives **10p** (Scheme 10g) [56]. It is important to note that the moderate reaction yields are due to the tendency of the produced 2,5-dihydro-1,2,4-oxadiazoles **10p** to undergo ring opening via N-O bond cleavage under standard reaction conditions. In 2022, Rastogi successfully synthesized a series of 3-imidazoline derivatives **10r** by employing *N*-tosyl benzylidene imines **10q** as dipolarophiles (Scheme 10h) [57].

1,3-Dipolar cycloaddition of 2*H*-azirines **11a** with 2,4,6-triarylpyrylium salts **11b** under blue-light irradiation for the synthesis of tetrasubstituted pyrroles **11c** was achieved by Rastogi et al. in 2021 (Scheme 11) [58]. Interestingly, the 2,4,6-triarylpyrylium salt **11b** was found to serve as both a dipolarophile and a photosensitizer in this reaction. Mechanistic investigations confirmed that the single electron oxidation of 2*H*-azirines **11a** was initiated by the photoexcited 2,4,6-triarylpyrylium

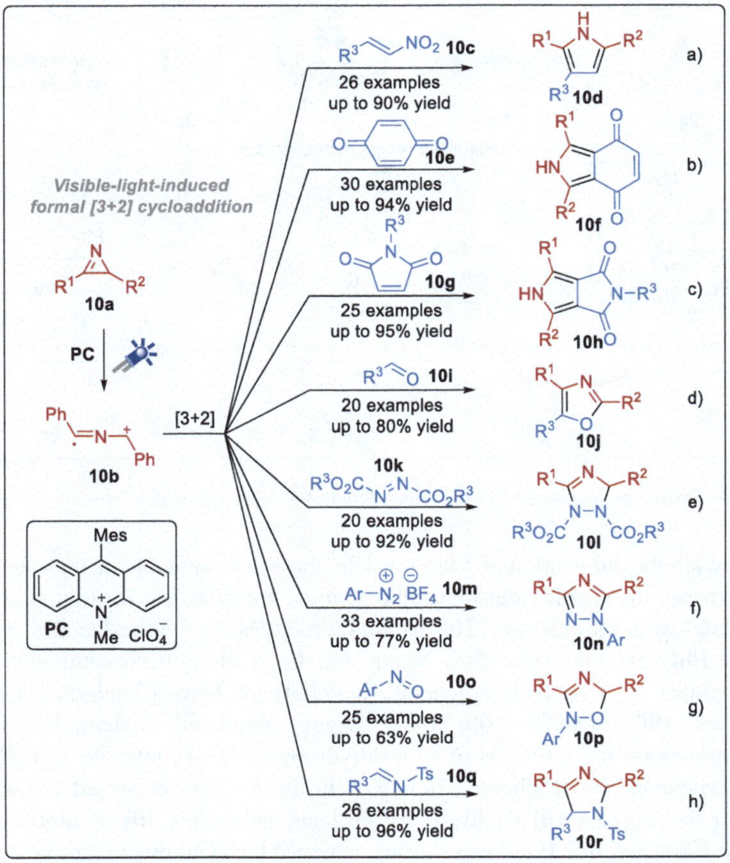

Scheme 10 Visible light-induced [3+2] cycloaddition of 2*H*-azirines with different dipolephiles. a) Synthesis of trisubstituted pyrroles using nitroalkenes as dipolarophiles; b) Synthesis of benzo[*f*] isoindole-4,9-dione derivatives using quinones as dipolarophiles; c) Synthesis of 4,6a-dihydropyrrolo[3,4-c]pyrrole derivatives using maleimides as dipolarophiles; d) Synthesis of oxazole derivatives using aldehydes as dipolarophiles; e) Synthesis of 1,2,4-triazolines using azodicarboxylates as dipolarophiles; f) Synthesis of 1,2,4-triazoles using aryldiazonium salts as dipolarophiles; g) Synthesis of 2,5-dihydro-1,2,4-oxadiazoles using nitrosoarenes as free radical acceptors; h) Synthesis of 3-imidazoline derivatives using *N*-tosyl benzylidene imines as dipolarophiles

salt **11b** under visible-light irradiation, leading to the subsequent coupling between a radical cation **11f** and a triarylpyranyl radical **11b**.

In 2016, the Wang group developed a novel three-component cyclization method involving 2*H*-azirines **12a**, alkynyl bromides **12b**, and O_2 under visible-light photoredox catalysis for the synthesis of substituted oxazole derivatives **12c** (Scheme 12) [59]. Unlike other 2*H*-azirines that cleave bonds under photocatalytic conditions, the 2*H*-azirines in this work undergo ring opening via C–N bonds rather

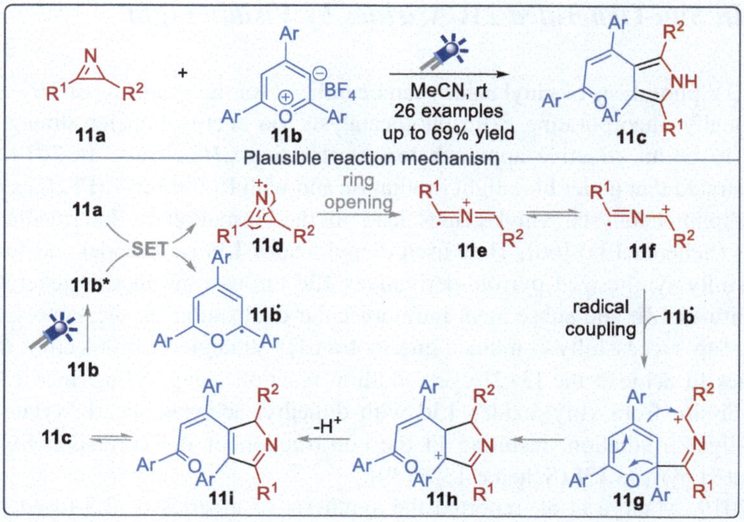

Scheme 11 Synthesis of tetrasubstituted pyrroles from 2*H*-azirines with triarylpyrylium salts

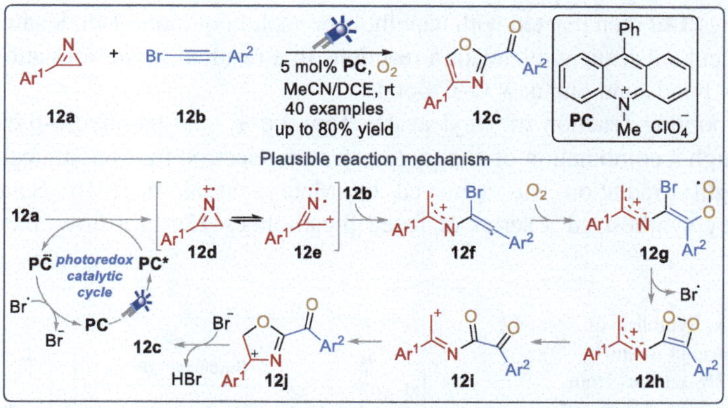

Scheme 12 Synthesis of oxazole through three-component cyclization of 2*H*-azirines

than C–C bonds. Isotope labeling experiments confirmed that the oxygen in the oxazole **12c** is derived from O_2 rather than H_2O. The authors also proposed potential reaction mechanisms. Initially, under visible-light irradiation, 2*H*-azirines **12a** are oxidized by the excited-state 9-phenyl-10-methylacridinium perchlorate **PC*** to generate radical cation **12d**. Subsequently, the addition of **12d** to alkynyl bromide **12b** provided intermediate **12f**. The formed **12f** reacted with O_2 to afford peroxyl radical **12g**, followed by intramolecular cyclization to generate intermediate **12h** along with the formation of Br radical. Finally, the ring opening of **12h** generated **12i**, followed by another intermolecular cyclization and β-H elimination to yield the desired oxazole **12c**.

3.2 *In Situ-Generated 2*H*-Azirines by Visible Light*

Direct UV photolysis of vinyl azides can be utilized for the synthesis of 2*H*-azirines. Additionally, incorporating extra photocatalysts via energy transfer strategies has proven to be an effective approach for constructing 2*H*-azirines. In 2014, Yoon demonstrated that under blue-light irradiation and with $Ru(dtbbpy)_3(PF_6)_2$ as energy transfer photocatalysts, vinyl azides lead to the formation of intermediate 2*H*-azirines (Scheme 13a) [60]. They used dienyl azides **13a** as a model substrate and successfully synthesized pyrrole derivatives **13c** through the in situ generation of vinylazirines **13b** and subsequent intramolecular cyclization. In the same year, the Xiao group successfully combined energy transfer strategies with electron transfer strategies to achieve the [3+2] cycloaddition reaction using 2*H*-azirines **13e** produced in situ from vinyl azides **13d** with dimethyl acetylenedicarboxylate under visible-light irradiation, resulting in the construction of the corresponding tetra-substituted pyrrole **13f** (Scheme 13b) [49].

In 2019, Maurya et al. reported the synthesis of a series of 2,3-fused pyrrole derivatives **14c** by coupling α-azidochalcones **14a** with naphthols or naphthoquinone **14b** under visible-light irradiation (Scheme 14) [61]. The process involved the photosensitized decomposition of α-azidochalcones **14a** into 2*H*-azirines **14d** using $Ru(bpy)_3(PF_6)_2$ as an energy transfer photocatalyst. The formed 2*H*-azirines **14d** then reacted with naphthols or naphthoquinone **14b**, leading to ring opening and subsequent cyclization reactions that resulted in the formation of two new C–N bonds and one new C–C bond.

The coupling reaction of vinyl azides **15a** and 1,2,3,4-tetrahydro-β-carbolines **15b** through a combination of energy transfer and electron transfer strategies under visible-light irradiation was achieved by Maurya et al. in 2016 (Scheme 15) [62]. They synthesized a series of fused β-carbolines **15c** in a flow microreactor

Scheme 13 Synthesis of pyrroles through in situ-generated 2*H*-azirines from vinyl azides. a) Synthesis of pyrrole derivatives via intramolecular [3+2] cycloaddition of vinyl azides; b) Combined energy and electron transfer strategies for the [3+2] cycloaddition of vinyl azides with dimethyl acetylenedicarboxylate under visible light

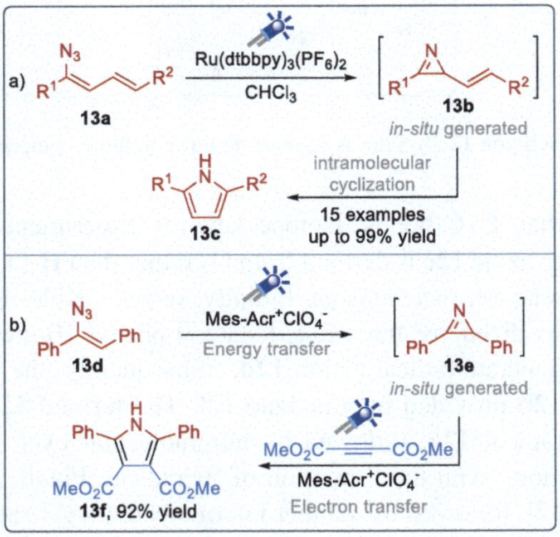

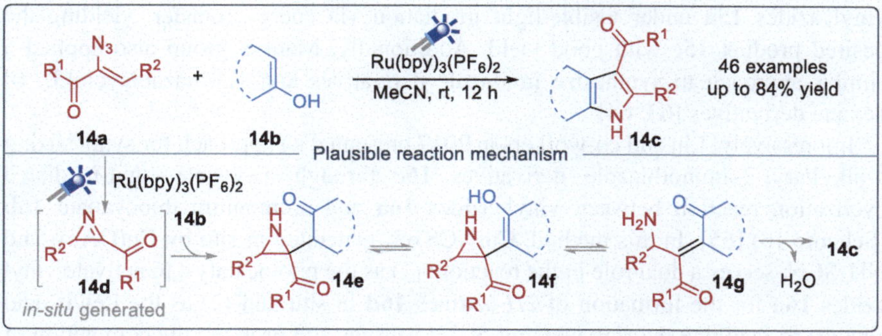

Scheme 14 Synthesis of 2,3-fused pyrroles through visible light-induced coupling of vinyl azides with naphthols or naphthoquinone

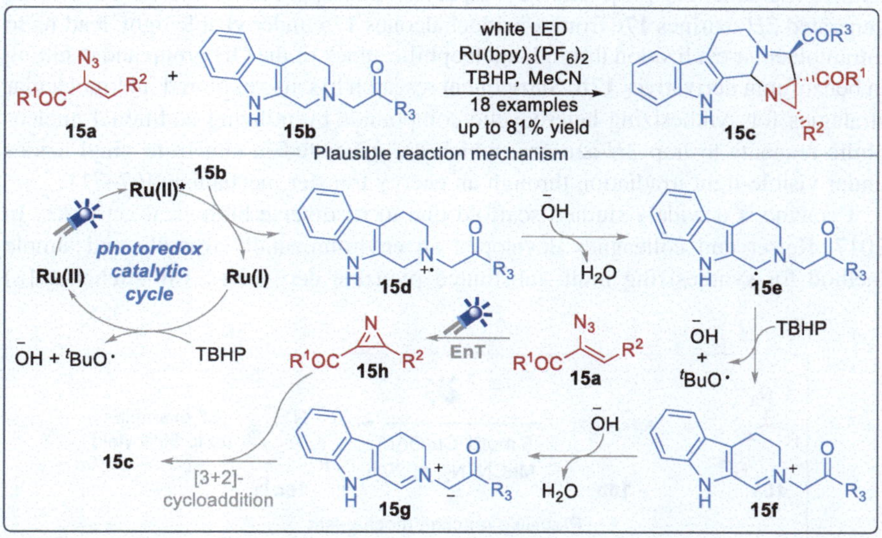

Scheme 15 Synthesis of β-carbolines through visible light-induced coupling of vinyl azides with 1,2,3,4-tetrahydro-β-carbolines

and proposed potential reaction mechanisms. Under visible-light irradiation, 1,2,3,4-tetrahydro-β-carbolines **15b** were oxidized by the excited $Ru(bpy)_3(PF_6)_2$* to form radical cations **15d**. Simultaneously, the reduced catalyst was oxidized by additional *tert*-butyl hydroperoxide (TBHP) to achieve a catalytic cycle, generating ^{t}BuO and OH^-. Subsequently, OH^- assisted in the deprotonation of **15d**, leading to the formation of α-amine radical **15e**. The oxidation of **15e** by TBHP resulted in the formation of imine ion **15f**, which upon further deprotonation gives the key 1,3-dipole intermediate **15g**. Finally, a [3+2] cycloaddition reaction occurred between 1,3-dipole intermediate **15g** and 2*H*-azirines **15h**, generated in situ from vinyl azides **15a** under visible-light irradiation via energy transfer, yielding the desired product **15c** with good yield. Additionally, Maurya group also applied a similar approach to synthesize imidazole derivatives and 1,3-diazabicyclo[3.1.0] hexane derivatives [63, 64].

Impressively, Liu and co-workers in 2017 presented an approach for synthesizing 4-alkyl/aryl-2-aminothiazole derivatives **16c** through a visible light-mediated cyclization reaction between vinyl azides **16a** and ammonium thiocyanate **16b** (Scheme 16) [65]. In this method, $Cu(NCS)_2^-$, generated in situ by $Cu(OAc)_2$ and NH_4SCN, serves a dual role in the reaction: (1) as the photocatalyst to activate vinyl azides **16a** for the formation of 2*H*-azirines **16d** in situ and (2) as the Lewis acid catalyst to facilitate the ring opening of 2*H*-azirines and promote the completion of the cycloaddition reaction with ammonium thiocyanate **16b**.

In 2017, Maurya and colleagues presented an effective approach for the synthesis of benzofuran derivatives **17b** from α-azidochalcones **17a** under visible-light irradiation (Scheme 17) [66]. The key aspect of this approach involves the in situ-generated 2*H*-azirines **17c** from α-azidochalcones **17a** under visible light, leading to intramolecular cyclization through nucleophilic attack of the OH group and resulting in benzofuran derivatives **17b**. Subsequent research has also explored intermolecular strategies for synthesizing heterocyclic compounds by utilizing additional nucleophilic reagents to trap 2*H*-azirines, which are generated in situ from vinyl azides under visible-light irradiation through an energy transfer mechanism [67–71].

Pyrazine is a widely studied scaffold due to its diverse biological activities. In 2017, Reiser and colleagues developed an environmentally friendly and simple method for synthesizing multi-substituted pyrazine derivatives **18b** (Scheme 18)

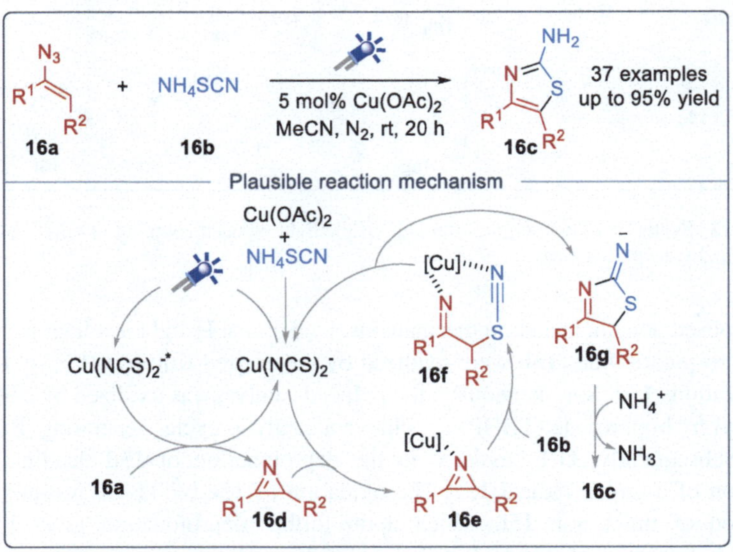

Scheme 16 Synthesis of 4-alkyl/aryl-2-aminothiazoles through visible light-induced cycloaddition of vinyl azides with ammonium thiocyanate

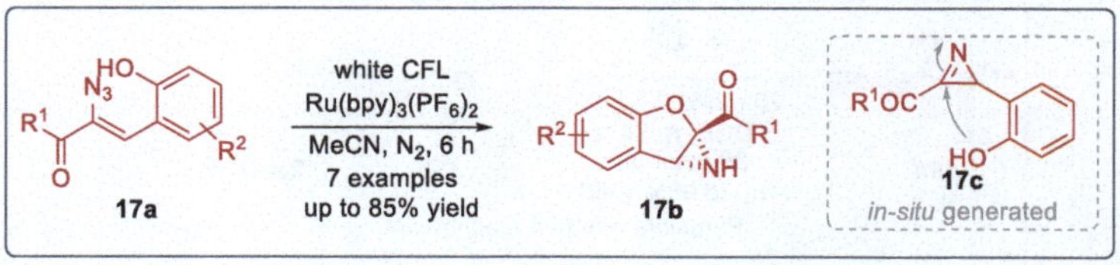

Scheme 17 Synthesis of benzofuran derivatives through visible light-induced intramolecular cycloaddition of α-azidochalcones

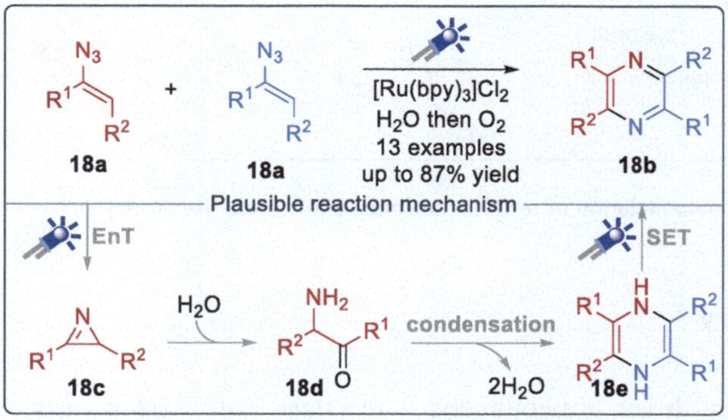

Scheme 18 Synthesis of pyrazines through visible light-induced dimerization of vinyl azides

[72]. This method involved the visible light-induced dimerization of vinyl azides **18a** using $Ru(bpy)_3Cl_2$ as a photocatalyst and atmospheric oxygen as the terminal oxidant. The proposed reaction mechanism suggested that the photoexcited Ru $(bpy)_3Cl_2$* catalyst transferred energy to the vinyl azide **18a**, leading to the formation of a highly reactive 2*H*-azirine intermediate **18c**. Water then acted as a nucleophilic reagent, facilitating the ring opening of 2*H*-azirines **18c**, followed by a condensation reaction to yield dihydropyrazine **18e**. Subsequent single-electron transfer (SET) and deprotonation of the dihydropyrazine **18e** ultimately led to the formation of the desired aromatized product **18b**.

In 2022, Maurya and co-workers reported a visible light-induced synthesis of 2,5-diaryloxazoles **19b** from α-azidochalcones **19a** (Scheme 19) [73]. Mechanistic studies revealed that 2,4-DNP functions as both a photocatalyst and an additive. Under visible-light irradiation, 2,4-DNP undergoes energy transfer process with α-azidochalcones **19a**, leading to the formation of a crucial 2*H*-azirine intermediate **19c**. Subsequently, 2,4-DNP interacts with 2*H*-azirine intermediate to form intermediate **19d**. Furthermore, under visible-light irradiation, **19d** is excited to its excited state, followed by ring opening to generate intermediate **19e**. This intermediate then undergoes intramolecular cyclization, leading to the release of 2,4-dinitrophenol and the formation of the desired product **19b**.

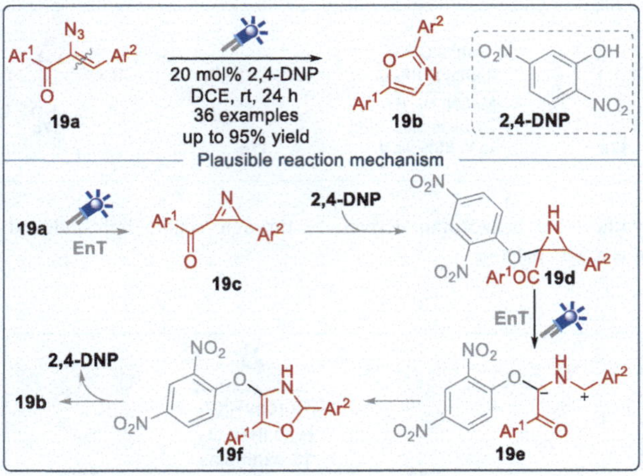

Scheme 19 Photoannulation of α-azidochalcones into 2,5-diaryloxazoles

3.3 Others

Visible light-driven photoredox reaction of 2*H*-azirines **20a** with tetrahydroisoquinoline **20b** for the synthesis of 5,6-dihydroimidazo[2,1-*a*] isoquinoline derivatives **20c** using a novel and photoactive EDA complex was established by Wang and Lu et al. in 2021 (Scheme 20) [74]. Interestingly, the resulting 5,6-dihydroimidazo[2,1-*a*]-isoquinoline **20c** exhibits strong photoluminescence in both solutions and powders, making it suitable for application in the production of blue OLED devices. Control experiments demonstrate that **20a** and **20b** initially form an EDA complex, which is then oxidized by the excited-state photocatalyst to produce the radical cation intermediate **20f**. Simultaneously, under photocatalytic conditions and in the presence of O_2, singlet oxygen is generated and subsequently reduced by the photocatalyst to yield O_2^{-}. Following a hydrogen atom transfer (HAT) from the radical cation **20f** results in the formation of the iminium intermediate **20g**. Ultimately, the desired product **20c** is obtained through intramolecular cyclization, deprotonation, and aromatization.

In 2022, He and co-workers described a visible light-promoted (3+3) cycloaddition of 2*H*-azirines **21a** with in situ-generated Huisgen zwitterions **21d** from azodicarboxylates **21b** and $P(NMe_2)_3$, resulting in the synthesis of 1,2,4-triazines **21c** with good yield (Scheme 21) [75]. The mechanistic verification experiment demonstrated that this (3+3) cycloaddition reaction proceeds through a cascade sequence involving nucleophilic addition to form EDA complex, visible light-driven 1,3-ester migration, and intramolecular cyclization.

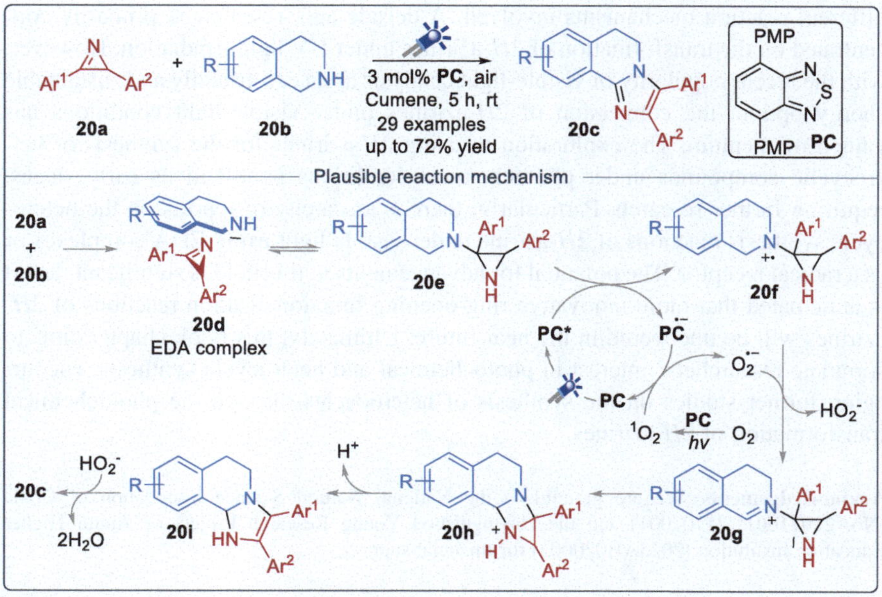

Scheme 20 Visible light-induced [3+2] cycloaddition of 2*H*-azirines with tetrahydroisoquinoline

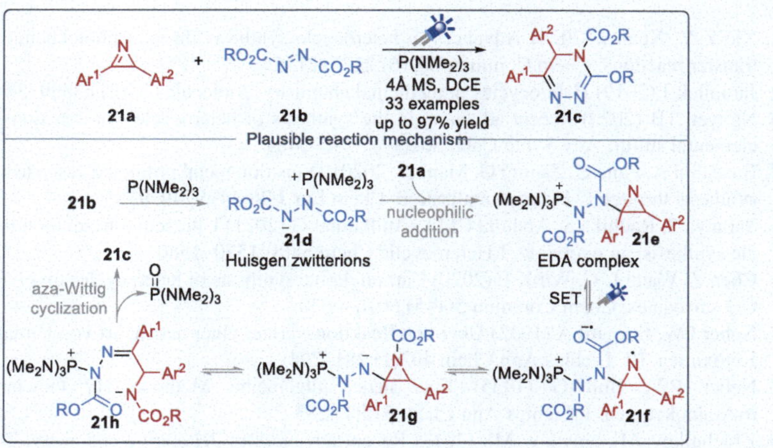

Scheme 21 Synthesis of 1,2,4-triazines through 2*H*-azirines with Huisgen zwitterions

4 Conclusion

This book chapter provides a detailed summary of the synthesis of heterocyclic compounds through the photochemical transformation of 2*H*-azirines. The discussion focuses on the light sources used, such as ultraviolet and visible light, and the different reaction mechanisms involved. A decade ago, researchers primarily concentrated on the transformation of 2*H*-azirines under UV light irradiation. However, with the recent popularity of visible-light catalysis as an eco-friendly and sustainable energy option, the conversion of 2*H*-azirines under visible-light conditions has gained momentum. The exploration of using 2*H*-azirines for the synthesis of heterocyclic compounds under photochemical conditions is still in its early stages, requiring further research. Particularly, there is a scarcity of reports on the heterocycle synthesis reactions of 2*H*-azirine under visible light using EDA complexes or as a radical receptor. The potential for advancements in this field is significant, and it is anticipated that more innovative ring-opening functionalization reactions of 2*H*-azirines will be uncovered in the near future. Ultimately, this book chapter aims to stimulate researchers' interest in photochemical and heterocycle synthesis, encouraging further studies on the synthesis of heterocycles through the photochemical transformation of 2*H*-azirines.

Acknowledgements We are grateful to the National Natural Science Foundation of China (No. 21971001, 21702001) and the Distinguished Young Research Project of Anhui Higher Education Institution (2023AH020003) for financial support.

References

1. Xie Z-Y, Xuan J (2024) Advances in heterocycle synthesis through photochemical carbene transfer reactions. Chem Commun 60(16):2125–2136
2. Jampilek J (2019) Heterocycles in medicinal chemistry. Molecules 24(21):3839–3842
3. Nguyen TB (2020) Recent advances in the synthesis of heterocycles via reactions involving elemental sulfur. Adv Synth Catal 362(17):3448–3484
4. Baccalini A, Faita G, Zanoni G, Maiti D (2020) Transition metal promoted cascade heterocycle synthesis through C-H functionalization. Chem Eur J 26(44):9749–9783
5. Saranya S, Radhika S, Abdulla CMA, Anilkumar G (2021) Ultrasound irradiation in heterocycle synthesis: an overview. J Heterocyclic Chem 58(8):1570–1580
6. Chen Z, Wang L-C, Wu X-F (2020) Carbonylative synthesis of heterocycles involving diverse CO surrogates. Chem Commun 56(45):6016–6030
7. Neber PW, Burgard A (1932) Über den Reaktionsverlauf einer neuen Art von Umlagerung bei Ketoximen. III. Liebigs Ann Chem 493(1):281–294
8. Neber PW, Huh G (1935) Eine neue, allgemeine Methode zur Gewinnung von α-Aminoketonen. I. Liebigs Ann Chem 515(1):283–296
9. Khlebnikov AF, Novikov MS (2013) Recent advances in 2H-azirine chemistry. Tetrahedron 69(16):3363–3401
10. Khlebnikov AF, Novikov MS, Rostovskii NV (2019) Advances in 2H-azirine chemistry: a seven-year update. Tetrahedron 75(18):2555–2624
11. Xu F, Zeng F-W, Luo W-J, Zhang S-Y, Huo J-Q, Li Y-P (2024) 2*H*-azirines: recent progress in synthesis and applications. Eur J Org Chem 27(11):e202301292
12. Lin S, Chu Y, Deng J, Xie F, Zhao L (2024) Light-induced synthesis of 2H-azirines and its applications in organic chemistry. J Saudi Chem Soc 28(2):101831–101849
13. Xuan J, Xiao W-J (2012) Visible-light photoredox catalysis. Angew Chem Int Ed 51(28): 6828–6838

14. Prier CK, Rankic DA, MacMillan DWC (2013) Visible light photoredox catalysis with transition metal complexes: applications in organic synthesis. Chem Rev 113(7):5322–5363
15. Chen Y, Lu L-Q, Yu D-G, Zhu C-J, Xiao W-J (2019) Visible light-driven organic photochemical synthesis in China. Sci China Chem 62(1):24–57
16. Cai B-G, Xuan J, Xiao W-J (2019) Visible light-mediated C–P bond formation reactions. Sci Bull 64(5):337–350
17. Wei Y, Zhou Q-Q, Tan F, Lu L-Q, Xiao W-J (2019) Visible-light-driven organic photochemical reactions in the absence of external photocatalysts. Synthesis 51(16):3021–3054
18. Xuan J, He X-K, Xiao W-J (2020) Visible light-promoted ring-opening functionalization of three-membered carbo- and heterocycles. Chem Soc Rev 49(9):2546–2556
19. Yu X-Y, Chen J-R, Xiao W-J (2021) Visible light-driven radical-mediated C–C bond cleavage/functionalization in organic synthesis. Chem Rev 121(1):506–561
20. Wang P-Z, Xiao W-J, Chen J-R (2023) Light-empowered contra-thermodynamic stereochemical editing. Nat Rev Chem 7:35–50
21. Padwa A, Smolanoff J (1971) Photocycloaddition of arylazirenes with electron-deficient olefins. J Am Chem Soc 93(2):548–550
22. Padwa A, Dharan M, SmoIanofi J, Wetmore SI (1973) Observations on the scope of the photoinduced 1,3-dipolar addition reactions of arylazirines. J Am Chem Soc 95(6):1945–1954
23. Lim RKV, Lin Q (2010) Azirine ligation: fast and selective protein conjugation via photoinduced azirine–alkene cycloaddition. Chem Commun 46(42):7993–7995
24. Padwa A, Dharan M, SmoIanoff J, Wetmore SI (1973) Photochemical transformations of small ring heterocyclic compounds. XLVII. Electronic details of the photocycloaddition of arylazirines. J Am Chem Soc 95(6):1954–1961
25. Dietliker K, Heimgartner H (1983) Photochemisch induzierte Reaktionen von 3-Amino-2H-azirinen. Helv Chim Acta 66(1):262–295
26. Padwa A, Smolanoff J, Tremper A (1975) Intramolecular cycloaddition reactions of vinyl-substituted 2H-azirines. J Am Chem Soc 97(16):4682–4691
27. Padwa A, Stengel T (2004) Transition metal catalyzed ring opening reactions of 2-phenyl-3-vinyl substituted 2H-azirines. Tetrahedron Lett 45(31):5991–5993
28. Weragoda GK, Das A, Sarkar SK, Sriyarathne HDM, Zhang X, Ault BS, Gudmundsdottir AD (2017) Singlet photoreactivity of 3-methyl-2-phenyl-2H-azirine. Aust J Chem 70(4):413–420
29. Muller F, Mattay J (1991) [3 + 2] Cycloadditions with azirine radical cations: a new synthesis of N-substituted imidazoles. Angew Chem Int Ed 30(10):1336–1337
30. Muller F, Mattay J (1992) A new synthesis for imidazolo- and pyrrolophanes by [3+2] cycloaddition with azaallenyl radical cations. Angew Chem Int Ed 31(2):209–210
31. Muller F, Mattay J, Steenken S (1993) Radical cation [3 + 2] cycloadditions of 2H-azirines. Mechanistic studies concerning the intermediate radical cation. J Org Chem 58(16):4462–4464
32. Albrecht E, Averdung J, Bischof EW, Heidbreder A, Kirschberg T, Müller F, Mattay J (1994) Photoinduced electron transfer (PET) in organic synthesis. [3 + 2]-type cycloaddition, cyclization and C-C bond cleavage reactions. J Photochem Photobiol A Chem 82:219–232
33. Horner L, Christmann A, Gross A (1963) Über Lichtreaktionen, XIV Bildung und Chemische Eigenschaften photolytisch erzeugter Arylimene. Chem Ber 96(2):399–406
34. Woerner FP, Reimlinger H, Arnold DR (1968) Formation of an azabicyclopentane derivative on photolysis of α-azidostyrene. Angew Chem Int Ed 7(2):130–131
35. Woerner FP, Reimlinger H (1970) 1.5-Dipolare Cyclisierungen, II. *ν*-Triazole aus Vinylaziden sowie durch Addition des Azid-Ions an die CC-Dreifachbindung. Chem Ber 103(6):1908–1917
36. Banert K, Ihle A, Kuhtz A, Penk E, Saha B, Wurthwein E-U (2013) Generation of highly strained 2,3-bridged 2H-azirines via cycloaddition reactions of 2-azidobuta-1,3-dienes and photolysis of the resulting cyclic vinyl azides. Tetrahedron 69(11):2501–2508
37. Nguyen TK, Titov GD, Khoroshilova OV, Kinzhalov MA, Rostovskii NV (2020) Light-induced one-pot synthesis of pyrimidine derivatives from vinyl azides. Org Biomol Chem 18(26):4971–4982

38. Cludius-Brandt S, Kupracz L, Kirschning A (2013) [3 + 2]-Cycloadditions of nitrile ylides after photoactivation of vinyl azides under flow conditions. Beilstein J Org Chem 9:1745–1750
39. Ullman EF, Singh B (1966) Photochemical transposition of ring atoms in five-membered heterocycles. The photorearrangement of 3,5-diphenylisoxazole. J Am Chem Soc 88(8): 1844–1845
40. Singh B, Ullman EF (1967) Photochemical transposition of ring atoms in 3,5-diarylisoxazoles. Unusual example of wavelength control in a photochemical reaction of azirines. J Am Chem Soc 89(26):6911–6916
41. Wang Q, Zhang Z, Zhang X, Zhang J, Kang Y, Peng J (2015) Synthesis of 7a-phenyl-1a,7a-dihydro-benzopyrano[2,3-b]azirin-7-ones via photoisomerization reaction. RSC Adv 5(7): 4788–4794
42. Paternoga J, Opatz T (2019) A copper-catalyzed synthesis of pyrroles through photochemically generated acylazirines. Eur J Org Chem 2019(42):7067–7078
43. Pusch S, Opatz T (2014) A photochemical one-pot three-component synthesis of tetrasubstituted imidazoles. Org Lett 16(20):5430–5433
44. Morita T, Fuse S, Nakamura H (2020) Photochemical conversion of isoxazoles to 5-hydroxyimidazolines. Org Lett 22(9):3460–3463
45. Pusch S, Kowalczyk D, Opatz T (2016) A photoinduced cobalt-catalyzed synthesis of pyrroles through in situ-generated acylazirines. J Org Chem 81(10):4710–4718
46. Pusch S, Schollmeyer D, Opatz T (2016) A light-induced vinylogous Nazarov-type cyclization. Org Lett 18(13):3043–3045
47. Qi X, Xu X, Park C-M (2012) Facile synthesis of 2-alkyl/aryloxy-2H-azirines and their application in the synthesis of pyrroles. Chem Commun 48(33):3996–3998
48. Filippov IP, Agafonova AV, Titov GD, Smetanin IA, Rostovskii NV, Khlebnikov AF, Novikov MS (2022) Synthesis of imidazo[1,2-a]pyridines via near UV light-induced cyclization of azirinylpyridinium salts. J Org Chem 87(9):6514–6519
49. Xuan J, Xia X-D, Zeng T-T, Feng Z-J, Chen J-R, Lu L-Q (2014) Visible-light-induced formal [3+2] cycloaddition for pyrrole synthesis under metal-free conditions. Angew Chem Int Ed 53(22):5653–5656
50. Karki BS, Devi L, Pokhriyal A, Kant R, Rastogi N (2019) Visible light-induced, metal-free denitrative [3+2] cycloaddition for trisubstituted pyrrole synthesis. Chem Asian J 14(24): 4793–4797
51. Wang L, Liu C, Li L, Wang X, Sun R, Zhou M-D, Wang H (2022) Visible-light-promoted [3 + 2] cycloaddition of 2H-azirines with quinones: access to substituted benzo[f]isoindole-4,9-diones. Chin J Chem 40(6):719–724
52. Guo H, Wang H, Zhao H, Zhang D (2023) Visible light-promoted synthesis of 4,6a-dihydropyrrolo[3,4-c]pyrrole-1,3(2H,3aH)-diones via [3+2] cycloaddition reaction of 2H-azirines with maleimides. New J Chem 47(12):5634–5638
53. Zeng T-T, Xuan J, Ding W, Wang K, Lu L-Q, Xiao W-J (2015) [3 + 2] Cycloaddition/oxidative aromatization sequence via photoredox catalysis: one-pot synthesis of oxazoles from 2H-azirines and aldehydes. Org Lett 17(16):4070–4073
54. Wang H, Ren Y, Wang K, Man Y, Xiang Y, Li N, Tang B (2017) Visible light-induced cyclization reactions for the synthesis of 1,2,4-triazolines and 1,2,4-triazoles. Chem Commun 53(69):9644–9647
55. Mishra P, Kumar P, Srivastava OS, Rastogi N (2023) Organophotoredox-mediated formal [3+2]-cycloaddition of 2H-azirines with aryldiazonium salts: direct access to trisubstituted 1,2,4-triazoles. Chem Asian J 18(9):e202300007
56. Cai B-G, Chen Z-L, Xu G-Y, Xuan J, Xiao W-J (2019) [3 + 2]-Cycloaddition of 2H-azirines with nitrosoarenes: visible-light-promoted synthesis of 2,5-dihydro-1,2,4-oxadiazoles. Org Lett 21(11):4234–4238
57. Mishra P, Shruti I, Kant R, Thakur TS, Kumar A, Rastogi N (2022) Visible light organo-photocatalytic synthesis of 3-imidazolines. Eur J Org Chem 2022(44):e202201079

58. Pokhriyal A, Karki BS, Kant R, Rastogi N (2021) Redox-neutral 1,3-dipolar cycloaddition of 2H-azirines with 2,4,6-triarylpyrylium salts under visible light irradiation. J Org Chem 86(6): 4661–4670
59. Chen L, Li H, Li P, Wang L (2016) Visible-light photoredox catalyzed three-component cyclization of 2H-azirines, alkynyl bromides, and molecular oxygen to oxazole skeleton. Org Lett 18(15):3646–3649
60. Farney EP, Yoon TP (2014) Visible-light sensitization of vinyl azides by transition-metal photocatalysis. Angew Chem Int Ed 53(3):793–797
61. Borra S, Chandrasekhar D, Newar UD, Maurya RA (2019) Access to 2,3-fused pyrroles via visible light driven coupling of α-azidochalcones with 1/2-naphthols, or 2-hydroxy-1,4-naphthoquinone. J Org Chem 84(2):1042–1052
62. Chandrasekhar D, Borra S, Nanubolu JB, Maurya RA (2016) Visible light driven photocascade catalysis: $Ru(bpy)_3(PF6)_2$/TBHP-mediated synthesis of fused β-carbolines in batch and flow microreactors. Org Lett 18(12):2974–2977
63. Tiwari DK, Maurya RA, Nanubolu JB (2016) Visible-light/photoredox-mediated sp^3 C-H functionalization and coupling of secondary amines with vinyl azides in flow microreactors. Chem Eur J 22(2):526–530
64. Borra S, Chandrasekhar D, Adhikary S, Rasala S, Gokulnath S, Maurya RA (2017) Visible-light driven photocascade catalysis: union of N,N-dimethylanilines and α-azidochalcones in flow microreactors. J Org Chem 82(4):2249–2256
65. Lei WD, Wang T, Feng KW, Wu LZ, Liu Q (2017) Visible-light-driven synthesis of 4-alkyl/aryl-2-aminothiazoles promoted by in situ generated copper photocatalyst. ACS Catal 7(11): 7941–7945
66. Borra S, Chandrasekhar D, Khound S, Maurya BA (2017) Access to 1a,6b-dihydro-1H-benzofuro[2,3-b]azirines and benzofuran-2-amines via visible light triggered decomposition of α-azidochalcones. Org Lett 19(19):5364–5367
67. Adiyala PR, Sastry KNV, Kovvuri J, Nagarajan A, Reddy VG, Sayeed IB, Nayak VL, Maurya RA, Kamal A (2017) Visible light driven coupling of 2-aminopyridines and α-keto vinyl azides for the synthesis of imidazo[1, 2-a]pyridines and their cytotoxicity. ChemistrySelect 2(26): 8158–8161
68. Man L, Copley RCB, Handlon AL (2019) Thermal and photochemical annulation of vinyl azides to 2-aminoimidazoles. Org Biomol Chem 17(27):6566–6569
69. Borra S, Borkotoky L, Newar UD, Das B, Maurya RA (2020) Photocatalyst-free visible-light enabled synthesis of substituted pyrroles from α-keto vinyl azides. Adv Synth Catal 362(16): 3364–3368
70. Adiyala PR, Jang S, Vishwakarma NK, Hwang Y-H, Kim D-P (2020) Continuous-flow photo-induced decarboxylative annulative access to fused imidazole derivatives via a microreactor containing immobilized ruthenium. Green Chem 22(5):1565–1571
71. Borkotoky L, Bora K, Maurya RA (2023) Visible-light-induced multicomponent synthesis of tetrasubstituted imidazoles. Asian J Org Chem 12(5):e202300146
72. Hossain A, Pagire SK, Reiser O (2017) Visible-light-mediated synthesis of pyrazines from vinyl azides utilizing a photocascade process. Synlett 28(14):1707–1714
73. Newar UD, Borra S, Maurya RA (2022) Visible-light 2,4-dinitrophenol-mediated photoannulation of α-azidochalcones into 2,5-diaryloxazoles. Org Lett 24(24):4454–4458
74. Wang Z, Li H, Wang Z, Suleman M, Wang Y, Lu P (2021) Photocatalytic approach for construction of 5,6-dihydroimidazo[2,1-a]isoquinolines and their luminescent properties. J Org Chem 86(12):8101–8111
75. Cao S, Tang T, Li J, He Z (2022) Visible light-driven [3 + 3] annulation reaction of 2H-azirines with Huisgen zwitterions and synthesis of 1,2,4-triazines. Org Chem Front 9(12):3342–3347

Top Heterocycl Chem (2025) 62: 161–200
https://doi.org/10.1007/7081_2025_90

Published online: 30 April 2025

Photo-Driven Dearomatization of Heteroarenes via Energy Transfer

Yuan-Zheng Cheng, Chao Zheng, and Shu-Li You

Contents

Y.-Z. Cheng (✉), C. Zheng (✉), and S.-L. You (✉)
New Cornerstone Science Laboratory, State Key Laboratory of Organometallic Chemistry, Shanghai Institute of Organic Chemistry, University of Chinese Academy of Sciences, Chinese Academy of Sciences, Shanghai, China
e-mail: chengyz@sioc.ac.cn; zhengchao@sioc.ac.cn; slyou@sioc.ac.cn

Abstract Dearomatization of heteroarenes has been proven to be a powerful tool for the construction of heterocycles bearing spiro-, fused, and bridged skeletons with increased degrees of molecular complexity and diversity, which are prevalent in natural products, pharmaceuticals, and functional materials. During the past decade, visible light-enabled dearomatization reactions have witnessed great progresses fueled by energy transfer photocatalysis. A variety of (hetero)arenes underwent dearomative cycloaddition with various biradical acceptors, such as alkene, alkyne, imine, and vinylcyclopropane. In this chapter, an overview of the state of the art of such synthetic methods and related mechanistic studies is provided. First, the general consideration of energy transfer-enabled dearomatization is introduced. Then, the main content of this chapter is categorized according to the types of heteroarenes, (1) reactions of indole/pyrrole derivatives, (2) reactions of pyridine/quinoline derivatives, and (3) reactions of (benzo)furan/(benzo)thiophene derivatives. Moreover, comprehensive mechanistic insights, involving those obtained by experiments and DFT calculations, are also discussed, aiming to give the readers a clear understanding of photo-driven dearomatization via energy transfer mechanism. Finally, this chapter is closed with a perspective on the current challenges and future directions in this field.

Keywords Dearomatization · Energy transfer · (Hetero)arene · Photocatalysis · Saturated cycles

1 Introduction

Dearomatization reactions of arenes have recently received considerable attention from synthetic chemists since such reactions provided a convenient avenue towards complex three-dimensional molecules from readily available aromatic feedstocks [1–5]. For instance, dearomatization of indole derivatives allowed facile synthesis of natural products including (−)-flustramine B, mollenine A (with revised structural assignment), pseudophrynaminol, (−)-chimonanthine, (−)-psychotriasine, and many others (Scheme 1a) [6, 7]. Thanks to the fast development in the past decade, several typical reaction modes for the dearomatization of indoles have been formulated. Indoles react with (pendant) electrophiles, nucleophiles, and radicals in regulated sequence allowing the highly efficient assembly of versatile polycyclic indoline scaffolds (Scheme 1b) [8–11]. When various organic or organometallic catalysts are applied, the dearomatization reaction can be made enantioselectively, resulting in the forging of quaternary stereogenic center(s) during the reactions.

Despite these progresses, the majority of known catalytic dearomatization reactions are carried out under thermal conditions (ground state, gs). Therefore, the thermodynamic restriction cannot be violated in these reactions (Scheme 2). In other words, the Gibbs free energy of the product should be lower than that of the substrate ($\Delta G^{\circ}_{gs} < 0$). It is well known that aromatic compounds possess extraordinary stability and aromaticity [12–14], due to the delocalization of π-electrons. In this regard, breaking aromaticity in dearomatization reactions often faces thermodynamical

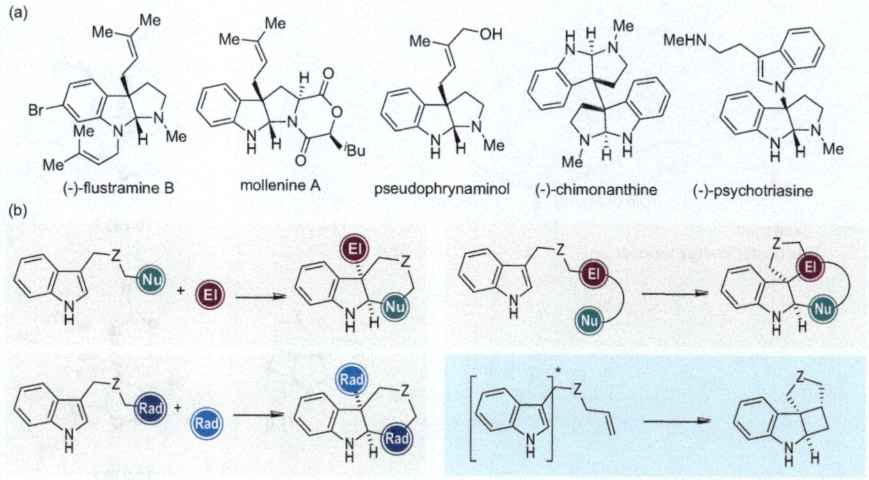

Scheme 1 (**a**) Selected natural products prepared using dearomatization of indole as the key step. (**b**) Typical reaction modes for the dearomatization of indoles under the ground state (in orange background) or excited state (in blue background)

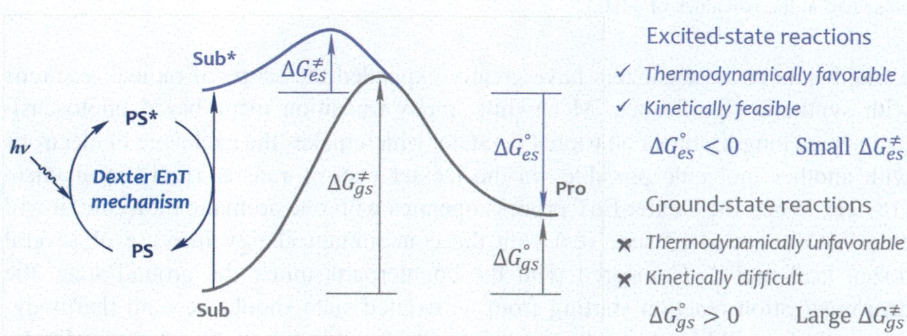

Scheme 2 Schematic energy diagram and energetic features of ground-state and excited-state reactions

obstacles besides kinetic difficulty (large $\Delta G^{\neq}_{gs}$). One way to circumvent this scenario is to couple the dearomatization with some thermodynamical advantages, for example, the formation of strong C–C or C–X bonds. However, the inherent thermodynamical limit in dearomatization reactions poses great challenges for preparing the dearomatized molecules of higher Gibbs free energy than that of the aromatic substrates ($\Delta G^{\circ}_{gs} > 0$).

Recent renaissance of visible-light catalysis brings great opportunities in organic synthesis [15–17]. Excited-state photosensitizers usually exhibit stronger electron transfer (ET) ability due to the population of one electron in high-energy orbitals. A large array of reactions initiated with the oxidative or reductive quenching of

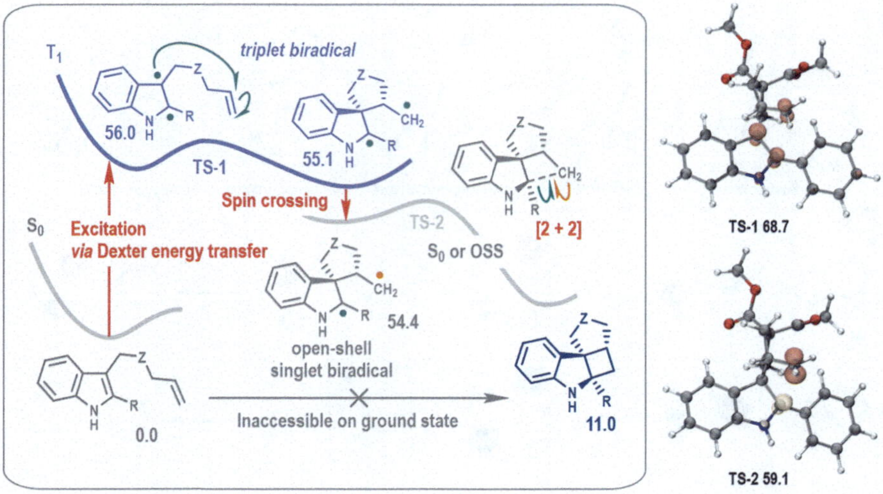

Scheme 3 Calculated energy profile of intramolecular dearomative [2+2] cycloaddition of indoles with alkenes (left) and structures and relative Gibbs free energies of key transition states (**TS-1** and **TS-2**) in this reaction. (U)DFT calculations were performed at the (U)B3LYP/6-31G(d,p)/SMD (DCM) level of theory. Energy values are in kcal mol^{-1}. Spin density at certain positions is presented at the isovalues of ±0.03

excited-state photosensitizers have greatly expanded the scope of radical reactions with synthetic significance. Meanwhile, many transition metal-based photosensitizers have long lifetimes on triple (T_1) state, which makes the exchange of electrons with another molecule possible via the Dexter energy transfer (EnT) mechanism [18–21]. When the Dexter EnT process operates with one aromatic molecule, it will enter into one excited state (es) with the concomitant energy increase of several dozen kcal mol^{-1}. Compared with the counterparts under the ground state, the dearomatization reaction starting from an excited state should be both thermodynamically favorable ($\Delta G^{\circ}_{es} < 0$) and kinetically feasible (small $\Delta G^{\neq}_{es}$). Accordingly, the photo-driven dearomatization reactions via energy transfer mechanism provide powerful complements to the synthesis of high-energy dearomatized products [22–25].

Mechanistically, the photo-driven dearomatization reactions involve several transient species on different electronic states (Scheme 3). Taking the intramolecular dearomative [2+2] cycloaddition of indoles with alkenes as an example (see Sect. 2.1 for details) [26], triplet indole ring can be roughly regarded to have two spin-paralleled electrons at the C2 and C3 positions. An intramolecular radical cyclization to the pendant olefin moiety furnishes a new 1,4-biradical intermediate bearing a spiro[4.4] bicyclic structure. This reaction on the T_1 state is quite facile, which is exemplified by a low energy barrier (**TS-1**, 12.7 kcal mol^{-1}). Subsequently, this triplet 1,4-biradical intermediate undergoes spin crossing, entering into the open-shell singlet state, and the biradical recombination occurs very fast (**TS-2**, 4.7 kcal mol^{-1}), delivering the highly strained angular tetracyclic spiroindoline product. Notably, the Gibbs free energy of such a product is 11.0 kcal mol^{-1} higher than that of the indole substrate, indicating the priority of this photo-driven dearomatization via energy transfer mechanism. Besides (unrestricted) DFT calculations, Stern–Volmer plots, quenching experiments with triplet quencher, and other experimental methods have been widely employed in the mechanistic studies of photo-driven dearomatization reactions.

In this chapter, representative examples of inter- and intramolecular dearomative photocycloaddition of aromatic rings via energy transfer process will be discussed, including the [2+2], [3+2], and [4+2] cycloaddition of indoles, pyrroles, pyridines, quinolines, (benzo)furans, (benzo)thiophenes, benzenes, or naphthalenes with various biradical acceptors such as alkenes, alkynes, allenes, oximes, and vinyl cyclopropanes. These reactions have provided efficient access to structurally complex polycyclic architectures. Notably, some of these reactions have been applied in the total synthesis of heteroatom-containing natural products and pharmaceutical candidates. Meanwhile, some arenes, such as naphthalenes and (benzo)thiophenes, also served as the triplet biradical acceptors to facilitate double-dearomative photocycloaddition. Besides, another mechanistically distinctive reaction type regarding the activation of bifunctional reagents via energy transfer was summarized. The activated bifunctional reagents underwent C–O bond cleavage to generate two different radicals, which thus reacted with arenes to realize difunctionalized dearomatization.

2 Reactions of Indoles/Pyrroles

2.1 *Cycloaddition with Alkenes*

In 2019, the You group reported a visible light-promoted intramolecular dearomative [2+2] cycloaddition of indoles with alkenes (Scheme 4) [26]. In the presence of $Ir(dFCF_3ppy)_2(dtbbpy)PF_6$ ($E_T = 60.8$ kcal mol^{-1}), the indole core was activated to its triplet state via energy transfer upon irradiation by blue LEDs. The energy transfer process was supported in that the introduction of a triplet quencher to this reaction significantly decreased the yield of the dearomatized product. On the basis of spin density population analysis, the phenyl group at the C2 position of indole stabilized the two spin-aligned electrons in the T_1 state, which lowered the T_1–S_0 energy gap of substrate **1**. A variety of functional groups, as well as *N*-tethered side chains, were all compatible with this protocol, affording cyclobutane-fused angular tetracyclic spiroindolines in excellent yields (up to 99%) with >20:1 dr (**2a–2g**). While the reaction efficiency of multi-substituted alkenes was slightly decreased (**2h**, **2i**), the substrate bearing electron-withdrawing group on alkene moiety was not amenable to this reaction (**2j**). In the next year, Oderinde, Dhar, and coworkers disclosed an intramolecular [2+2] cycloaddition of indoles, benzofurans, and benzothiophenes with an olefin tethered to the aromatic core at the C2 or C3 position (Scheme 5) [27]. Zhang, Wei, Fu, and coworkers developed an

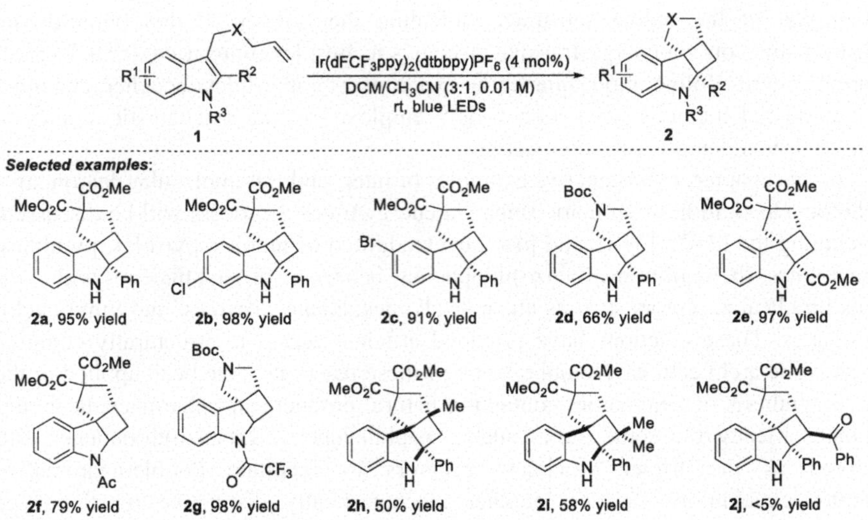

Scheme 4 Intramolecular dearomative [2+2] cycloaddition of indoles with a tethered olefin at the C3 position

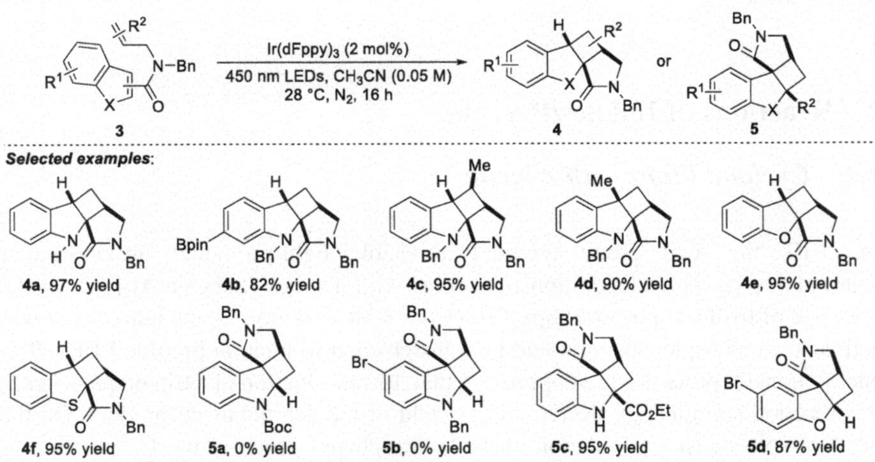

Scheme 5 Intramolecular dearomative [2+2] cycloaddition of indoles, benzofurans, and benzothiophenes with a tethered olefin at the C2 or C3 position

intramolecular dearomative [2+2] cycloaddition of indole decorating an alkene side chain at N1 atom (Scheme 6) [28].

In 1984, the Ikeda group disclosed an UV-promoted regio- and stereoselective [2+2] photocycloaddition of indoles with olefins in the presence of acetophenone as the triplet photosensitizer [29]. To broaden the application scope of this reaction, a visible light-enabled intermolecular dearomative [2+2] cycloaddition of heteroarenes with alkenes was developed by the Oderinde group in 2021 (Scheme

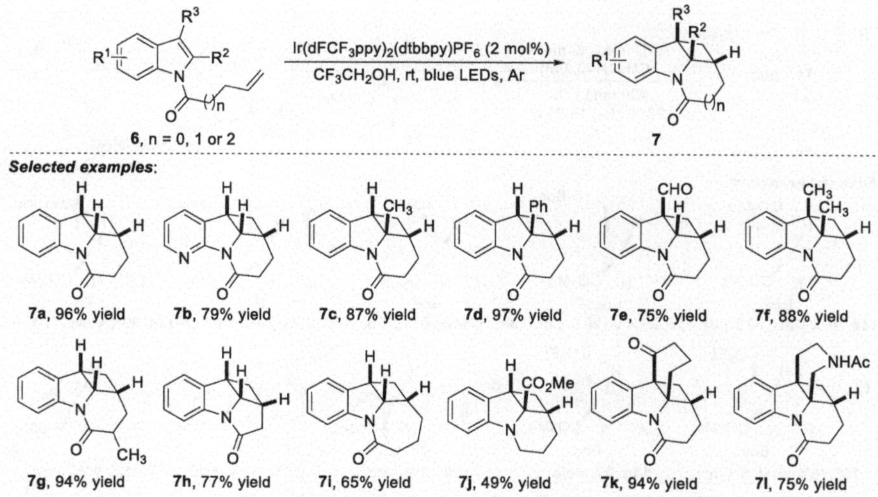

Scheme 6 Intramolecular dearomative [2+2] cycloaddition of indoles with a tethered olefin at the N1 position

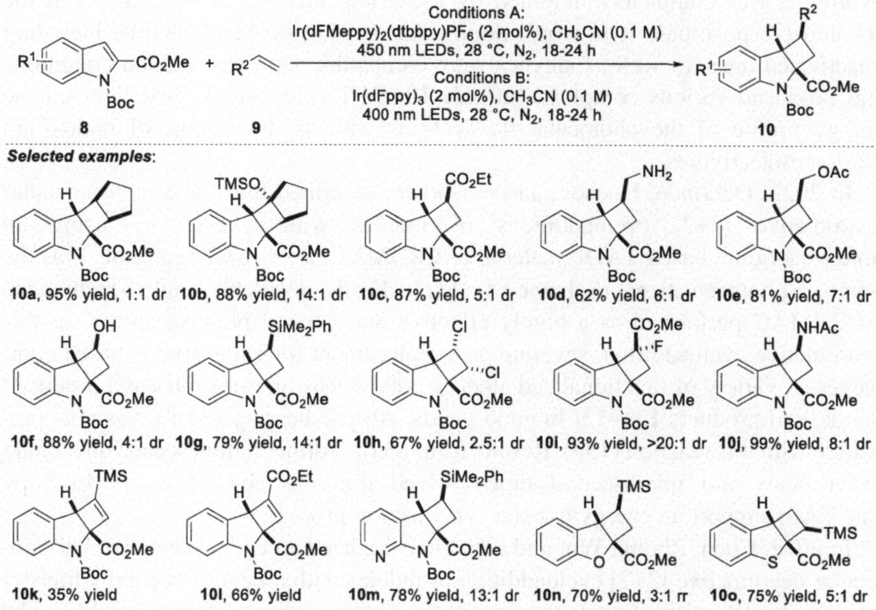

Scheme 7 Intermolecular dearomative [2+2] cycloaddition of heteroarenes with alkenes

7) [30]. This reaction was initiated by energy transfer process from excited Ir(dFMeppy)$_2$(dtbbpy)PF$_6$ (E_T = 62.9 kcal mol^{-1}) or Ir(dFppy)$_3$ (E_T = 63.5 kcal mol^{-1}) to arenes. When the reactivity of C2-substituted indoles

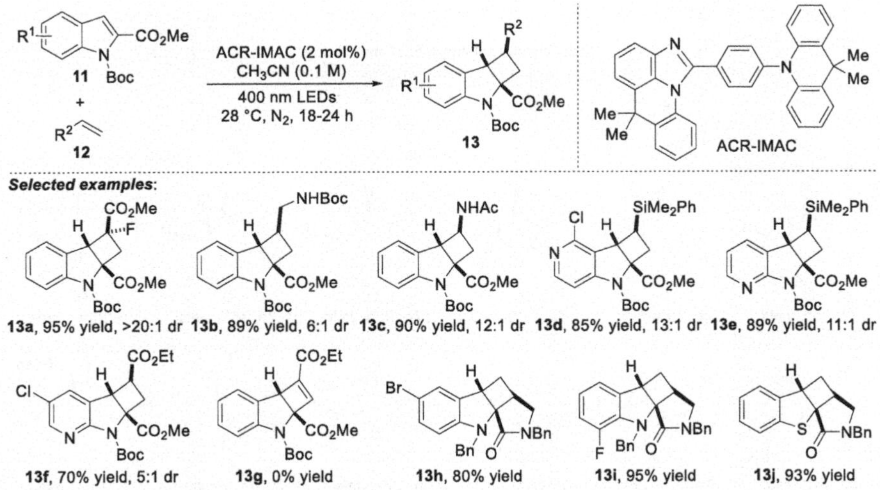

Scheme 8 TADF material catalyzed dearomative [2+2] cycloaddition of indoles with alkenes

was evaluated with vinylsilane, it was found that the undesired homodimer byproduct was completely inhibited by increasing the size of substituents at the N1 and C2 positions. It is worth noting that a wide range of olefins, including unactivated ones, as well as alkynes were compatible with the reaction conditions and produced various complex scaffolds **10**. DFT calculations shed light on the energy profile of the photocatalytic cycle as well as the origins of regio- and diastereoselectivities.

In 2022, Oderinde, Hudson, and coworkers described intra- and intermolecular dearomative [2+2] cycloadditions of indoles with alkenes by using an imidazoacridine-based TADF material (ACR-IMAC, $E_T = 63.7$ kcal mol^{-1}) as the organic photosensitizer (Scheme 8) [31]. Easily available and bench-stable ACR-IMAC performed as a highly effective and general photosensitizer for this dearomative cycloaddition, severing as a replacement for expensive Ir-based complexes. A variety of functionalized alkenes proved feasible substrates and produced the desired products **13a–13f** in good yields. Alkyne-bearing CO_2Et was not compatible with this reaction (**13g**). Results from Stern-Volmer luminescence quenching experiments and microsecond time-resolved transient absorption spectroscopy (ms-TAS) support an energy transfer sensitization process.

In 2022, Chen, Zhong, Wu, and coworkers realized an enantioselective intramolecular dearomative [2+2] cycloaddition of indoles with alkenes via energy transfer by using triplet artificial photoenzymes as the photosensitizers (Scheme 9) [32]. The photoenzymes were developed by encoding photosensitive unnatural amino acids to introduce a new photocatalytic center in the platform protein LmrR and optimized through directed evolution. Upon irradiation by an UV light, these photoenzymes decorated with a benzophenone group (E_T = 69.1 kcal mol^{-1}) were excited and transferred the energy to indoles. Since the resulting triplet indoles are associated with a chiral environment via multiple interactions with the surrounding amino acid

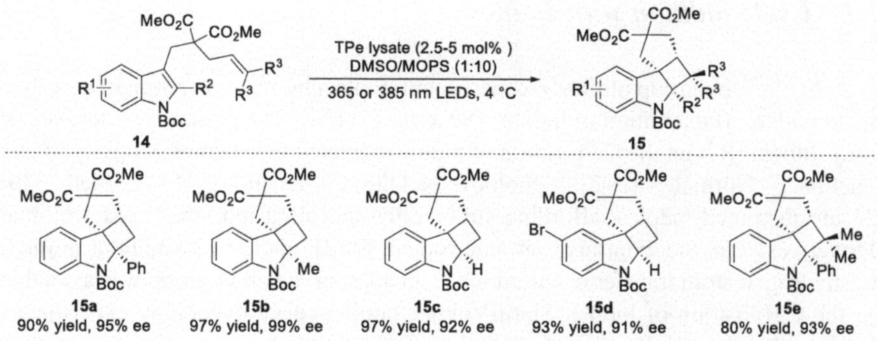

Scheme 9 Photoenzyme-catalyzed enantioselective intramolecular dearomative [2+2] cycloaddition of indoles with alkenes

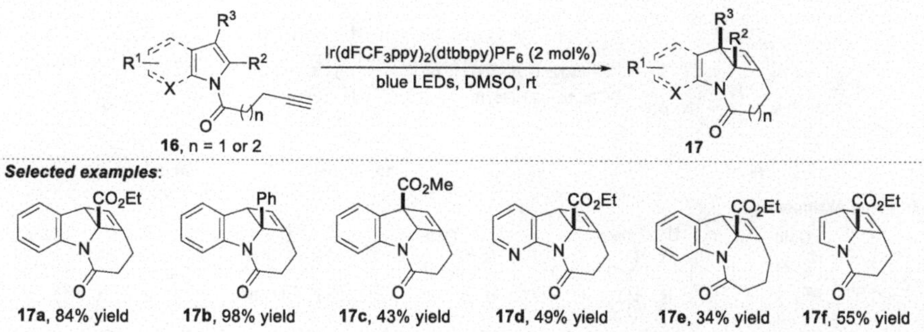

Scheme 10 Intramolecular dearomative [2+2] cycloaddition of indoles/pyrroles with alkynes

residues, the subsequent radical addition, ISC, and radical recombination afforded the cyclobutane-containing products with high enantioselectivity (up to 99% ee). Mechanistically, multiple interactions of the substrate with the chiral environment, which induced the enantioselectivity, were elucidated by the crystal structure of triplet photoenzymes.

2.2 *Cycloaddition with Alkynes*

In 2020, the You group employed alkynes as biradical acceptors in an energy transfer-enabled intramolecular dearomative [2+2] cycloaddition of indoles and pyrroles (Scheme 10) [33]. Due to the diminished reactivity of alkynes relative to alkenes, and the stronger ring strain of cyclobutenes than cyclobutanes, alkynes were much more challenging partners in visible light-promoted dearomative cycloaddition of indoles. In this reaction, a series of cyclobutene-fused indolizidines inaccessible under thermal conditions were afforded in moderate to excellent yields (21–98%). DFT calculations suggested the energy profile of this reaction.

2.3 Cycloaddition with Imines

In 2020, the You group utilized oximes as intramolecular reaction partners to realize dearomative cycloaddition of indoles (Scheme 11) [34]. The presence or absence of a substituent (R^3) on the C3 position of indole regulated the chemoselectivity of this reaction. Normal [2+2] photocycloaddition reactions proceeded for C3-unsubstituted indoles, affording structurally appealing indoline-fused azetidine derivatives. On the contrary, an interrupted [2+2] photocycloaddition through 1,5-hydrogen atom transfer occurred when an appropriate alkyl group was available on the C3 position of indole. Stern-Volmer luminescence quenching experiments revealed that the indole moiety, rather than C=N double bond in oximes, quenched the excited photosensitizer through energy transfer. The energy profile of visible light-promoted [2+2] cycloaddition or 1,5-hydrogen atom transfer was investigated

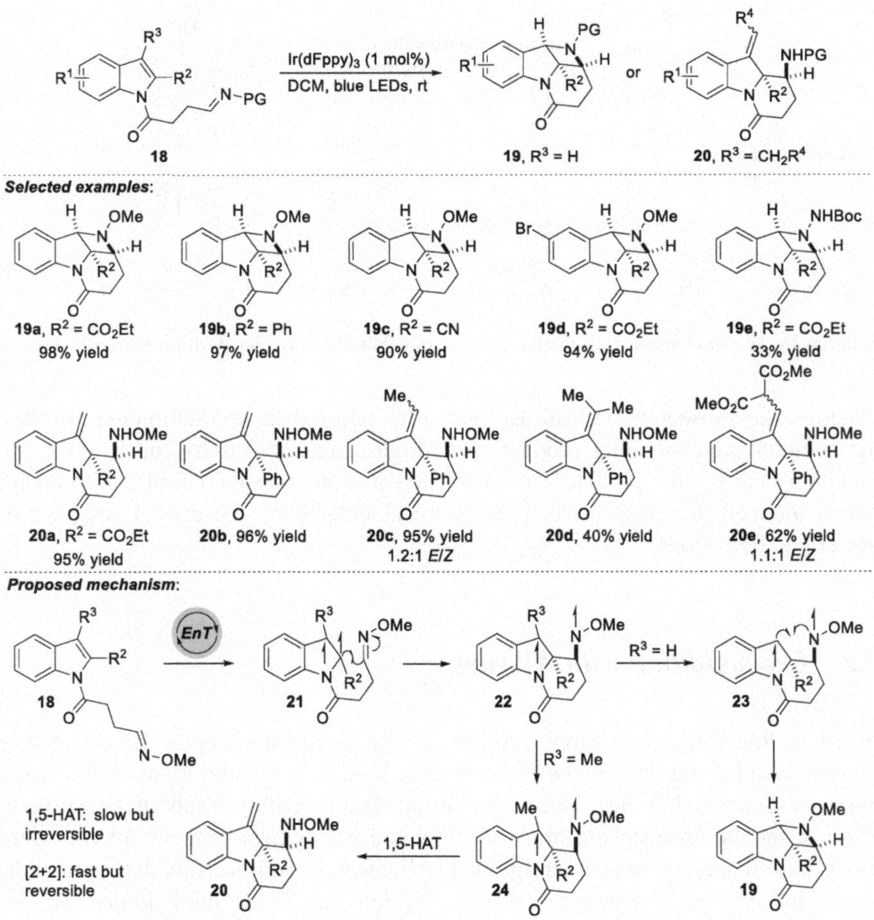

Scheme 11 Intramolecular dearomative (interrupted) [2+2] cycloaddition of indoles with oximes

by a series of experimental and computational studies. These results suggested that kinetically favored dearomative [2+2] cycloaddition products were formed at the early stage in the reactions of C3-alkyl-substituted substrates, but were later transformed into thermodynamically more stable 1,5-hydrogen atom transfer products.

2.4 Cycloaddition with Arenes

In 2021, the You group described a visible light-induced intramolecular double-dearomative cycloaddition of indoles with naphthalenes (Scheme 12) [35]. It provided an array of bridged polycyclic indolines in good yields. Interestingly, mechanistic insights were obtained by monitoring the reaction progress by ^{1}H NMR. Dearomative [2+2] adduct of naphthalene was observed in the early stage along with the formation of [4+2] adduct, but disappeared as the reaction proceeded. Moreover, the isolated [2+2] adduct could be converted to [4+2] adduct under standard conditions. These results, combined with DFT calculations, suggested a mechanism involving parallel reversible [2+2] and irreversible [4+2] pathways of transient open-

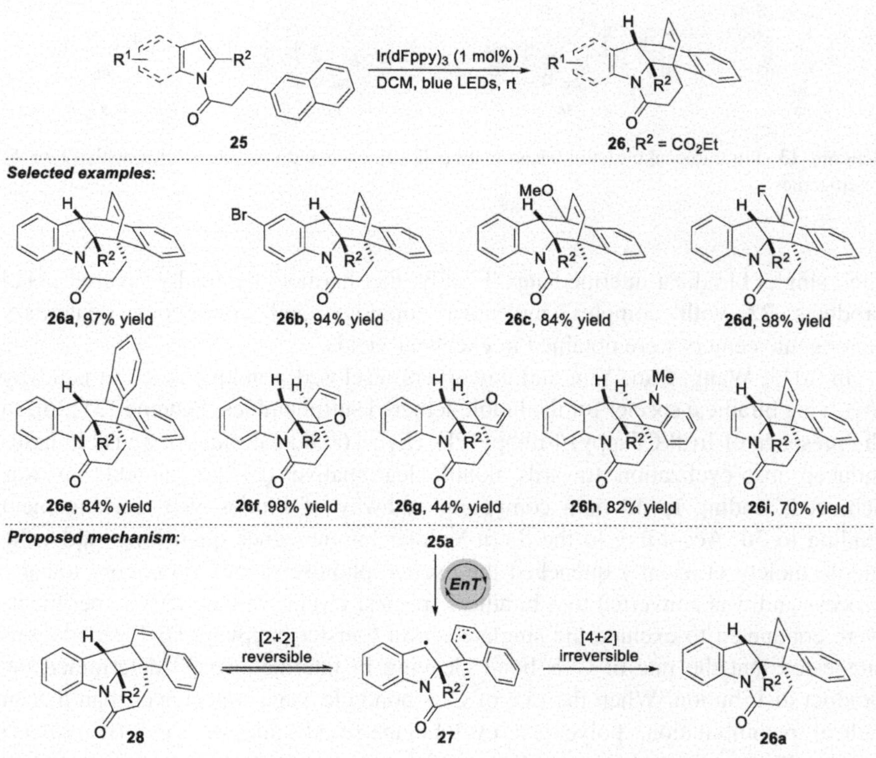

Scheme 12 Intramolecular double-dearomative cycloaddition of indoles with naphthalenes

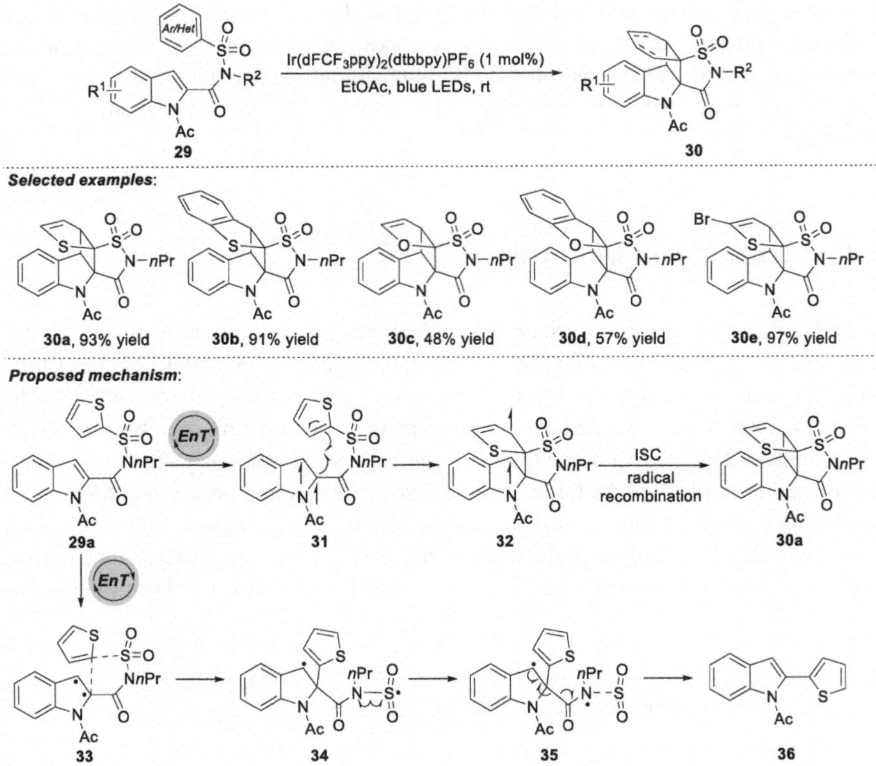

Scheme 13 Intramolecular double-dearomative [2+2] cycloaddition of indoles with *O* or *S*-heteroarenes

shell singlet biradical intermediates. Finally, the thermodynamically favored [4+2] products **22** with complex molecular topology and contiguous quaternary stereogenic centers were obtained in excellent yields.

In 2023, Wang, Cao, Yin, and coworkers disclosed a unique reaction pathway involving biradical species for the indole-tethered sulfonamides (Scheme 13) [36]. In the presence of Ir(dFCF$_3$ppy)$_2$(dtbbpy)PF$_6$ ($E_T = 60.8$ kcal mol^{-1}), a visible light-induced ipso-cyclization towards double-dearomative [2+2] cycloaddition was achieved, leading to **30** or a competing pathway of Smiles-type rearrangement leading to **36**. According to the Stern-Volmer luminescence quenching study, the indole moiety efficiently quenched the excited photosensitizer via energy transfer process and was converted to a biradical species. Cyclic voltammetry experiments were conducted to exclude the single-electron transfer pathway. DFT calculations suggested that the rate of C–S bond cleaving in intermediate **33** determined the product distribution. When the rate of C–S bond cleavage was slower than that of radical recombination, polycyclic cyclobutane-fused indolines and (hetero)aryl indoles **30** were formed in up to 98% yields.

2.5 Cycloaddition with Vinylcyclopropanes

In 2021, the You group disclosed a visible light-promoted intramolecular dearomative [5+2], [2+2], and [5+4] cycloaddition of indoles or pyrroles by using vinylcyclopropanes as the biradical acceptors (Scheme 14) [37]. The chemoselectivity was regulated by engineering substrate structures and tuning reaction conditions. Mechanistically, substrates **37** were activated to their triplet state via energy transfer by the excited photosensitizer. When no substituent was present at the C3 position (R^3 = H), 1,4-biradical intermediate **41** underwent ring

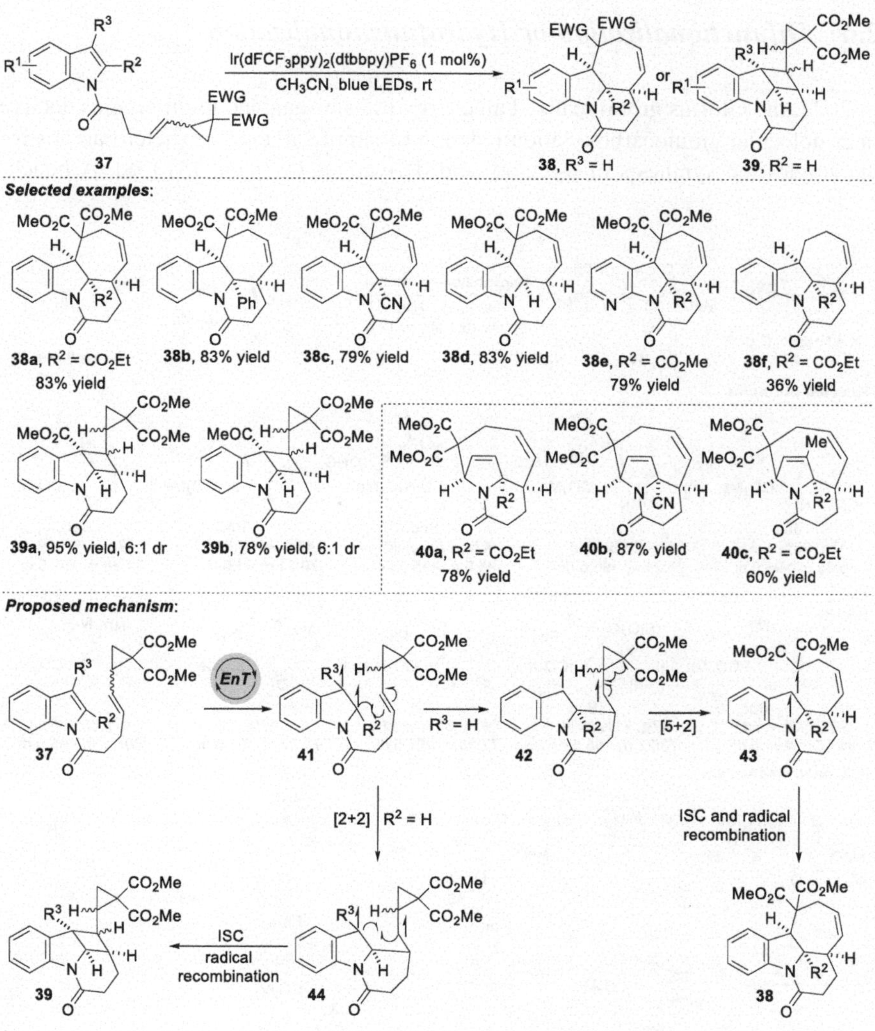

Scheme 14 Intramolecular dearomative [5+2], [2+2], and [5+4] cycloaddition of indoles or pyrroles with vinylcyclopropanes

opening of vinyl cyclopropane to form 1,7-biradical intermediate **43**, followed by radical recombination to afford [5+2] cycloaddition products **38**. Increasing the steric encumbrance at the C3 position of indole ($R^2 = H$, $R^3 \neq H$) significantly suppressed the [5+2] cycloaddition pathway, and [2+2] cycloaddition without ring-opening process of vinylcyclopropane took place, generating cyclobutane-fused polycyclic scaffold **39**. When pyrrole-derived vinyl cyclopropanes were employed in this reaction, dearomative [5+4] cycloaddition occurred instead of [5+2] cycloaddition **40**. The energy profile for this dearomative cycloaddition was investigated by DFT calculations.

2.6 Difunctionalization or Hydrofunctionalization

In 2022, the Glorius group realized an energy transfer-enabled highly regioselective intermolecular aminocarboxylation reaction of simple alkenes or (hetero)arenes for the single-step synthesis of β-amino acid derivatives (Scheme 15) [38]. A bench-

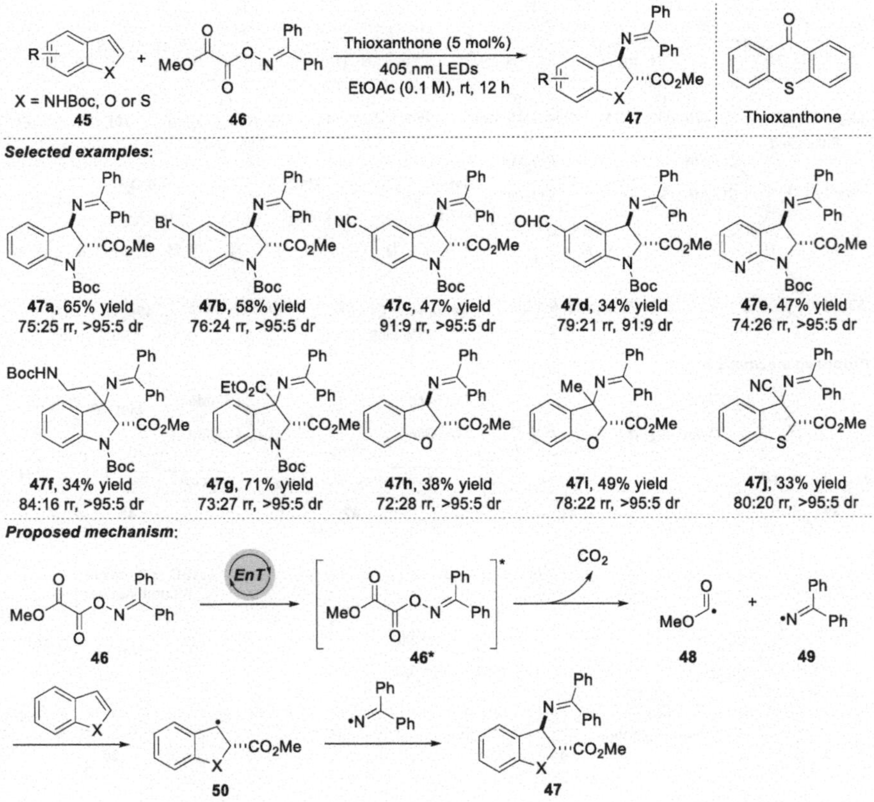

Scheme 15 Intermolecular aminocarboxylation reaction of (hetero)arenes

stable bifunctional oxime oxalate ester **46** was employed as the source of amine and ester functionalities for the reaction. Upon irradiation by a blue LED light ($\lambda_{max} = 405$ nm), this bifunctional reagent was excited in the presence of commercially available thioxanthone ($E_T = 60–65$ kcal mol^{-1}) as the photosensitizer, followed by C–O bond cleavage to give a C-centered ester radical **48** along with an N-centered iminyl radical **49**. The TEMPO quenching and radical clock experiments revealed that both of these two radical species were involved in this reaction. Moreover, yields of the 2,3-*trans*-difunctionalized indolines **47** were correlated with the triplet energies of photocatalysts, rather than their redox potentials. Combining with Stern-Volmer luminescence studies and cyclic voltammetry measurement, it was the energy transfer process that resulted in the generation of both two radicals and hence the downstream difunctionalization reactions. According to the DFT calculations, the difunctionalization of alkene was initiated with a fast and regioselective addition of C-center radical, leading to a stabilized benzylic radical which then underwent a radical-radical cross coupling with a long-lived N-center radical.

In the same year, the Glorius group developed an energy transfer-enabled dearomative unsymmetrical diamination of various electron-rich (hetero)arenes with **52** as the diamination reagent, affording previously inaccessible vicinal diamines **53** with excellent regio- and diastereoselectivities (Scheme 16) [39]. A series of mechanistic investigations were conducted, including TEMPO quenching experiment, radical clock experiments, UV-vis absorption spectroscopy, Stern-Volmer analysis, as well as cyclic voltammetry measurement. These studies collectively suggested that energy transfer from excited photosensitizer to diamination reagent led to the triplet state of the latter. The subsequent fragmentation of excited-state **52** via N–O bond cleavage resulted in the generation of a transient amidyl radical **54** and a persistent iminyl radical **55**, along with the release of CO_2 and acetone. Based on the persistent radical effect, the electrophilic amidyl radical **54** underwent radical addition to indoles, benzofurans, and benzothiophenes at the C2 position, affording transient C-centered radical **56**. The radical-radical coupling of **56** and persistent iminyl radical **55** produced the desired products. The less steric hindrance of C2 position and the formation of stabilized benzyl radical were considered as the origin of excellent regioselectivity. The difunctionalization across the C=C double bond occurred in an exclusively *trans* fashion, such that the steric repulsion between two bulky amine groups could be minimized.

In 2022, the Bach group achieved a dearomatization reaction of indoles via a visible light-promoted hydrogen atom abstraction/cyclization cascade (Scheme 17) [40]. In previous works, the excited state indoles underwent cycloaddition with alkenes to construct cyclobutane-fused polycyclic scaffold. In this reaction, the triplet reactivity of indoles was unlocked for an unprecedented dearomative hydrogen atom transfer/cyclization reaction. Following energy transfer-enabled excitation of indoles, one of the radical centers abstracted a hydrogen from the tethered side chain to afford 1,4-biradical **63**. Then, the radical recombination of **63** was facilitated

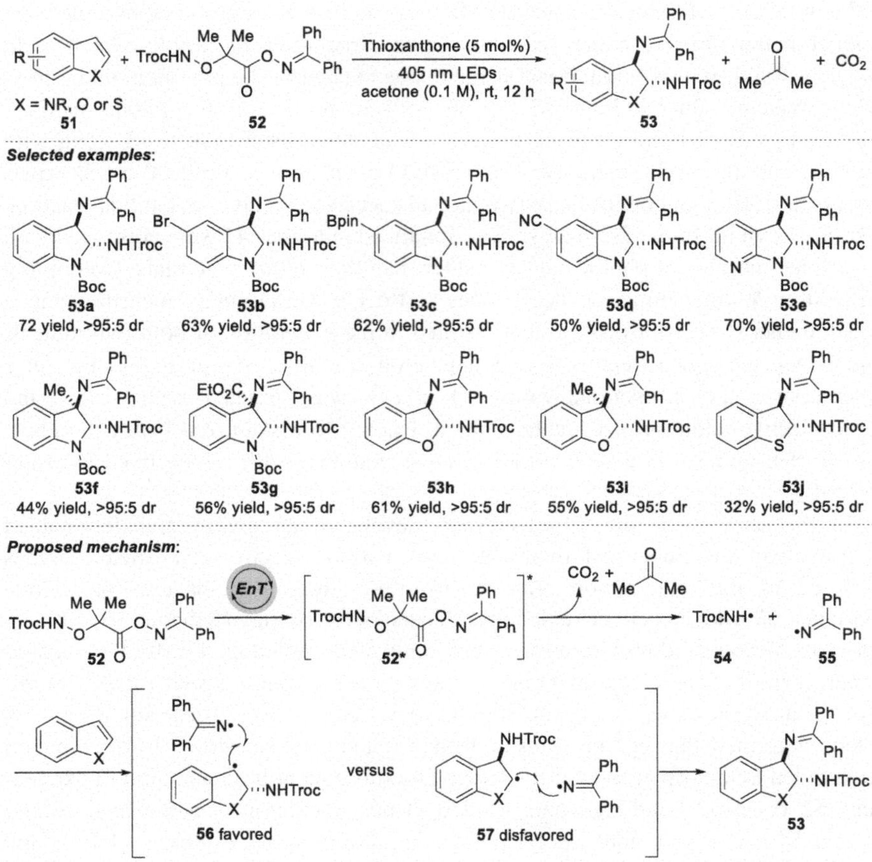

Scheme 16 Dearomative unsymmetrical diamination of various electron-rich (hetero)arenes

to form the spiro[azetidine-3,3′-indoline] **60** and 1,3-dihydrospiro[indene-2,3′-indoline] skeleton **61**. Although the triplet energy of photosensitizer was slightly lower than that of substrate **58** or **59**, it was speculated that energy transfer should be possible due to the long lifetime of triplet photosensitizer. The deuterium labeling experiments demonstrated that hydrogen atom transfer occurred at the C2 position of indole. When indole **58e** tethering an *N*-isopropyl amide side chain was subjected to the standard reaction conditions, dealkylated carboxamide **60e** was obtained in 47% yield via intermediates **66** and **67**, which further supported the occurrence of 1,5-HAT process in excited state **58** or **59**.

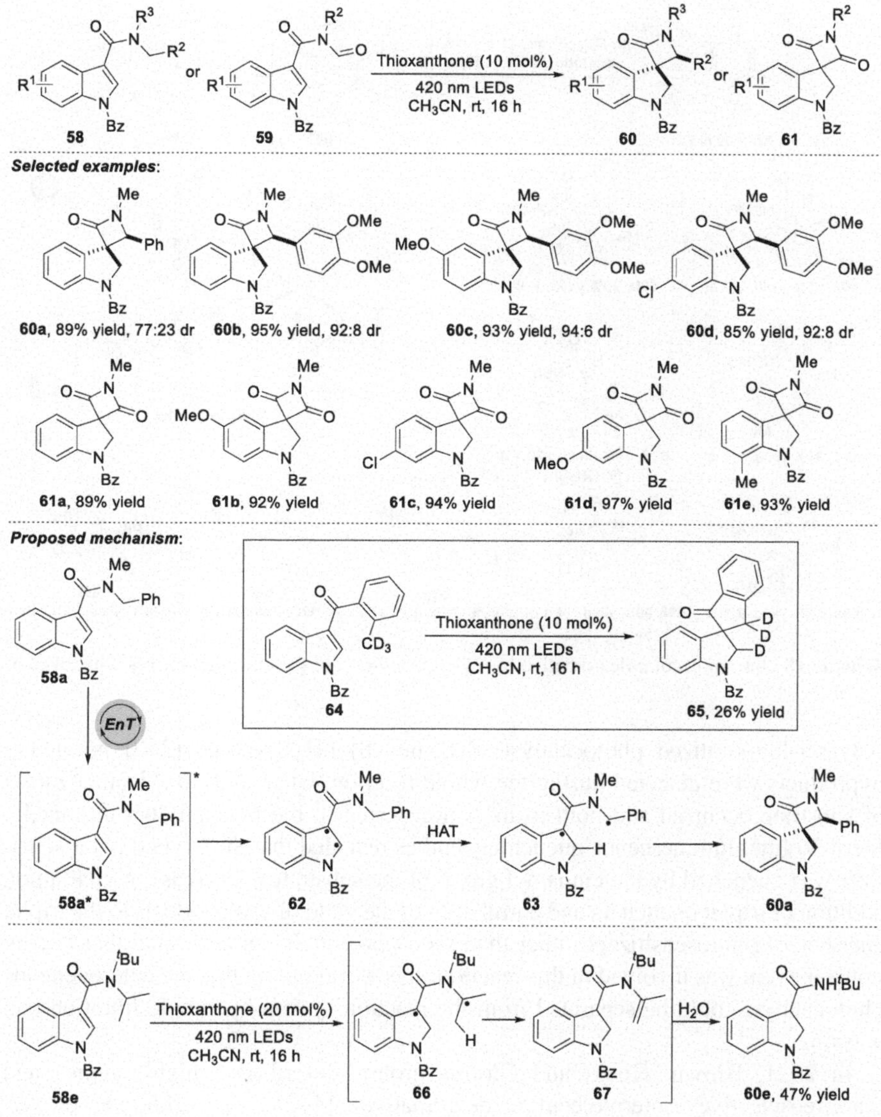

Scheme 17 Visible light-promoted dearomatization of indoles via hydrogen atom abstraction/cyclization

3 Reactions of Pyridines/Quinolines

3.1 Cycloaddition with Alkenes

In 2019, the Glorius group developed an intramolecular dearomative [4+2] cycloaddition of pyridine with alkene in the presence of a recyclable and sustainable

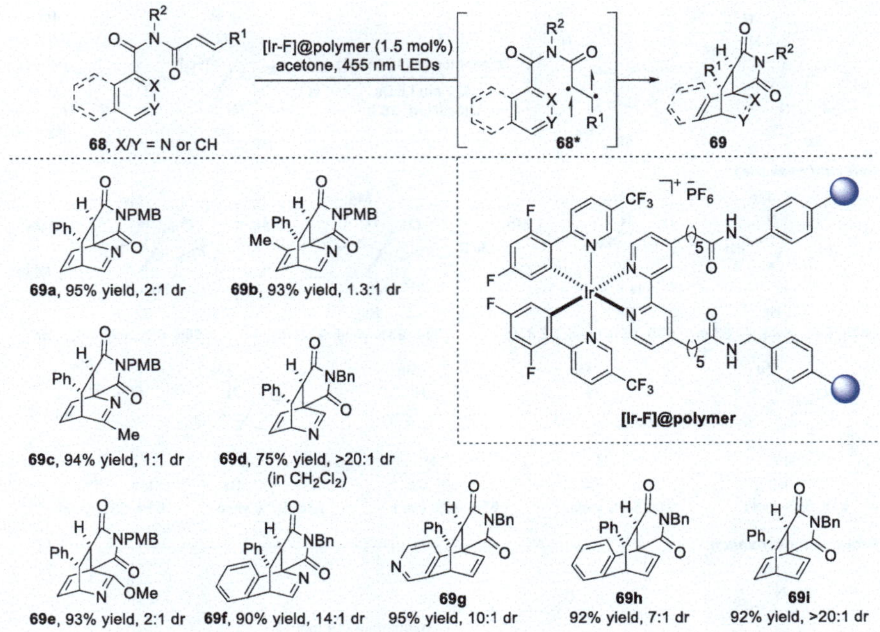

Scheme 18 Intramolecular dearomative [4+2] cycloaddition of pyridine derivatives with alkenes

polymer-immobilized photocatalyst (Scheme 18) [41]. Neither [2+2] nor [3+2] byproducts were detected during the whole reaction, whereas *E* to *Z* isomerization of substrate occurred and both isomers transferred to the final product ultimately. Stern-Volmer luminescence quenching studies revealed that the excited photosensitizer was quenched by the cinnamyl group of the substrates. Decrease in yield upon addition of triplet quenchers and correlation of the yield of cycloadducts to the triplet energies of photosensitizers rather than redox potentials demonstrated that energy transfer event was involved in this reaction. It is worth noting that the heterogeneous photocatalyst could be separated from the reaction system by simple filtration and washing.

In 2021, Brown, Houk, and Glorius groups described a highly regio- and diastereoselective intermolecular dearomative [4+2] cycloaddition of *N*-heteroarenes with alkenes via energy transfer catalysis in the presence of Lewis or Brønsted acid (Scheme 19) [42]. Mechanistic studies including ^{1}H NMR titration, nuclear Overhauser effect spectroscopy, and Job's plot analysis were performed to support the formation of hydrogen bonding interaction between alcoholic proton of 1,1,1,3,3,3-hexafluoro-2-propanol (HFIP) and the nitrogen atom of substrates **70**. Meanwhile, a dual hydrogen bonding interaction was directly observed in the cocrystal structure of quinoline and HFIP in the solid state. DFT calculations suggested hydrogen bonding or Lewis acid-base interaction with $BF_3{\cdot}OEt_2$ as the Lewis acid decreased the singlet-triplet energy gap. Cyclic voltammetry study and correlation of reaction rate to triplet energies rather than redox potentials of the

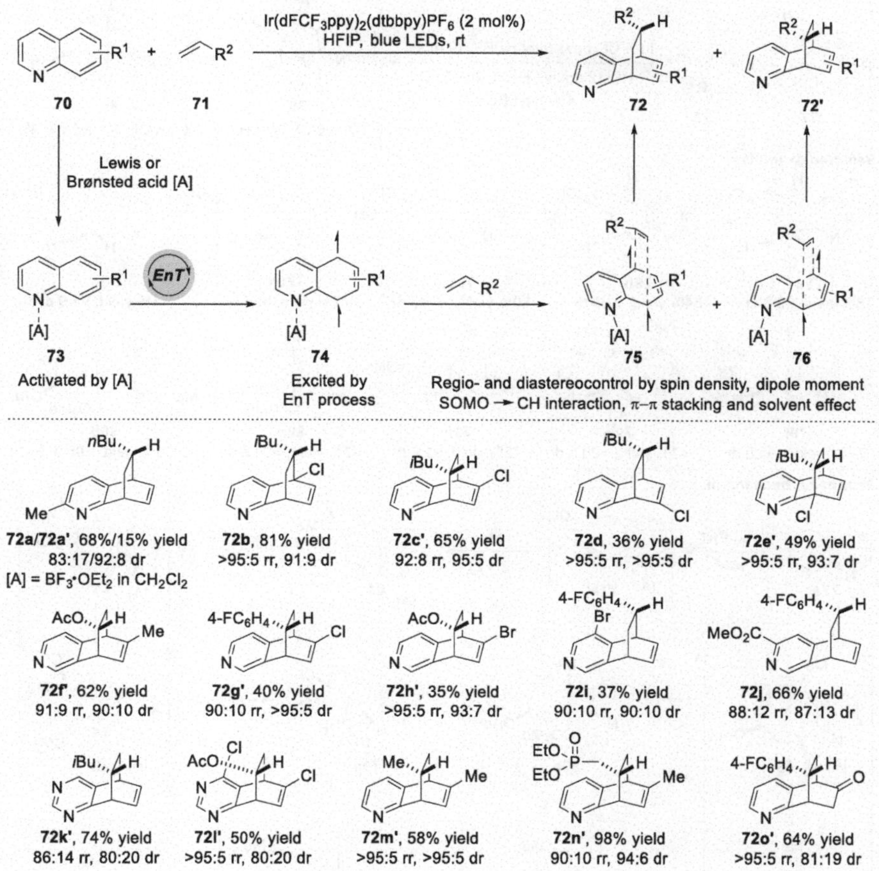

Scheme 19 Intermolecular dearomative [4+2] cycloaddition of *N*-heteroarenes with alkenes

photocatalysts excluded the single-electron transfer process. Transition-state polarity and solvent stabilization exhibited a pivotal role in regulating regio- and diastereoselectivities. This method was compatible with various kinds of olefins, including styrenes, unactivated aliphatic olefins, vinyl esters, and so on. Almost at the same time, the Kano group reported a dearomative [4+2] photocycloaddition of quinolines with olefins via protonation-enhanced reactivity of triplet state [43]. In the next year, the Houk, Glorius, Chen, and Brown groups further investigated the detailed mechanism of this dearomative [4+2] cycloaddition of *N*-heteroarenes with alkenes using a combined experimental and computational approach [44]. The results revealed that processes of [2+2] cycloaddition and sequential rearrangement to [4+2] adduct were involved in this transformation.

In 2022, Houk and Glorius disclosed an intermolecular cascade dearomative [2+2] cycloaddition/rearrangement reaction of quinolines with alkenes, providing a facile avenue to fused 2D/3D cyclic architectures (Scheme 20) [45]. This reaction commenced with dearomative [2+2] cycloaddition, in which the regioselectivity was

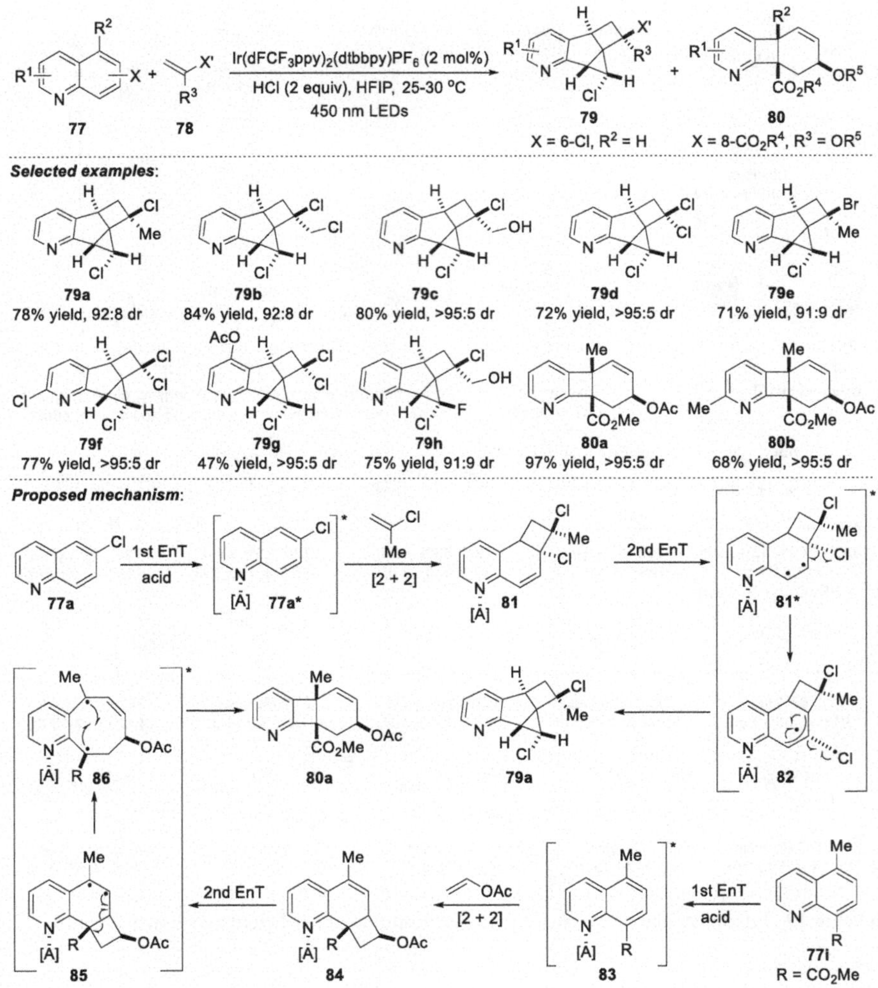

Scheme 20 Intermolecular dearomative [2+2] cycloaddition/rearrangement of *N*-heteroarenes with alkenes

controlled by decorating chloro or ester substitutions on the benzenoid rings. Then, the resulting cinnamyl chloride analogue intermediate **81** was further excited by a second energy transfer. When the substituent on benzenoid ring was 6-Cl, a homolytic bond dissociation of C–Cl and the following C–C and C–Cl bond formations occurred to afford structurally unique pyridine-fused 6-5-4-3 ring system **79** in a highly diastereoselective fashion. When 8-CO_2Me-substituted quinolines were employed, fused 6-4-6 rings **80** were achieved through ring-opening/reforming event triggered by a second energy transfer process. DFT calculations were also performed to support this mechanistic proposal.

3.2 Cycloaddition with Allenes

In the abovementioned work by the Houk, Glorius, Chen, and Brown groups in 2022, allenes were also employed in the dearomative [4+2] cycloaddition with $BF_3{\cdot}OEt_2$ as a Lewis acidic promoter (Scheme 21) [44]. 1,1-Disubstituted and internal allenes reacted with quinoline smoothly, and the desired products were afforded in excellent yields with 1.8–5:1 *Z*/*E* ratios (**89a**–**89d**). Moderate yield was obtained for the reaction of 1,3-disubstituted allene possibly because of its large steric hindrance (**89e**).

3.3 Cycloaddition with BCBs

In 2023, the Houk and Glorius groups disclosed an ortho-selective dearomative [2π+2σ] photocycloaddition of bicyclic *N*-heteroarenes to bicyclo[1.1.0]butanes (BCBs) (Scheme 22) [46]. BCBs have emerged as highly efficient building blocks for synthesizing the potential bioisosteres of phenyl ring due to their unique bicyclic structures. The release of the ring strain facilitated this dearomative cycloaddition, affording the corresponding adducts in up to 89% yield. Compared with previously reported cycloaddition of quinolines with alkenes [45], the isomerization triggered by the second photoexcitation was suppressed. Mechanistic studies suggested two possible reaction pathways. On the one hand, the complex **93*** of excited-state substrate with Lewis acid could undergo stepwise [2π+2σ] cycloaddition with BCBs to construct the desired products **92** via biradical intermediates **96**. On the other hand, a single-electron transfer with excited-state photocatalyst Ir(III)*, either before or after the addition of **93*** with BCBs, led to radical cations **95**, which induced the formation of final products in a chain propagation process. The existence of radical cations **94** was confirmed by nucleophilic trapping with pyrazole. DFT calculations were also conducted, providing the energy profile and the rationale of the regioselectivity (*para-* vs. *ortho-*).

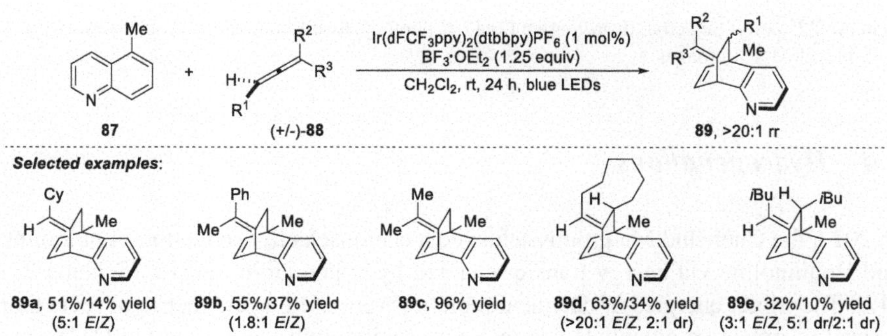

Scheme 21 Intermolecular dearomative [4+2] cycloaddition of 5-methyl quinoline with allenes

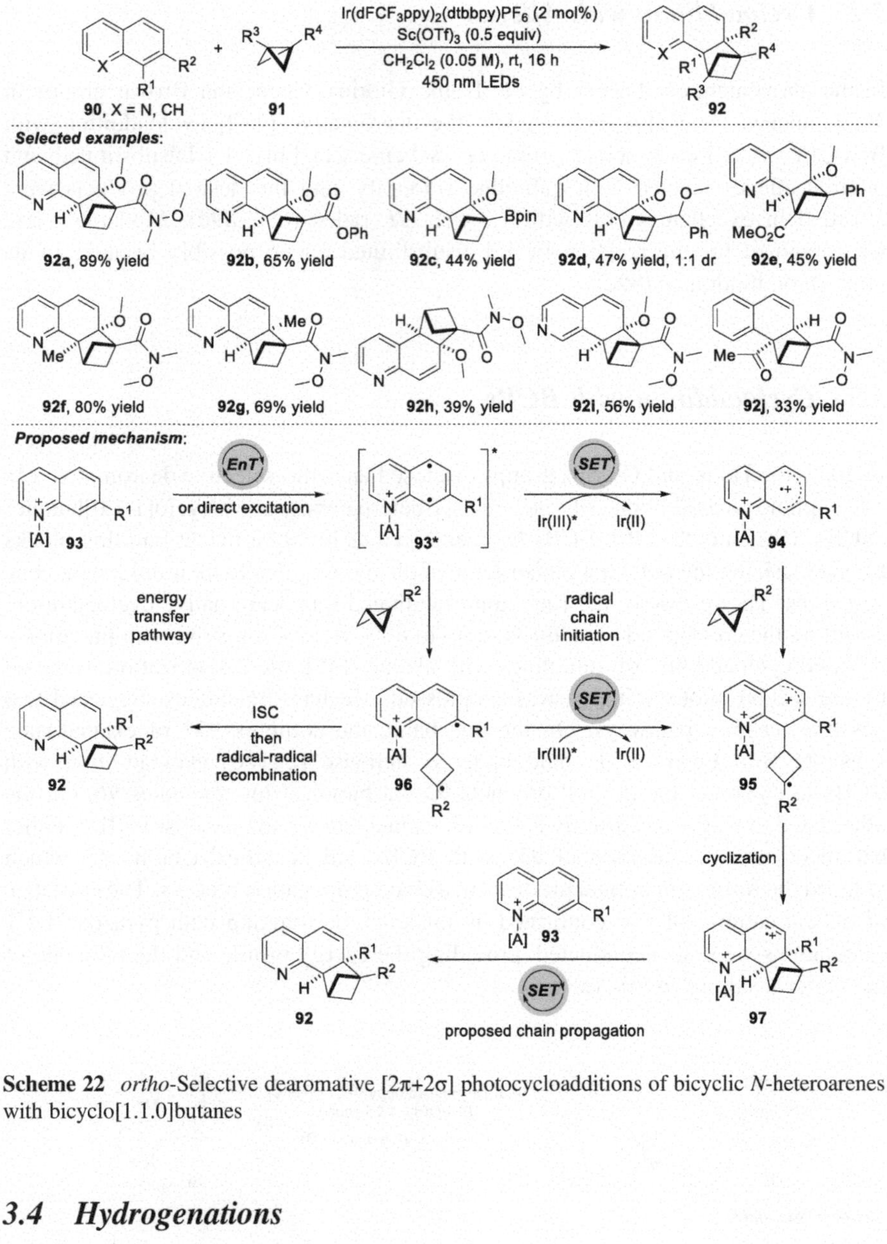

Scheme 22 *ortho*-Selective dearomative [2π+2σ] photocycloadditions of bicyclic *N*-heteroarenes with bicyclo[1.1.0]butanes

3.4 Hydrogenations

In 2023, the Chen and Ma groups achieved a chemoselective reduction of quinoline and isoquinoline via energy transfer-enabled hydrogen atom transfer (Scheme 23) [47]. The triplet energies of chemical moieties were used as the metric of reduction instead of reductive potentials. Upon complexation with Lewis acid $Gd(OTf)_3$, the triplet energy of *N*-heteroarene moieties became lower than that of the pendant groups, such as α,β-unsaturated amide and *N*-Boc indole, resulting in selective

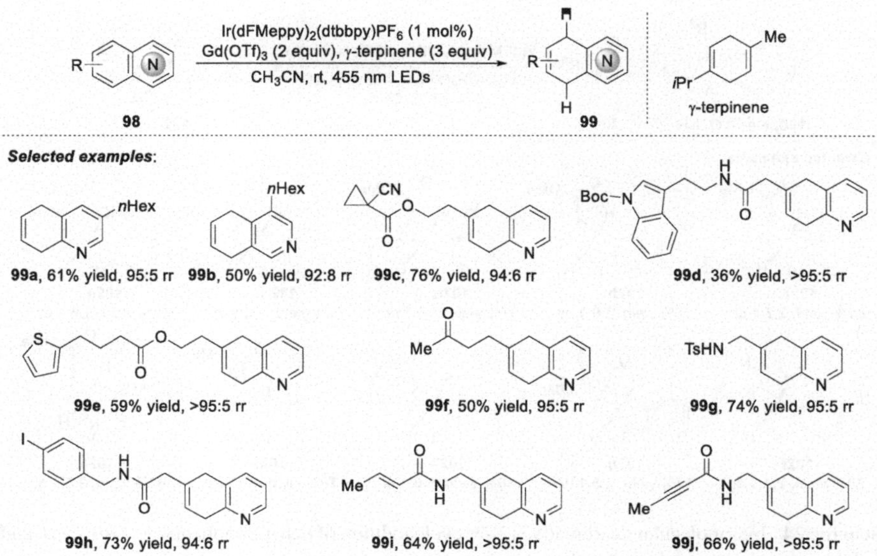

Scheme 23 Energy transfer-enabled Birch-type reduction of quinolines and isoquinolines

energy transfer from the excited photosensitizer to aromatic cores. The subsequent hydrogen abstraction transfer from terpinene to the biradical species led to Birch-type products with excellent chemoselectivity. Some unsaturated functional groups, which were unselectively reduced with thermal methods, remained intact under these photoreduction conditions and gave the arene-reduced products **99**. DFT calculations were conducted to elucidate the origins of excellent chemoselectivity in this photo-Birch-type reduction reaction.

4 Reactions of (Benzo)Furans/(Benzo)Thiophenes

4.1 Cycloaddition with Alkenes

In 2019, the Guldi and Glorius groups developed an intermolecular dearomative [2+2] cycloaddition of benzothiophenes, benzofurans, and indoles by using a combining "mechanism-based" and "reaction-based" two-dimensional screening protocol (Scheme 24) [48]. During the optimization of reaction conditions, nanosecond and femtosecond transient absorption spectroscopy, pulse radiolytic studies, as well as spectro-electrochemistry were performed to shed light on the mechanism. Based on the detailed spectroscopic corroboration, the photosensitizer with higher triplet energy but lower reductive potential was employed to suppress the undesired electron transfer. Upon irradiation with 455 nm LEDs, energy transfer from excited Ir(dFMeppy)$_2$(dtbbpy)PF$_6$ to heteroarenes was facilitated, followed by radical addition to α,β-unsaturated olefins. This intermolecular energy transfer-enabled

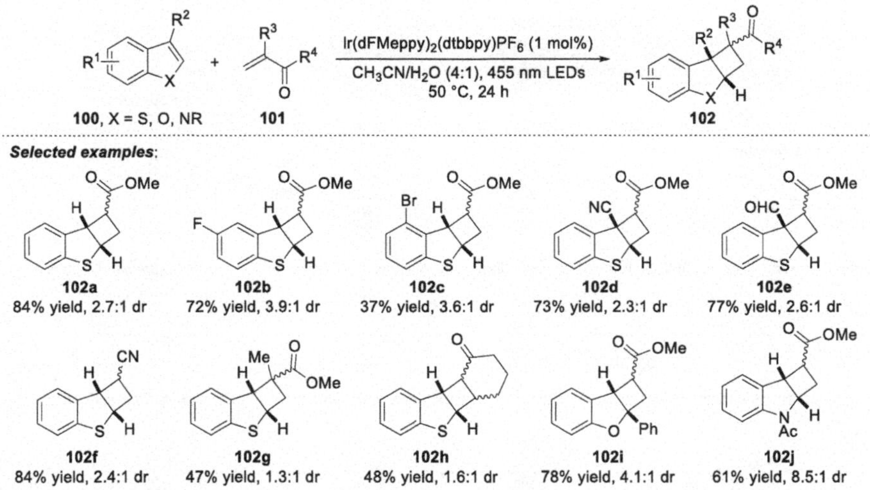

Scheme 24 Intermolecular dearomative [2+2] cycloaddition of benzothiophenes, benzofurans, and indoles with electron-deficient olefins

dearomative [2+2] cycloaddition afforded cycloadducts **102** in moderate to good yields with moderate diastereoselectivity.

In 2023, the Wang, Cao, and Yin groups reported energy transfer-promoted intramolecular dearomative [2+2] or [4+2] cycloaddition reactions of heteroarenes, naphthalenes, and benzenes in the presence of 4CzIPN or thioxanthone, allowing for the direct construction of fused/bridged three-dimensional polycyclic scaffolds (Scheme 25) [49]. Based on the experimental mechanistic studies and previous reports, the cinnamyl moiety was excited to its triplet state by excited photosensitizer. The subsequent addition of biradical to C=C bond, ISC, and radical recombination occurred to give the dearomative products **104**. In the reactions of naphthalenes, monitoring the reaction progress revealed the initial formation of dearomative [2+2] cycloadducts, followed by the second energy transfer-triggered rearrangement, giving the final [4+2] adducts. On the contrary, [2+2] adducts were not observed in the reactions of benzene derivatives.

In 2024, Zhong, Liang, and coworkers described an energy transfer-enabled dearomative [2+2] cycloaddition of benzofurans and benzothiophenes with maleimides (Scheme 26) [50]. In the reaction of benzofuran, maleimide was excited to triplet state via energy transfer from excited-state photosensitizer and then underwent cycloaddition reaction with ground-state benzofuran. On the contrary, the triplet photosensitizer was quenched by benzothiophene in its reaction with maleimide on the basis of Stern-Volmer luminescence quenching studies. The UV-visible absorption experiments excluded the possibility of direct excitation of benzofuran, benzothiophene, or maleimide under the reaction conditions (390 nm LEDs). Introduction of large substituents on arenes or reaction partners significantly improved the diastereoselectivity to 20:1 dr (**107d**, 2-Et; **107e**, OMe; **107h**, 2-Me).

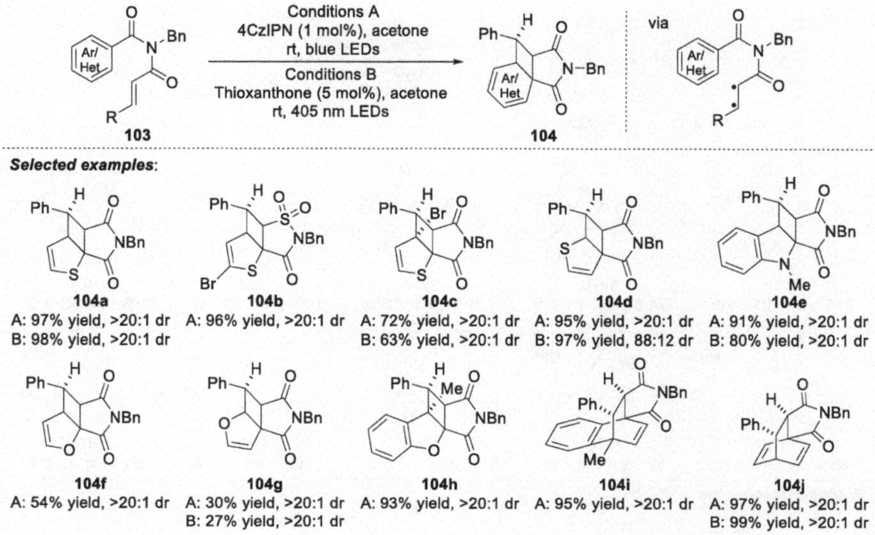

Scheme 25 Intramolecular dearomative [2+2] or [4+2] cycloaddition reactions of heteroarenes, naphthalenes, and benzenes

4.2 Cycloaddition with Allenes

In 2024, the Maestri group disclosed intramolecular dearomatization reactions of *N*- or *S*-heteroarenes via sequential allene-alkene [2+2] photocycloaddition/selective retro [2+2] processes (Scheme 27) [51]. This dearomatization method was not accessible with *N*-substituted indole and pyrrole derivatives probably because of the steric hindrance stemming from the substituents on nitrogen. DFT calculations were performed to rationalize the mechanism of this skeletal rearrangement/dearomatization reaction. The calculated results suggested that allenamide moiety, rather than cinnamamide group, was activated by excited photosensitizer via energy transfer. The subsequent [2+2] cycloaddition generated [3.2.0] bicyclic intermediate **112**, which was characterized by a significant steric strain. Then, this intermediate facilitated either the second energy transfer event to form triplet intermediate **113** or C–C cleavage to afford open-shell singlet intermediate **116**. Both intermediates **115** and **116** were engaged in the formation of desired products.

5 Reactions of Benzenes and Naphthalenes

The dearomative cycloaddition reactions of benzenes and naphthalenes with alkenes have been extensively investigated in 1950s. In most cases, high-energy ultraviolet was used to directly excite arenes to their singlet excited state, which will not be included in this review. Recently, the utilization of triplet sensitizers allowed the

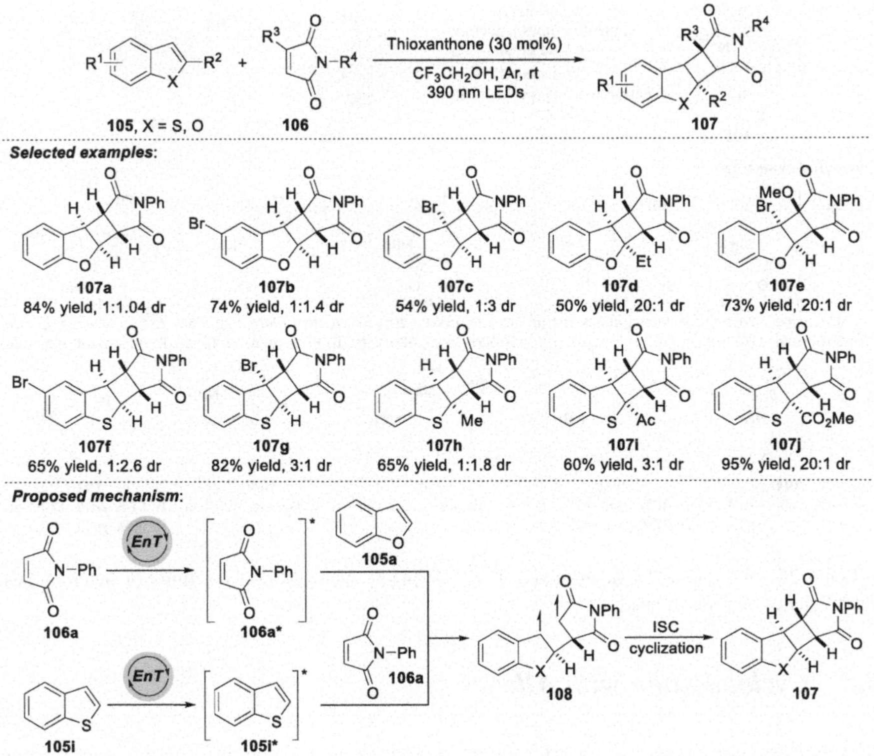

Scheme 26 Substrate-dependent dearomative [2+2] cycloaddition of benzofurans and benzothiophenes with maleimides

renaissance of dearomative cycloaddition of benzenes/naphthalenes by taking advantages of triplet excited-state aromatics through energy transfer processes.

5.1 Cycloaddition with Alkenes

In 1960, the Steinmetz group reported a benzophenone-photosensitized cycloaddition of benzene with maleic anhydride (Scheme 28) [52]. Two years later, the Lodge group suggested that this reaction involved a triplet intermediate on the basis of the inhibition of this reaction by oxygen [53]. In 1967, Hammond and coworkers further studied the mechanism of this sensitized transformation [54]. It was suggested that triplet benzophenone excited maleic anhydride-benzene complex to its triplet state via energy transfer. The collapse of the excited complex delivered a [2+2] monoadduct, which subsequently reacted with another molecule of maleic anhydride in a thermal [4+2] reaction to generate the final product.

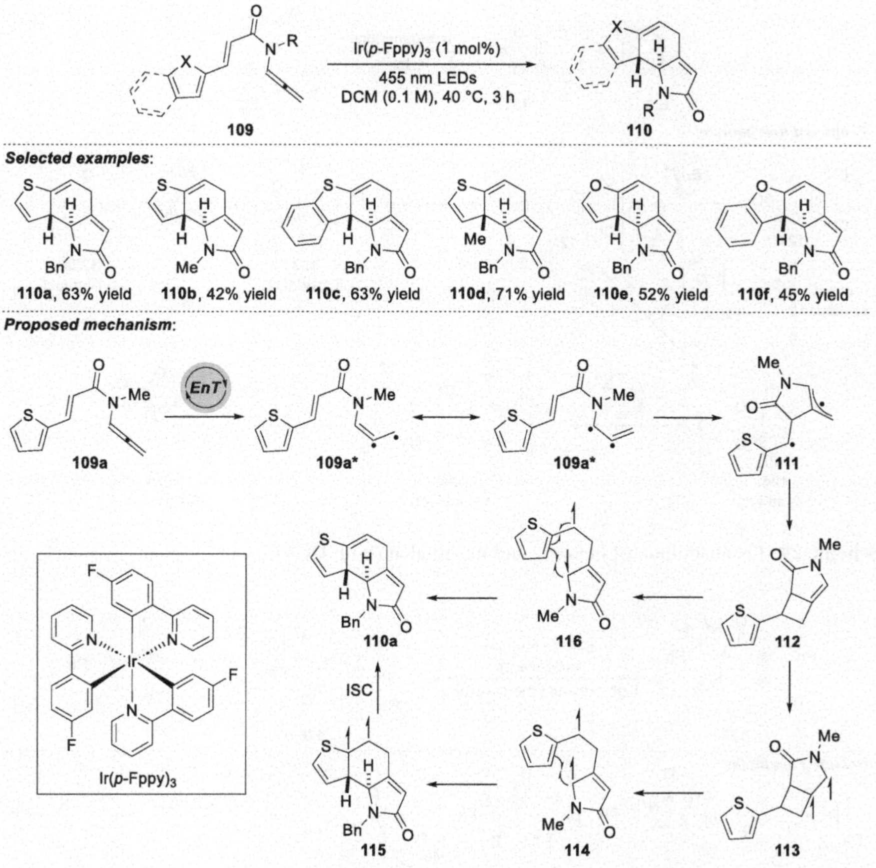

Scheme 27 Sequential skeletal rearrangement/dearomatization of *N*- or *S*-heteroarenes

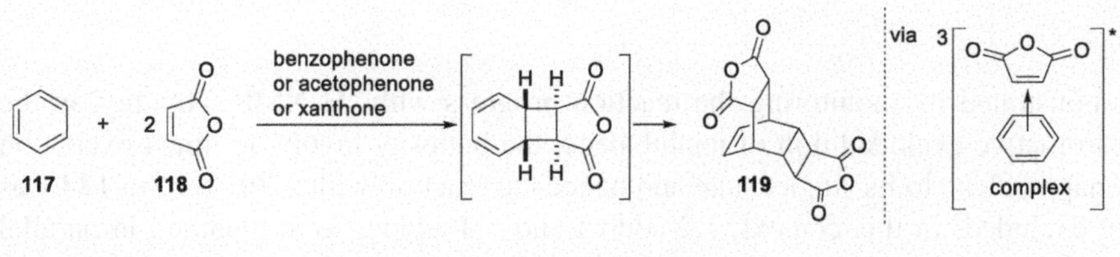

Scheme 28 Cycloaddition of benzene with maleic anhydride

In 1975, the Aretz group described the mechanism of dearomative cycloaddition reaction of benzene and naphthalene with dichlorovinylene carbonate **121** (DCVC) by using acetophenone as the photosensitizer (Scheme 29) [55]. Energy transfer from excited acetophenone to DCVC led to triplet DCVC*, which reacted with benzene or naphthalene to generate 1,2-adduct **123**. Subsequently, 1,2-adduct **123** rearranged to 1,4-adduct **122a** under the reaction conditions, which was

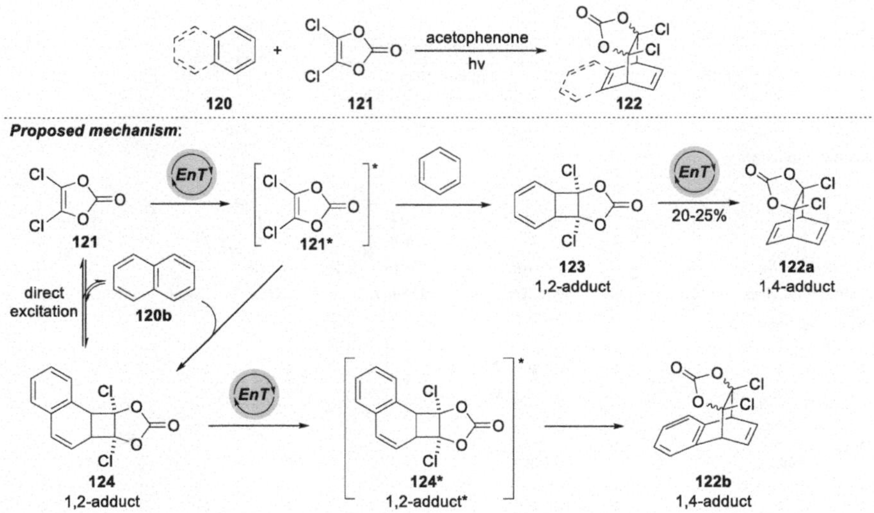

Scheme 29 Cycloaddition of benzene and naphthalene with DCVC

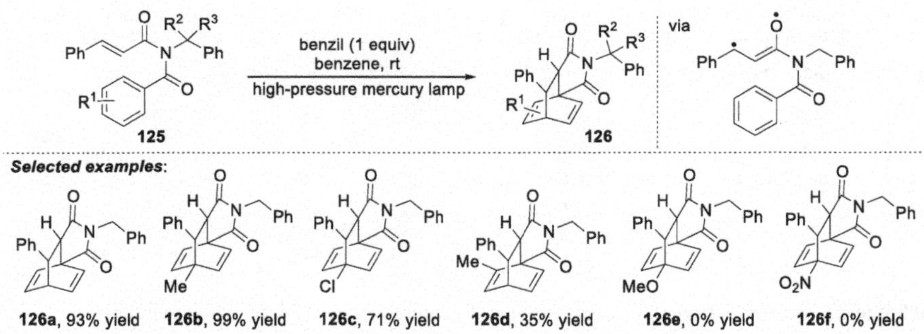

Scheme 30 Intramolecular photocycloaddition of benzenes with α,β-unsaturated amides

demonstrated by monitoring the reaction progress with ^{1}H NMR. Notably, in the dearomative cycloaddition of naphthalene, the pathway involving direct excitation of naphthalene to its singlet state and hence the reaction with **120b** to form **124** was not excluded. In this context, 1,2-adduct and 1,4-adduct were obtained in parallel reactions via a common intermediate **124**.

In 1997, the Kishikawa group disclosed an intramolecular [4+2] photocycloaddition of benzene with α,β-unsaturated amide in the presence of benzyl ($E_T = 53$ kcal mol^{-1}) and irradiation by a high-pressure mercury lamp (Scheme 30) [56]. Initially, the cinnamyl group of **125** was excited to reactive biradical species by the triplet photosensitizer via energy transfer. Then, excited cinnamyl moiety preferentially reacted with the benzoyl group but not with the benzyl group. This result indicated that electron density distribution of the benzoyl group might control the reactivity. The pathway involving charge transfer complex composed of phenyl ring

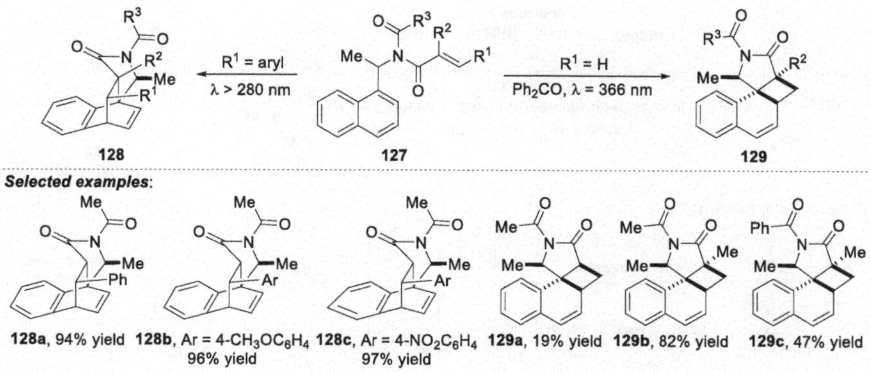

Scheme 31 Highly peri- and stereoselective intramolecular photocycloaddition of naphthalenes with α,β-unsaturated amides

and cinnamyl moiety was excluded [57], since the face-to-face interaction of these two groups for charge transfer was difficult because of the rigid junction. Therefore, [2+2] photocycloaddition of substrates **125** was suppressed significantly, and [4+2] adducts became the major products (**126a–126d**). **125e** bearing an OMe group was not reactive because an electron-rich benzene ring was an unsuitable acceptor of radical. The complete suppression of reactivity by NO_2 was explained in terms of that the biradical was stabilized by the electron-withdrawing group.

In 2001, the Kohmoto group developed a highly peri- and stereoselective intramolecular photocycloaddition of *N*-(1-naphthylethyl)prop-2-enamides (Scheme 31) [58]. The substituents on α,β-unsaturated amide moiety decided the reaction pathway, which induced the formation of [2+2] or [4+2] adducts, respectively. When cinnamyl acid-derived substrates **127** (R^1 = Ar) were used, [4+2] adducts **128** with the phenyl group *anti* to adjacent carbonyl were obtained stereoselectively. The possibility of rearrangement from [2+2] adducts to [4+2] adducts was ruled out because the [2+2] adducts were stable under irradiation in the presence of sensitizer. On the contrary, [2+2] adducts **129** were generated exclusively when acryloyl-substituted **127** (R^1 = H) were employed. In this pathway, the naphthalene group was suggested to be excited by the triplet benzophenone.

In 2018, the Glorius group realized a visible light-promoted intramolecular dearomative [2+2] cycloaddition of naphthol derivatives in the presence of Ir(ppy)$_2$(dtbbpy)PF$_6$ (E_T = 49.2 kcal mol^{-1}) or Ir(dFCF$_3$ppy)$_2$(dtbbpy)PF$_6$ (E_T = 60.1 kcal mol^{-1}), accessing to complex polycyclic molecular scaffolds through sensitizer-selective energy transfer (Scheme 32) [59]. Upon irradiation by a 455 nm LED light, naphthyl moiety was excited to triplet-state intermediate **130***. The subsequent [2+2] cycloaddition of naphthyl biradical **130*** to C=C double bond in the side chain gave cyclobutanes **131**. When Ir(ppy)$_2$(dtbbpy)PF$_6$ was used as the photosensitizer, **131** was afforded as the major product. If Ir(dFCF$_3$ppy)$_2$(dtbbpy)-PF$_6$ was used instead of Ir(ppy)$_2$(dtbbpy)PF$_6$ in this reaction, the remaining π system in product **131** could be efficiently excited to intermediate **131*** via a second energy

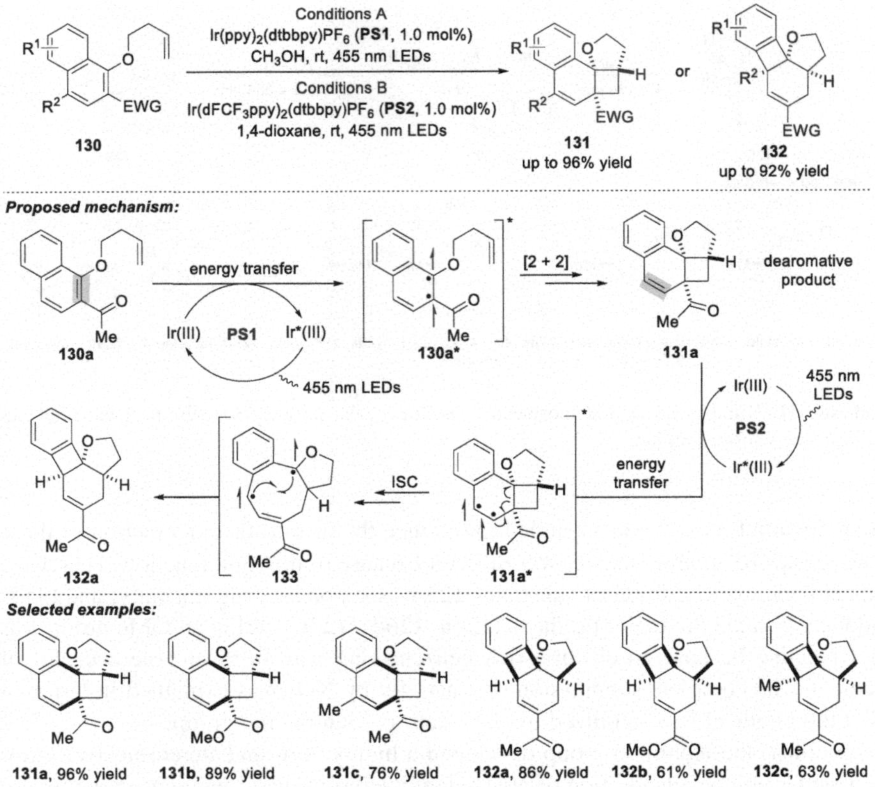

Scheme 32 Photosensitizer-dependent intramolecular dearomatization of naphthol derivatives with alkene

transfer. Then, intermediate **131*** engaged in a vinyl cyclobutane rearrangement to afford benzocyclobutene **132**.

In 2022, the Maji group reported an intermolecular dearomative [4+2] cycloaddition reaction of naphthalene derivatives with alkenes via visible light-mediated energy transfer mechanism (Scheme 33a) [60]. Stern-Volmer luminescence quenching studies, cyclic voltammetry studies, as well as experiments in the absence of photosensitizer or in the presence of triplet quenchers (oxygen and 2,5-dimethylhexa-2,4-diene) collectively indicated that this reaction proceeded through a pathway that involved energy transfer. In the kinetic experiments, the rate of formation of [4+2] adducts **136** was found to be equal to that of the disappearance of substrate **134**. This supported the direct formation of **136** without the generation of [2+2] adduct intermediate, which was different from the previously reported intramolecular variants. Almost at the same time, the Brown group disclosed a similar reaction with Ir(p-CF_3ppy)$_3$ (E_T = 56.4 kcal mol^{-1}) as the photosensitizers, where acyl pyrazole was employed as the activating group (Scheme 33b) [61]. The competition reactions of naphthalenes substituted by ester

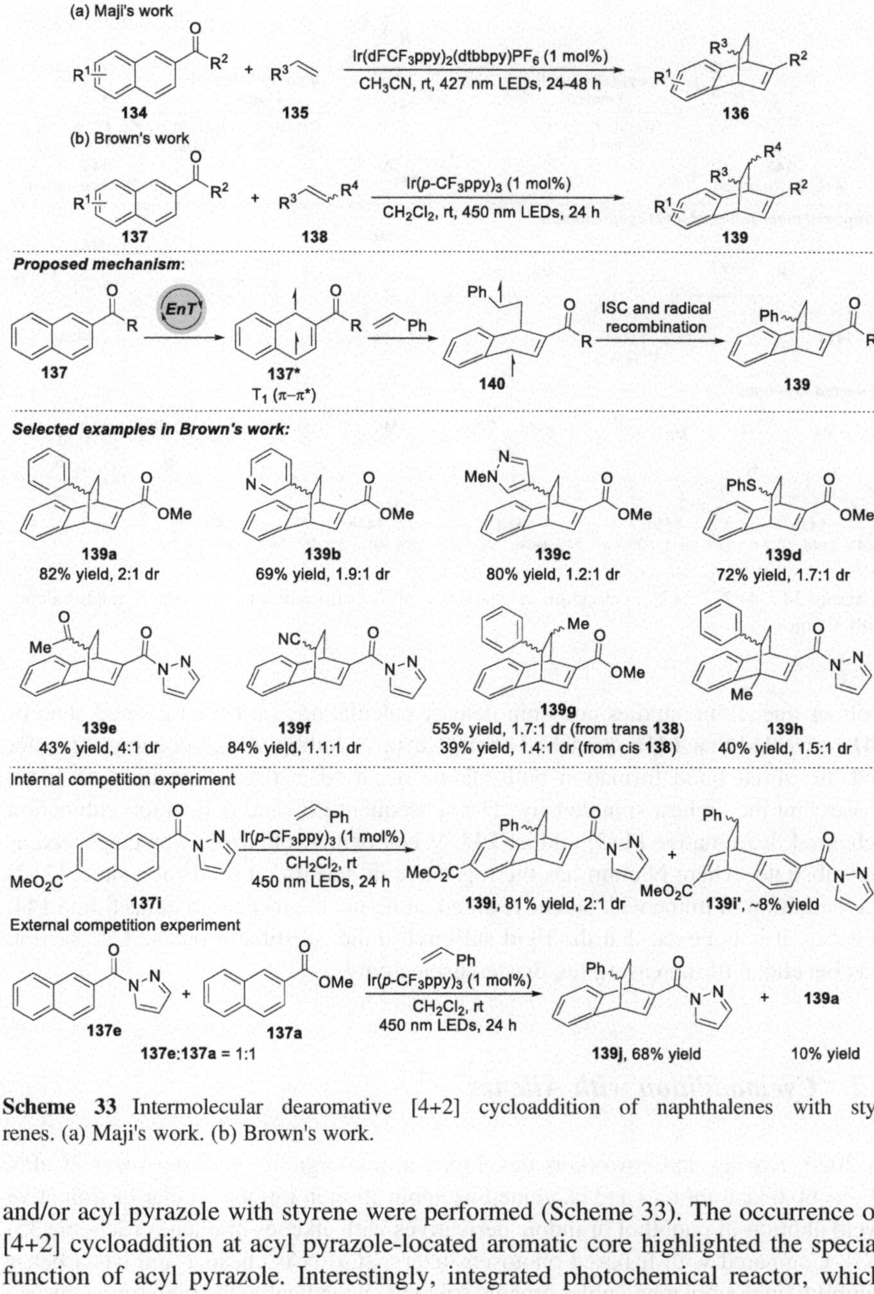

Scheme 33 Intermolecular dearomative [4+2] cycloaddition of naphthalenes with styrenes. (a) Maji's work. (b) Brown's work.

and/or acyl pyrazole with styrene were performed (Scheme 33). The occurrence of [4+2] cycloaddition at acyl pyrazole-located aromatic core highlighted the special function of acyl pyrazole. Interestingly, integrated photochemical reactor, which supplies a significantly higher photon flux, elevated the efficiency of the reaction of the corresponding ester substrate.

In 2023, the Brown group achieved selective dearomative [4+2] cycloaddition of naphthalenes and [2+2] cycloaddition of imines to alkenes by modulating the *N*-sulfonylimine group attached to the aromatic substrates (Scheme 34) [62]. Stern-

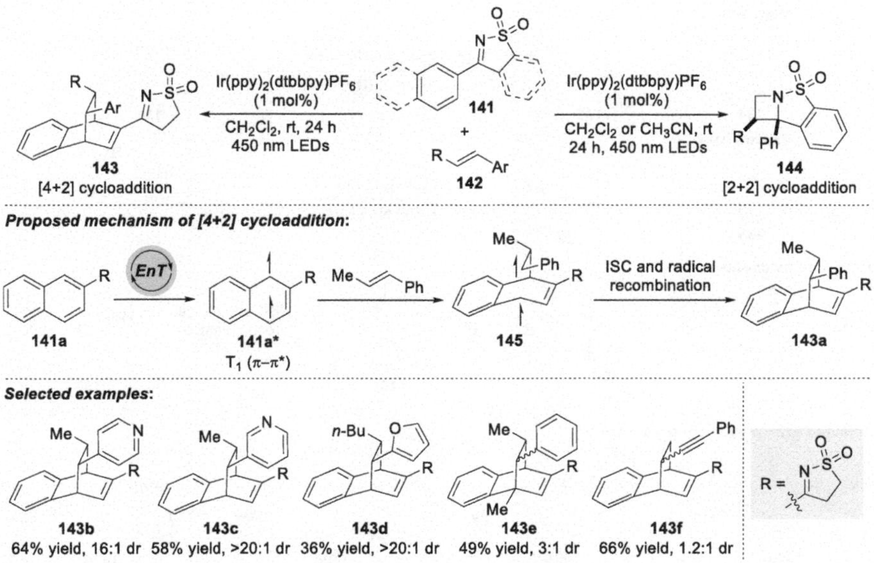

Scheme 34 [4+2]-/[2+2] cycloaddition reactions of *N*-sulfonylimine-substituted naphthalenes with alkenes

Volmer quenching studies and spin density calculations on triplet excited state of **141a** revealed that **141a** quenched excited $Ir(ppy)_2(dtbbpy)PF_6$ via energy transfer and the initial bond formation with alkene occurred at the C1 position of **141a**, possessing the highest spin density. The subsequent ISC and radical recombination delivered dearomative [4+2] adduct **143**. When benzoisothiazole was employed as the substrate, where N-atom has the highest spin density on the triplet state, [2+2] cycloaddition of imine with alkene resulted, allowing the formation of azetidine **144**. Notably, it is believed that the rigid sulfonyl imine substituent on the C2 position was beneficial of increasing the diastereoselectivity.

5.2 *Cycloaddition with Allenes*

In 2020, Koenig and coworkers developed a new organic photosensitizer 2CzPN ($E_T = 60.6$ kcal mol^{-1}) and examined its application in intramolecular dearomative cycloaddition of naphthol or indole derivatives with alkenes or allenes (Scheme 35) [63]. Compared with Ir-based photosensitizers, 2CzPN is cheaper and has a better solubility in nonpolar and polar organic solvents. Specifically, the high triplet energy of 2CzPN made it a preferred sensitizer for effectively activating aromatic substrates to their triplet state via energy transfer. In the presence of 2CzPN, allene worked as a biradical acceptor for the dearomative cycloaddition of naphthol derivative, giving rise to the dearomatized product with a complex polycyclic molecular framework.

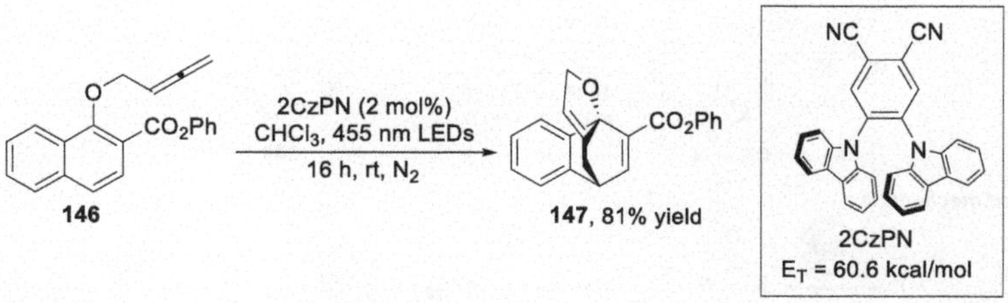

Scheme 35 Dearomative [4+2] cycloaddition of naphthalene with a tethered allene

In 2023, the Maestri group achieved a visible light-enabled intramolecular dearomative [4+2] cycloaddition of benzenes, naphthalenes, and *N*-heteroarenes with allenes in the presence of $Ir(ppy)_3$ (E_T = 57.8 kcal mol^{-1}) (Scheme 36) [64]. Initially, the reaction efficiency of benzene derivatives could not be improved by optimizing the reaction conditions, including solvents, photosensitizers, and higher catalyst loading. However, when the naphthalene derivative was used in the model reaction, its desired product **149b** was produced in a significantly faster rate and isolated in 99% yield. On the basis of this observation, the authors speculated that the π-system in naphthalene moiety could stabilize transient biradical. As expected, the addition of naphthalene (20 equiv.) as an additive efficiently promoted the reaction of benzene derivative to complete in 38 h, giving **149g** in 65% yield. DFT calculations were also performed to support this hypothesis. A series of functional groups at the *meta*- or *para*-positions of benzene were all compatible with this dearomative transformation. The intermolecular dearomative cycloaddition of simple naphthalene with allene **154** proceeded smoothly without any modification of the reaction conditions.

5.3 Cycloaddition with Vinylcyclopropanes

In 2024, the You group described a visible light-induced intramolecular dearomative [5+4] cycloaddition of naphthol derivatives bearing a vinyl cyclopropane side chain, allowing the construction of highly strained medium-sized ring (Scheme 37) [65]. In the presence of $Ir(ppy)_3$, a series of complex polycyclic architectures containing an *E*-alkene moiety, a challenging target via available methods, were afforded in good yields and *E*/*Z* ratios **157**. Energy transfer from excited photosensitizer to substrate **156** delivered the triplet excited naphthalene. Subsequent addition to the C=C bond in vinyl cyclopropane, C–C cleavage of cyclopropane, and radical recombination were facilitated to form the products. Detailed DFT calculations were performed to demonstrate the energy landscape of this reaction and the origin for the favorable formation of thermodynamically less stable medium-sized rings.

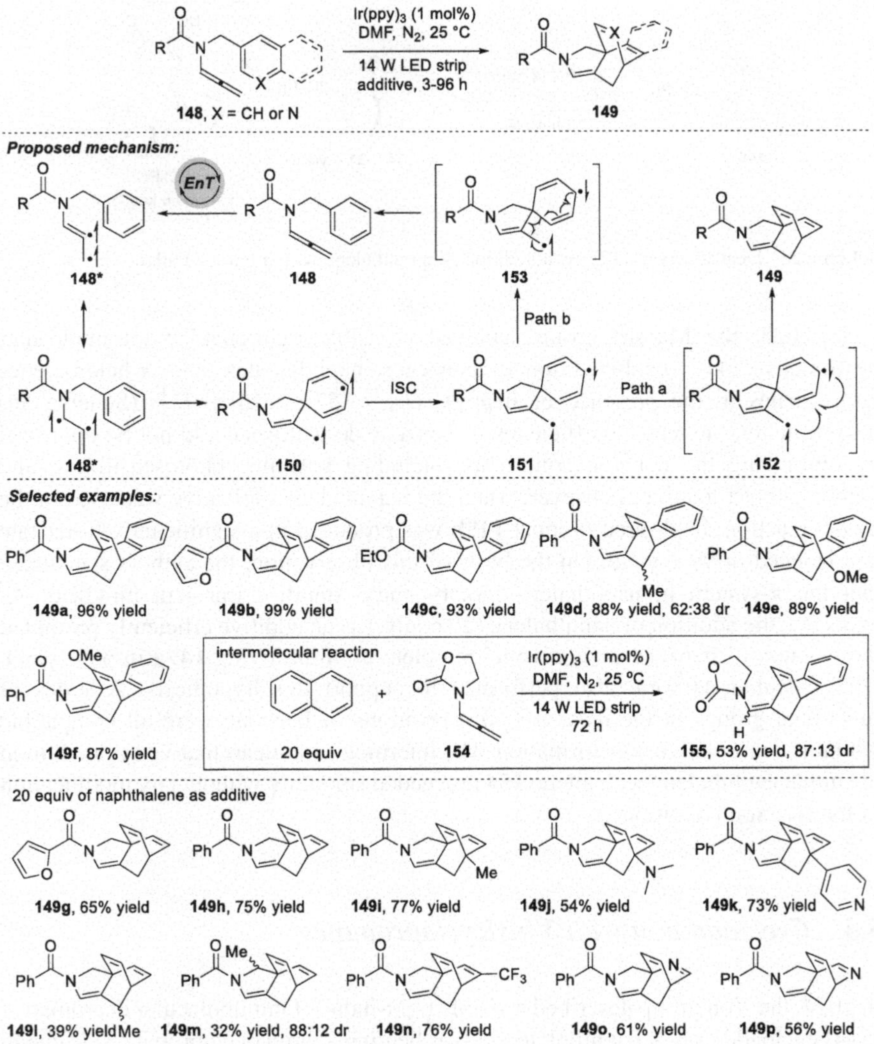

Scheme 36 Dearomative [4+2] cycloaddition of benzenes, naphthalenes, and pyridines with allenes

5.4 Hydrogenation

In 2019, the Koenig group realized a Birch-type photoreduction of (hetero)arenes by combining energy transfer and single-electron transfer processes (Scheme 38) [66]. In this reaction, it was found that plain arenes that are difficult to be reduced in ground state could be reduced easily in their excited state. Mechanistically, arenes were transferred to their excited state **158*** by triplet photosensitizer via energy

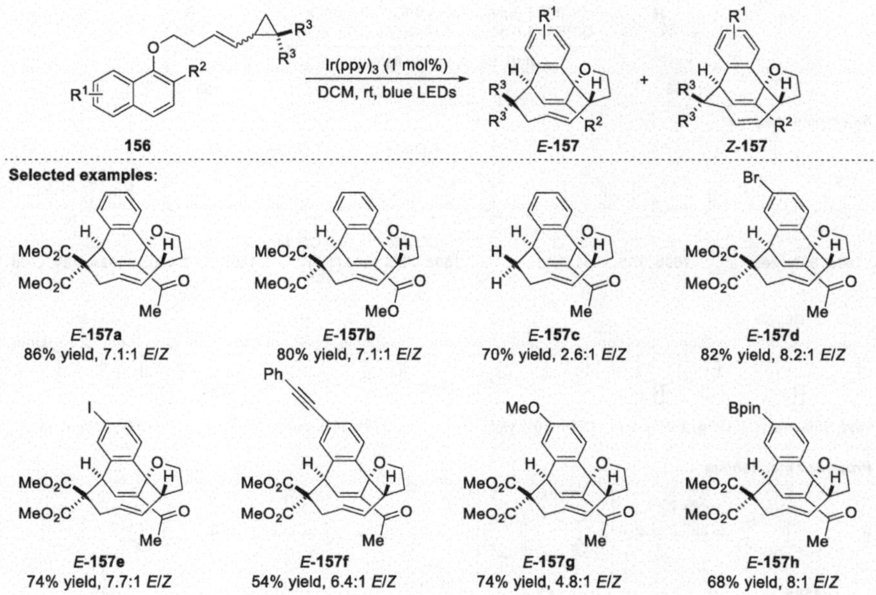

Scheme 37 Intramolecular dearomative [5+4] cycloaddition of naphthol derivatives with vinyl cyclopropanes

transfer. In a parallel process, the excited photosensitizer was reduced by an electron donor DIPEA. Then, excited-state arene **160** was reduced by Ir(II) to form a radical anion **161**. The following fast hydrogen atom transfer and protonation afforded the desired product. Due to the strong reduction ability of this reaction system, part of the dearomatized products **159b** (left) in the photoreduction of naphthalenes were further reduced to **159b** (right). Moreover, the regioselectivity of the reactions of unsymmetric naphthalene could not be controlled (**159c**). This reaction manifold unlocked an alternative strategy for dearomatization of plain aromatics.

5.5 Difunctionalization

In 2019, You, Cho, and coworkers developed an intramolecular dearomative double-functionalized reaction of benzenes enabled by energy transfer catalysis in the presence of molecular oxygen (Scheme 39) [67]. Mechanistically, the insertion of singlet oxygen to excited **162a*** at the C5 position gave hydroperoxy intermediate **164**, followed by homolytic N–O bond cleavage to generate the key iminyl radical **165** along with the hydroperoxide radical. This process represented an unprecedented approach for the generation of N radical. A series of controlled experiments and density functional theory studies supported this activation process.

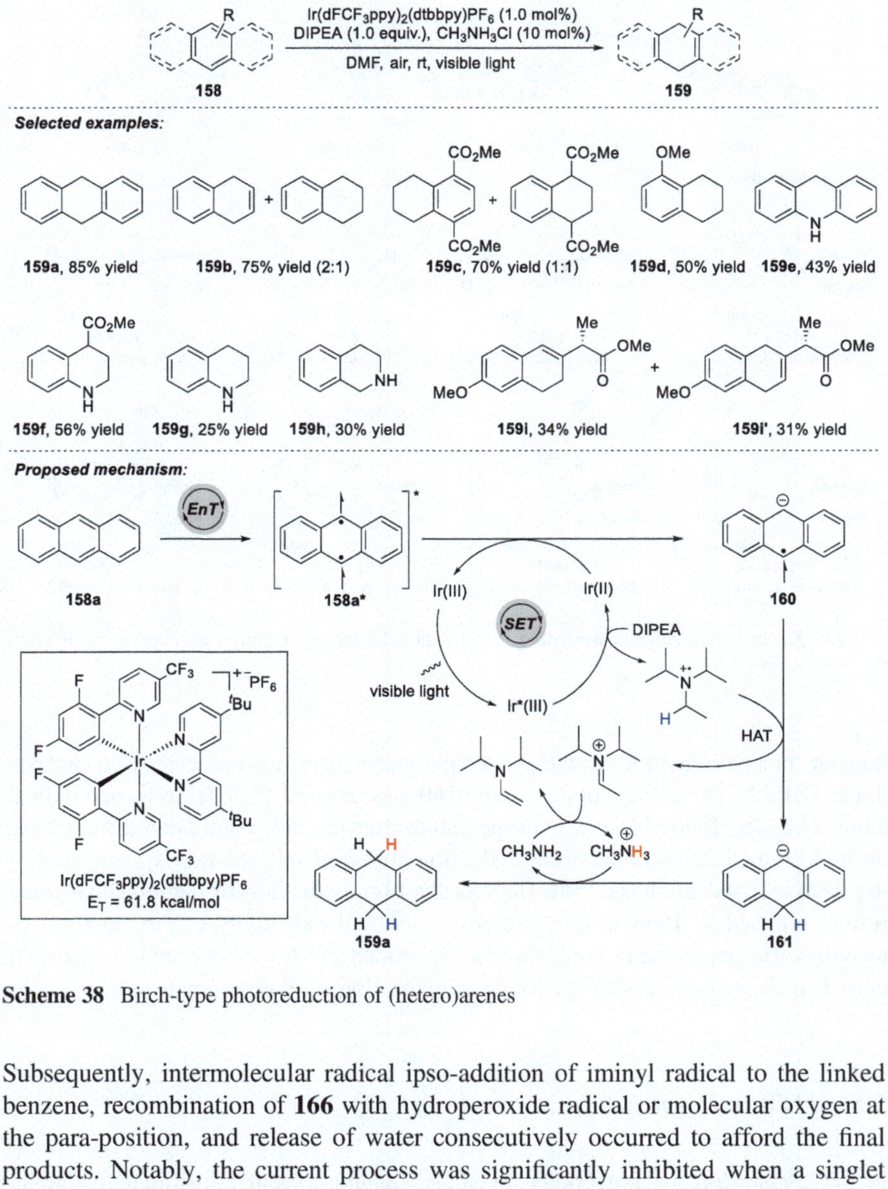

Scheme 38 Birch-type photoreduction of (hetero)arenes

Subsequently, intermolecular radical ipso-addition of iminyl radical to the linked benzene, recombination of **166** with hydroperoxide radical or molecular oxygen at the para-position, and release of water consecutively occurred to afford the final products. Notably, the current process was significantly inhibited when a singlet oxygen scavenger such as DABCO or NaN_3 was present, which suggested the participation of singlet-state oxygen in this reaction. The single-electron transfer from excited fac-Ir(ppy)$_3$* to oxadiazolines was excluded because of the unfavorable redox potentials.

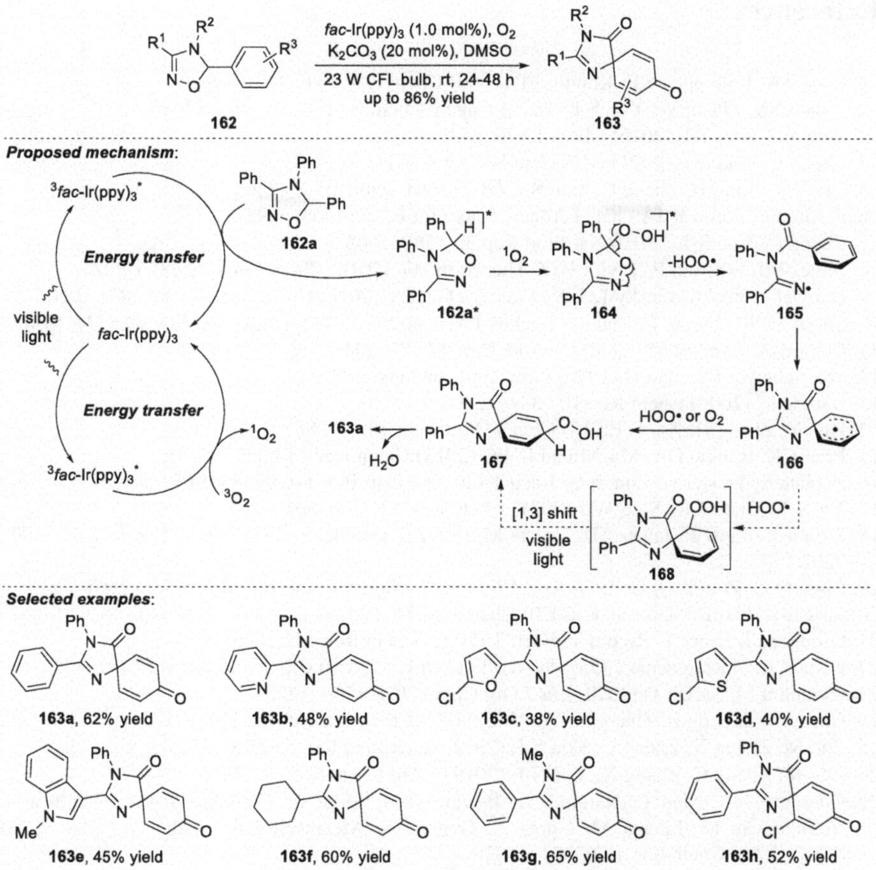

Scheme 39 Intramolecular dearomative double-functionalized reaction of benzenes

6 Conclusion and Perspectives

In this chapter, we have presented the recent progresses on photoinduced dearomatization reactions via energy transfer mechanism. These reactions cover a large array of aromatic compounds and coupling partners. Diverse cyclization modes including [2+2], [3+2], [4+2], [5+2], [5+4], etc. have been realized, allowing expedient synthesis of polycyclic molecules with complex topologies, some of which possess extremely high geometric strain and thus are not easily accessible under thermal conditions. Enantioselective reactions have also been proved possible when appropriate chiral chemical or enzymatic catalytic systems are applied. Further development in this area can be expected to fuel organic chemistry by more exciting synthetic methodologies expanding the chemical spaces of polycyclic molecules.

References

1. Pape AR, Kaliappan KP, Kündig EP (2000) Chem Rev 100:2917–2940
2. Zhuo C-X, Zhang W, You S-L (2012) Angew Chem Int Ed 51:12662–12686
3. Zheng C, You S-L (2016) Chem 1:830–857
4. Zheng C, You S-L (2021) ACS Cent Sci 7:432–444
5. Liu Y-Z, Song H, Zheng C, You S-L (2022) Nat Synth 1:203–216
6. Roche SP, Porco Jr JA (2011) Angew Chem Int Ed 50:4068–4093
7. Zheng C, You S-L (2019) Nat Prod Rep 36:1589–1605
8. James MJ, O'Brien P, Taylor RJK, Unsworth WP (2016) Chem Eur J 22:2856–2881
9. Bariwal J, Voskressensky LG, Van der Eycken EV (2018) Chem Soc Rev 47:3831–3848
10. Sharma UK, Ranjan P, Van der Eycken EV, You S-L (2020) Chem Soc Rev 49:8721–8748
11. Cerveri A, Bandini M (2020) Chin J Chem 38:287–294
12. von Schleyer PR, Jiao H (1996) Pure Appl Chem 68:209–218
13. Randić M (2003) Chem Rev 103:3449–3606
14. Hua Y, Zhang H, Xia H (2018) Chin J Org Chem 38:11–28
15. Prier CK, Rankic DA, MacMillan DWC (2013) Chem Rev 113:5322–5363
16. Poplata S, Tröster A, Zou Y-Q, Bach T (2016) Chem Rev 116:9748–9815
17. Yu X-Y, Chen J-R, Xiao W-J (2021) Chem Rev 121:506–561
18. Strieth-Kalthoff F, James MJ, Teders M, Pitzer L, Glorius F (2018) Chem Soc Rev 47:7190–7202
19. Zhou Q-Q, Zou Y-Q, Lu L-Q, Xiao W-J (2019) Angew Chem Int Ed 58:1586–1604
20. Strieth-Kalthoff F, Glorius F (2020) Chem 6:1888–1903
21. Großkopf J, Kratz T, Rigotti T, Bach T (2022) Chem Rev 122:1626–1653
22. Festa AA, Voskressensky LG, Van der Eycken EV (2019) Chem Soc Rev 48:4401–4423
23. Okumura M, Sarlah D (2020) Eur J Org Chem 2020:1259–1273
24. Cheng Y-Z, Feng Z, Zhang X, You S-L (2022) Chem Soc Rev 51:2145–2170
25. Zhu M, Zhang X, Zheng C, You S-L (2022) Acc Chem Res 55:2510–2525
26. Zhu M, Zheng C, Zhang X, You S-L (2019) J Am Chem Soc 141:2636–2644
27. Oderinde MS, Mao E, Ramirez A, Pawluczyk J, Jorge C, Cornelius LAM, Kempson J, Vetrichelvan M, Pitchai M, Gupta A, Gupta AK, Meanwell NA, Mathur A, Dhar TGM (2020) J Am Chem Soc 142:3094–3103
28. Zhang Z, Yi D, Zhang M, Wei J, Lu J, Yang L, Wang J, Hao N, Pan X, Zhang S, Wei S, Fu Q (2020) ACS Catal 10:10149–10156
29. Ikeda M, Ohno K, Mohri S-I, Takahashi M, Tamura Y (1984) J Chem Soc Perkin Trans 1:405–412
30. Oderinde MS, Ramirez A, Dhar TGM, Cornelius LAM, Jorge C, Aulakh D, Sandhu B, Pawluczyk J, Sarjeant AA, Meanwell NA, Mathur A, Kempson J (2021) J Org Chem 86:1730–1747
31. Sauvé ER, Mayder DM, Kamal S, Oderinde MS, Hudson ZM (2022) Chem Sci 13:2296–2302
32. Sun N, Huang J, Qian J, Zhou T-P, Guo J, Tang L, Zhang W, Deng Y, Zhao W, Wu G, Liao R-Z, Chen X, Zhong F, Wu Y (2022) Nature 611:715–720
33. Zhu M, Huang X-L, Xu H, Zhang X, Zheng C, You S-L (2020) CCS Chem 2:652–664
34. Zhu M, Zhang X, Zheng C, You S-L (2020) ACS Catal 10:12618–12626
35. Zhu M, Xu H, Zhang X, Zheng C, You S-L (2021) Angew Chem Int Ed 60:7036–7040
36. Zhen G, Zeng G, Jiang K, Wang F, Cao X, Yin B (2023) Chem Eur J 29:e202203217
37. Zhu M, Huang X-L, Sun S, Zheng C, You S-L (2021) J Am Chem Soc 143:13441–13449
38. Tan G, Das M, Keum H, Bellotti P, Daniliuc C, Glorius F (2022) Nat Chem 14:1174–1184
39. Tan G, Das M, Kleinmans R, Katzenburg F, Daniliuc C, Glorius F (2022) Nat Catal 5:1120–1130
40. Xiong Y, Großkopf J, Jandl C, Bach T (2022) Angew Chem Int Ed 61:e202200555
41. Ma J, Strieth-Kalthoff F, Dalton T, Freitag M, Schwarz JL, Bergander K, Daniliuc C, Glorius F (2019) Chem 5:2854–2864

42. Ma J, Chen S, Bellotti P, Guo R, Schäfer F, Heusler A, Zhang X, Daniliuc C, Brown MK, Houk K, Glorius F (2021) Science 371:1338–1345
43. Morofuji T, Nagai S, Chitose Y, Abe M, Kano N (2021) Org Lett 23:6257–6261
44. Guo R, Adak S, Bellotti P, Gao X, Smith WW, Le SN, Ma J, Houk KN, Glorius F, Chen S, Brown MK (2022) J Am Chem Soc 144:17680–17691
45. Ma J, Chen S, Bellotti P, Wagener T, Daniliuc C, Houk K, Glorius F (2022) Nat Catal 5:405–413
46. Kleinmans R, Dutta S, Ozols K, Shao H, Schäfer F, Thielemann RE, Chan HT, Daniliuc CG, Houk KN, Glorius F (2023) J Am Chem Soc 145:12324–12332
47. Liu D-H, Nagashima K, Liang H, Yue X-L, Chu Y-P, Chen S, Ma J (2023) Angew Chem Int Ed 62:e202312203
48. Strieth-Kalthoff F, Henkel C, Teders M, Kahnt A, Knolle W, Gómez-Suárez A, Dirian K, Alex W, Bergander K, Daniliuc CG, Abel B, Guldi DM, Glorius F (2019) Chem 5:2183–2194
49. Zhen G, Zeng G, Wang F, Cao X, Yin B (2023) Adv Synth Catal 365:43–52
50. Liao Z-Y, Gao F, Ye Y-H, Yu Q-H, Yang C, Luo Q-Y, Du F, Pan B, Zhong W-W, Liang W (2024) Chem Commun 60:4455–4458
51. Chiminelli M, Scarica G, Serafino A, Marchiò L, Viscardi R, Maestri G (2024) Molecules 29: 595
52. Schenck GO, Steinmetz R (1960) Tetrahedron Lett 1:1–8
53. Bryce-Smith D, Lodge JE (1962) J Chem Soc:2675–2680
54. Hardham WM, Hammond GS (1967) J Am Chem Soc 89:3200–3205
55. Scharf H-D, Leismann H, Erb W, Üaidetzka HW, Aretz J (1975) Pure Appl Chem 41:581–600
56. Kishikawa K, Akimoto S, Kohmoto S, Yamamoto M, Yamada K (1997) J Chem Soc Perkin Trans 1:77–84
57. Bradshaw JS (1966) J Org Chem 31:3974–3976
58. Kohmoto S, Miyaji Y, Tsuruoka M, Kishikawa K, Yamamoto M, Yamada K (2001) J Chem Soc Perkin Trans 1:2082–2088
59. James MJ, Schwarz JL, Strieth-Kalthoff F, Wibbeling B, Glorius F (2018) J Am Chem Soc 140: 8624–8628
60. Rai P, Maji K, Jana SK, Maji B (2022) Chem Sci 13:12503–12510
61. Wang W, Cai Y, Guo R, Brown MK (2022) Chem Sci 13:13582–13587
62. Wang W, Brown MK (2023) Angew Chem Int Ed 62:e202305622
63. Rolka AB, Koenig B (2020) Org Lett 22:5035–5040
64. Chiminelli M, Serafino A, Ruggeri D, Marchiò L, Bigi F, Maggi R, Malacria M, Maestri G (2023) Angew Chem Int Ed 62:e202216817
65. Zhu M, Gao Y-J, Huang X-L, Li M, Zheng C, You S-L (2024) Nat Commun 15:2462
66. Chatterjee A, König B (2019) Angew Chem Int Ed 58:14289–14294
67. Soni VK, Hwang HS, Moon KY, Park S-W, You Y, Cho EJ (2019) J Am Chem Soc 141: 10538–10545

Top Heterocycl Chem (2025) 62: 201–218
https://doi.org/10.1007/7081_2025_91

Published online: 17 April 2025

Arylation of Heterocycles Enabled by Photocatalysis

Kai Sun and Bing Yu

Contents

Abstract Arylated heterocycles are cornerstone structures in a wide array of chemical sciences, serving as critical components for numerous applications thanks to their unique structural qualities. The synthesis of these complex molecules represents a key endeavor in contemporary organic chemistry. The rise of photocatalysis marks a significant shift in the arylation techniques of heterocycles, positioning itself as an attractive alternative to the traditional transition metal-catalyzed cross-coupling methods. This modern tactic has spawned a plethora of protocols noted for their wide-ranging compatibility with different functional groups, their adaptability, and their ease of use. In this chapter, we aim to update readers on the cutting-edge progress in this rapidly advancing field, methodically categorized by the various aryl radical precursors employed. We will shine a light on breakthrough visible light-driven arylation methods, underscoring critical progressions in this research area. The discussion will explore the breadth and limitations of these processes, prioritizing a deeper understanding of the mechanisms that facilitate aryl radical creation.

Keywords Arylation · Heterocycles · Metal-free · Photocatalysis · Radical

K. Sun and B. Yu (✉)
College of Chemistry, Zhengzhou University, Zhengzhou, China
e-mail: bingyu@zzu.edu.cn

1 Introduction

Arylation of heterocycles stands as a pivotal technique with substantial impact across medicinal chemistry, synthesis of natural products, and development of functional materials (Scheme 1) [1–7]. This process, defined by the integration of aryl groups into heterocyclic structures, plays an essential role in significantly altering both the physicochemical and biological characteristics of these compounds. Recognizing its critical importance, there has been an evolution of diverse synthetic methods aimed at enhancing this transformation.

Traditionally, the production of arylated heterocycles has largely relied on transition metal-catalyzed cross-coupling reactions. Foremost among these are the Suzuki–Miyaura [8, 9], Negishi [10], Hiyama [11], and Stille reactions [12], which typically involve the employment of organic (pseudo)halides (I, Br, Cl, OTf, etc.) and organometallic reagents (Mg, Zn, B, Si, Sn, etc.) as coupling agents (Scheme 2a). Although these methods are effective, their reliance on specific transition metals and ligands for the C–H arylation processes has sparked concerns, especially regarding the potential contamination of the final products with metal residues. In response to these challenges and in the spirit of adopting more sustainable practices, radical-based C–H arylation has surfaced as an innovative alternative (Scheme 2b) [13, 14]. This method capitalizes on the generation of aryl radicals from a broad array of arylating agents, including aryl halides, aryl diazonium salts, aryl

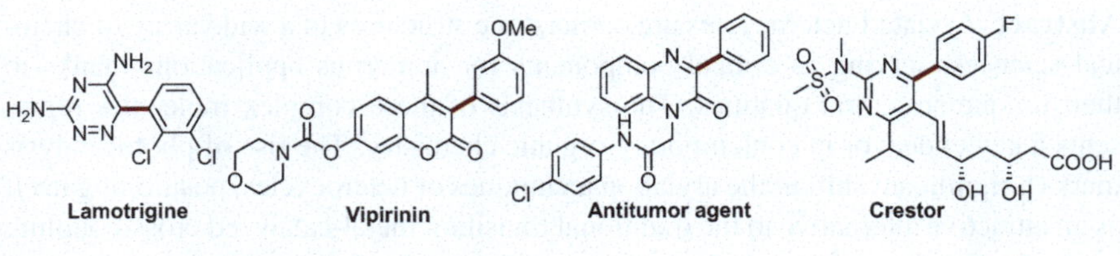

Scheme 1 Aryl-containing bioactive molecules

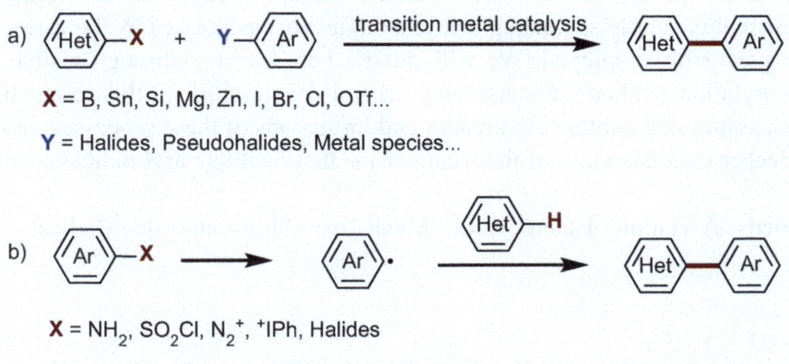

Scheme 2 Strategies for the construction of arylated heterocycles. (**a**) Transition-metal-catalyzed arylation; (**b**) Radical C-H arylation

carboxylic acids, and aryl boronic acids. It sidesteps the need for costly transition metals, stoichiometric redox reagents, and severe reaction conditions, all the while showing greater tolerance for various functional groups.

Driving the field toward unprecedented innovation, photocatalytic methods that utilize visible light have accelerated the development and advancement of organic synthesis [15–17]. These approaches enable the production of aryl radicals from sources such as aryl halides, aryldiazonium salts, and aryl sulfonium salts, facilitating their integration into heterocyclic molecules. Remarkably, this photocatalytic route operates under exceptionally mild conditions, dispensing with the need for specialized ligands and marking a move toward a more sustainable, efficient arylation process [18]. This significant progress signifies a major shift to environmentally friendly and versatile synthetic strategies, emphasizing the potential for upcoming breakthroughs in the construction of arylated heterocycles, with far-reaching implications for drug discovery and materials science.

2 Aryl Diazonium Salts as Aryl Radical Precursors

Diazonium salts have long been recognized as valuable sources of aryl radicals in organic synthesis. The Meerwein arylation reaction has inspired the development of numerous variations utilizing different reducing agents, notably transition metals [19]. The appeal of diazonium salts as aryl radical precursors lies in their versatility, stemming from the wide range of aniline precursors available for their synthesis, as well as their high reactivity attributed to a low reduction potential (0.16 V vs. SCE) [20]. With the emergence of novel redox systems, diazonium salts quickly found utility as reagents due to their favorable properties.

While the concept of photocatalyzed Pschorr cyclization dates back to 1984 [19], it was the recent advancements in photoredox catalysis that sparked interest in harnessing diazonium salts for more general photocatalytic reactions [21]. A notable milestone in this endeavor was achieved by Sanford et al. in 2011, who successfully combined photoredox and palladium catalysis for directed C–H arylation reactions (Scheme 3) [22]. This innovative strategy involved the generation of aryl radicals through single-electron transfer (SET) from a photoexcited ruthenium complex to diazonium salts, akin to the mechanism observed in Pschorr cyclization. During the process, the resulting aryl radicals **3d** underwent addition to a palladium (II) complex, leading to the formation of a Pd(III) species. Subsequent single-electron oxidation of the intermediate by the ruthenium complex or another molecule of diazonium salt yielded a palladium(IV) complex, which underwent reductive elimination to furnish the desired product **3c**. This elegant approach highlights the capability of photoredox chemistry to access aryl radicals under mild reaction conditions, showcasing its compatibility with diverse catalytic systems.

In 2012, König et al. disclosed a green light-activated arylation of heteroarenes catalyzed by eosin Y. This innovative protocol enables the C–H arylation of furans, pyrroles, and thiophenes utilizing a diverse array of diazonium salts with just 1 mol

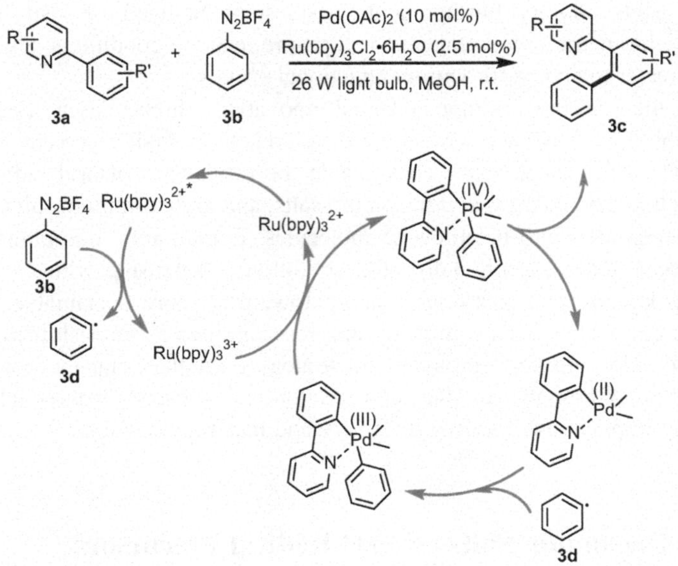

Scheme 3 Sanford's directed arylation using diazonium salts as arylating reagents

% of the photocatalyst (Scheme 4) [23]. Notably, diazonium salts featuring electron-withdrawing substituents exhibit superior performance in radical-mediated transformations. This observation was underscored in their investigation, where para-halo (**4d**, **4h**), nitro (**4e**), cyano (**4f**), and ester (**4g**) diazonium salts delivered high to excellent yields in furan reactions. In contrast, reactions with electron-rich salts like methoxy (**4i**), hydroxy (**4j**), and alkyl (**4k**) substituents yielded moderate product yields.

The mechanistic pathway of this process showcases oxidative quenching (Scheme 4). Upon excitation by 530 nm light, the excited eosin Y undergoes a single-electron transfer (SET) with diazonium salt **4a**, leading to the generation of an aryl radical **4q** along with nitrogen extrusion. The resulting aryl radical **4q**, possessing ample electrophilicity, selectively adds to electron-rich systems such as 5-membered heteroaromatics **4b**. After C–C bond formation, the resultant radical **4r** is oxidized through electron transfer to either the eosin Y radical cation or another diazonium salt. The latter scenario signifies a chain propagation step, a prevalent pathway in photoredox chemistry. The final deprotonation step restores aromaticity, culminating in product formation **4c**.

Xiao et al. developed a photocatalytic arylation technique for *N*-heteroarenes employing $[Ru(bpy)_3Cl_2]\cdot 6H_2O$ (Scheme 5) [24]. They achieved successful arylation of various pyridine hydrochlorides in aqueous solutions using aryldiazonium salts, resulting in compound **5c**. Furthermore, caffeine **5d** was also arylated under these photocatalytic conditions, though it required 88% aqueous formic acid as the solvent to yield compound **5e** in 42–84% yield. This method also demonstrated good to moderate yields when applied to other heterocycles such as quinoline, acridine, phthalazine, pyridine, and pyrazine.

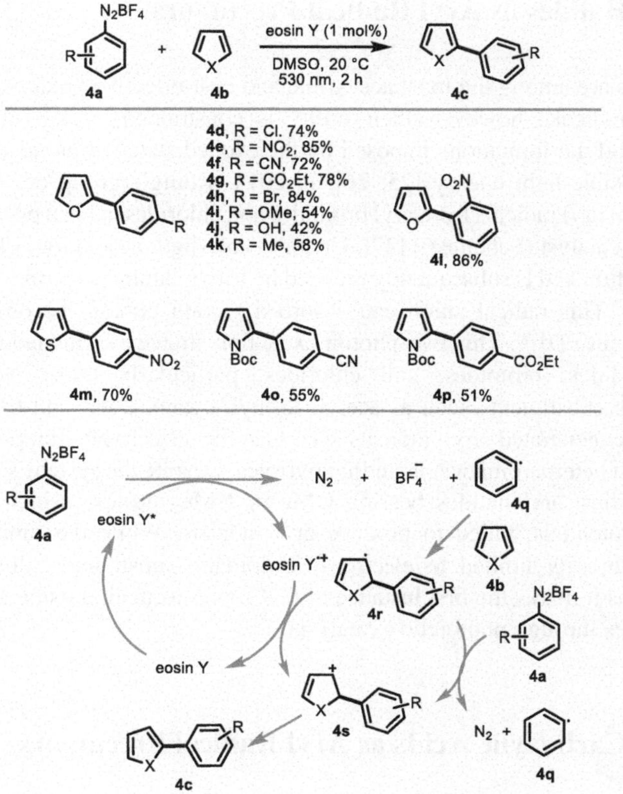

Scheme 4 König's eosin Y-photocatalyzed C–H arylation of heteroarenes

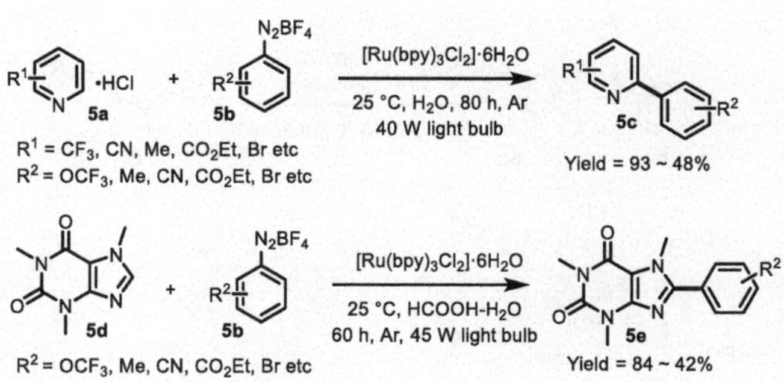

Scheme 5 Xiao's Ru-photocatalyzed arylation of heteroarenes

3 Aryl Halides as Aryl Radical Precursors

Aryl halides are among the most accessible and cost-effective precursors for generating aryl radicals; however, their utility is constrained by the extreme redox potentials and the limitations imposed by the excited states of photosensitizers and available visible-light energy [25, 26]. In 2014, König's group demonstrated the generation of aryl radicals from aryl bromides and chlorides using a perylene diimide (PDI) photocatalyst (Scheme 6) [27]. Under visible-light irradiation, PDI undergoes photoexcitation and is subsequently reduced by triethylamine to form a stable radical anion PDI^{-}. This radical anion can absorb a second photon, becoming a potent reducing agent PDI^{-*}. This two-photon excitation strategy facilitated the reduction of aryl iodides, bromides, and chlorides, particularly those with electron-withdrawing substituents such as acetyl, formyl, cyano, ester, and trifluoromethyl groups. The generated aryl radicals can then be effectively trapped by excess electron-rich heteroaromatics, including pyrroles. Despite the system's strong reducing capabilities, aryl halides bearing OMe or NMe_2 groups, which have higher reduction potentials, failed to produce aryl radicals. While the functional group scope is primarily limited to electron-withdrawing substituents, this protocol is significant as it marks the first instance of C–C bond formation using aryl bromides and chlorides through photoredox catalysis.

4 Aryl Carboxylic Acids as Aryl Radical Precursors

Aryl carboxylic acids have emerged as valuable precursors for generating aryl radicals, offering several significant advantages. Their chemical stability, compared to other aryl radical precursors like aryl bromides or iodides, ensures safer handling

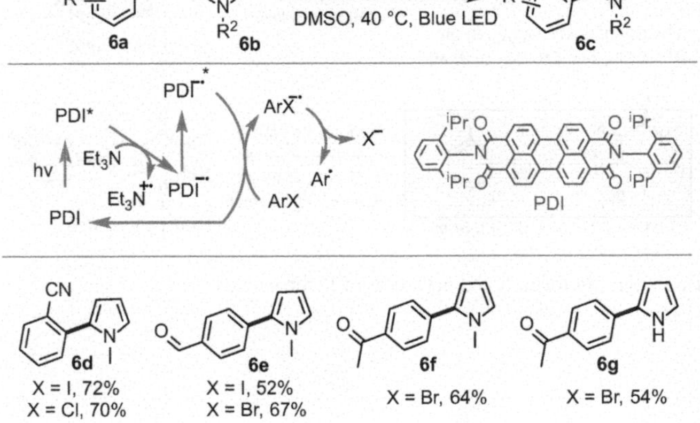

Scheme 6 König's PDI-photocatalyzed arylation of heteroarenes

and storage, reducing the risk of degradation and side reactions [28]. Furthermore, aryl carboxylic acids are readily available through cost-effective synthetic routes or natural sources, making them economically viable for radical generation processes. However, the direct formation of aryl radicals from aryl carboxylic acids remains a challenging task due to the typically low decarboxylation rates of benzoyl radicals ($k = 1.4 \times 10^6\ M^{-1}\ s^{-1}$), which are often outcompeted by faster reactions such as intermolecular hydrogen atom transfer ($k = 1.2 \times 10^7\ M^{-1}\ s^{-1}$) and addition to aromatic hydrocarbons ($k = 2.2 \times 10^8\ M^{-1}\ s^{-1}$) [29]. Overcoming these challenges requires innovative approaches to enhance decarboxylation efficiency, thereby improving the selectivity and yield of aryl radical production.

Recent advancements in the field have concentrated on the development of novel catalytic systems and tailored reaction conditions to confront the intrinsic challenges associated with the generation of aryl radicals from aryl carboxylic acids. In a seminal study conducted in 2017, Glorious et al. introduced a groundbreaking solution to the decarboxylation sluggishness by steering the reaction along an alternative trajectory (Scheme 7) [28]. They postulated that if the direct generation of aryl radicals from carboxyl radicals was kinetically infeasible, the intervention of an alternative intermediate might be requisite. Carboxyl radicals, known for their rapid reactivity with iodination reagents to yield benzoyl hypoiodites, inspired the researchers to hypothesize that a single-electron transfer (SET) to a bromo-analogue could precipitate a direct decarboxylative fragmentation, culminating in the formation of a halogen anion and an aryl radical. This innovative approach was empirically validated, demonstrating that by selecting an appropriate brominating reagent and employing an iridium-based photocatalyst, the in situ formation of carboxyl hypobromite could be achieved. Subsequent SET from the reduced photocatalyst to the bromite facilitated the generation of aryl radicals, which were then effectively utilized in the synthesis of biaryls. Significantly, the aryl radicals generated through this methodology exhibit the capacity to incorporate halogen functionalities at diverse positions and to accommodate a spectrum of neutral-to-donating functional groups. Although this approach has demonstrated significant efficacy, it is not devoid of limitations, particularly the diminished selectivity encountered in reactions involving nonsymmetrical arenes. This phenomenon highlights the fundamentally low-selective characteristics associated with radical additions.

Nonetheless, this research delineates a novel and auspicious avenue for the generation of aryl radicals from benzoic acids. Ongoing endeavors are directed towards the optimization of these methodologies, with the ultimate goal of enhancing efficiency and selectivity in aryl radical generation, thereby broadening their utility in organic synthesis and other applications.

5 Aryl Thianthrenium Salts as Aryl Radical Precursors

The direct C–H thianthrenation method has emerged as a robust synthetic technique for aromatic ring functionalization. Operating under mild conditions, it demonstrates exceptional tolerance towards a diverse array of functional groups. This method

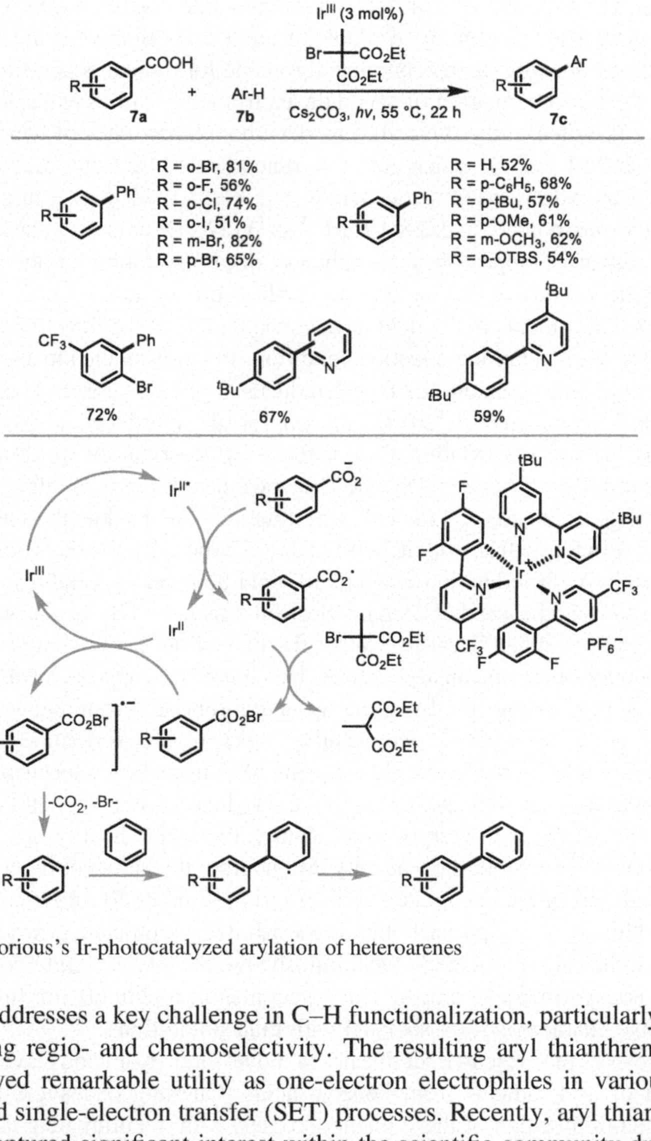

Scheme 7 Glorious's Ir-photocatalyzed arylation of heteroarenes

effectively addresses a key challenge in C–H functionalization, particularly in terms of controlling regio- and chemoselectivity. The resulting aryl thianthrenium salts have displayed remarkable utility as one-electron electrophiles in various photochemical and single-electron transfer (SET) processes. Recently, aryl thianthrenium salts have captured significant interest within the scientific community due to their remarkable reactivity and uniqueness [30, 31]. These compounds exhibit high versatility and value as electrophiles in a wide range of metal-catalyzed and photoredox-catalyzed reactions. Their versatility and reactivity highlight their importance in advancing synthetic methodologies and enriching the toolkit for intricate molecular synthesis.

In 2021, Ritter and coworkers presented a groundbreaking study on the late-stage heteroarylation of aryl thianthrenium salts utilizing an α-amino alkyl radical as the catalyst, as illustrated in Scheme 8 [32]. This innovative activation approach deviates from conventional halogen atom transfer (XAT) techniques, allowing for the

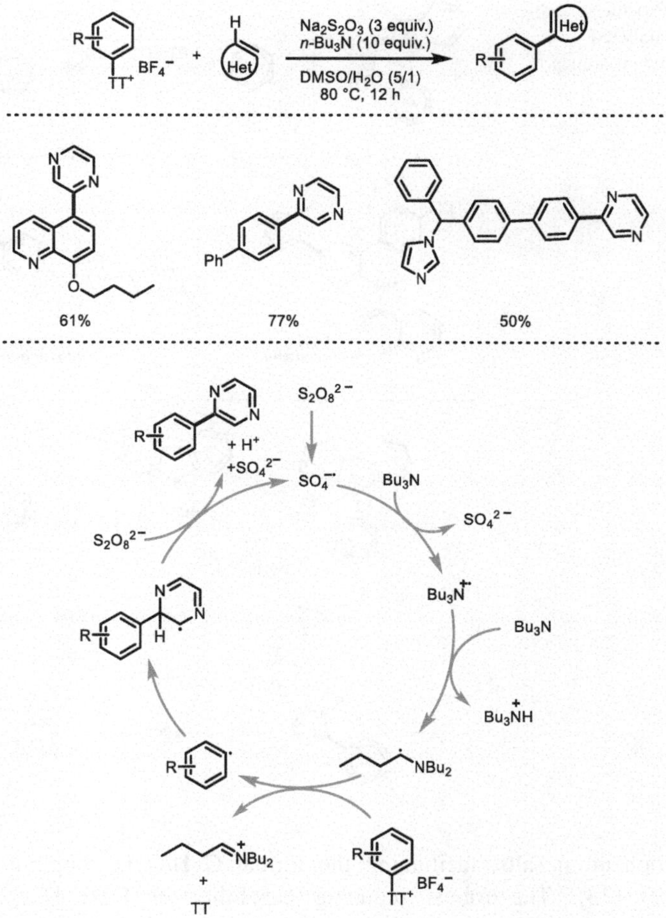

Scheme 8 Ritter's photocatalyzed arylation of heteroarenes

efficient synthesis of a diverse range of biaryls incorporating heteroarene motifs. The proposed mechanistic pathway involves the initial formation of an amine radical cation through single-electron oxidation with a sulfate radical anion, resulting in the generation of an α-aminoalkyl radical via deprotonation. The subsequent interaction between the α-amino alkyl radical and aryl thianthrenium salt leads to the formation of thianthrene and an aryl radical. The aryl radical then engages in an addition reaction with a heteroarene, followed by single-electron oxidation and deprotonation, ultimately affording the desired biaryl product. Importantly, this transformation proceeds under mild conditions without the need for additives, light exposure, or transition metal catalysts and demonstrates excellent compatibility with air and water.

In 2020, Procter and colleagues employed 10-phenylphenothiazine (PTH) as a highly effective photocatalyst for the activation of in situ-generated

Scheme 9 Procter's PTH-photocatalyzed arylation of heteroarenes

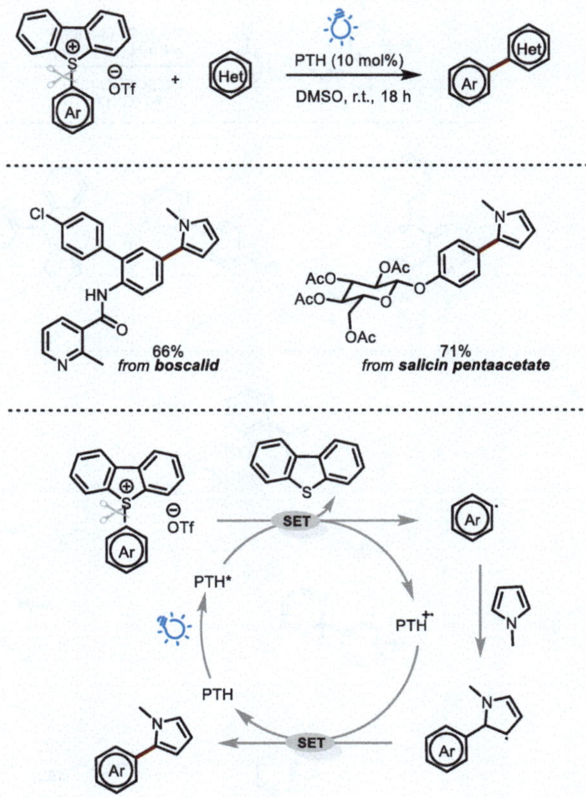

dibenzothiophenium salts, facilitating the formal C–H/C–H coupling of arenes (Scheme 9) [33]. The robust reducing capability of PTH [$E_{1/2}^{red*}$ = −2.1 V vs. SCE in dimethyl sulfoxide (DMSO)] was instrumental in ensuring the smooth progression of the reaction by effectively reducing the dibenzothiophenium salts. Stern–Volmer experiments confirmed the oxidative quenching of the excited state of the photocatalyst by the dibenzothiophenium salts. The aryl radical, produced through a single-electron transfer (SET) process between PTH* and dibenzothiophenium salts, underwent radical addition and single-electron oxidation, resulting in the desired products. This methodology facilitates the C–H/C–H coupling of arenes from aromatic compounds through a simple one-pot procedure.

Recently, our research group developed a novel and practical photoactivation strategy for aryl sulfonium salts, which enables the generation of aryl radicals through the formation of an electron donor-acceptor (EDA) complex between aryl sulfonium salts and 1,4-diazabicyclo[2.2.2]octane (DABCO) (Scheme 10) [34]. This method facilitated the synthesis of arylated *N*-heterocycles with moderate to good yields. Notably, these mild reaction conditions were successfully extended to the coupling of bioactive molecules with complex drugs or agricultural pharmaceuticals. Mechanistic studies, including radical trapping experiments, UV–vis absorbance

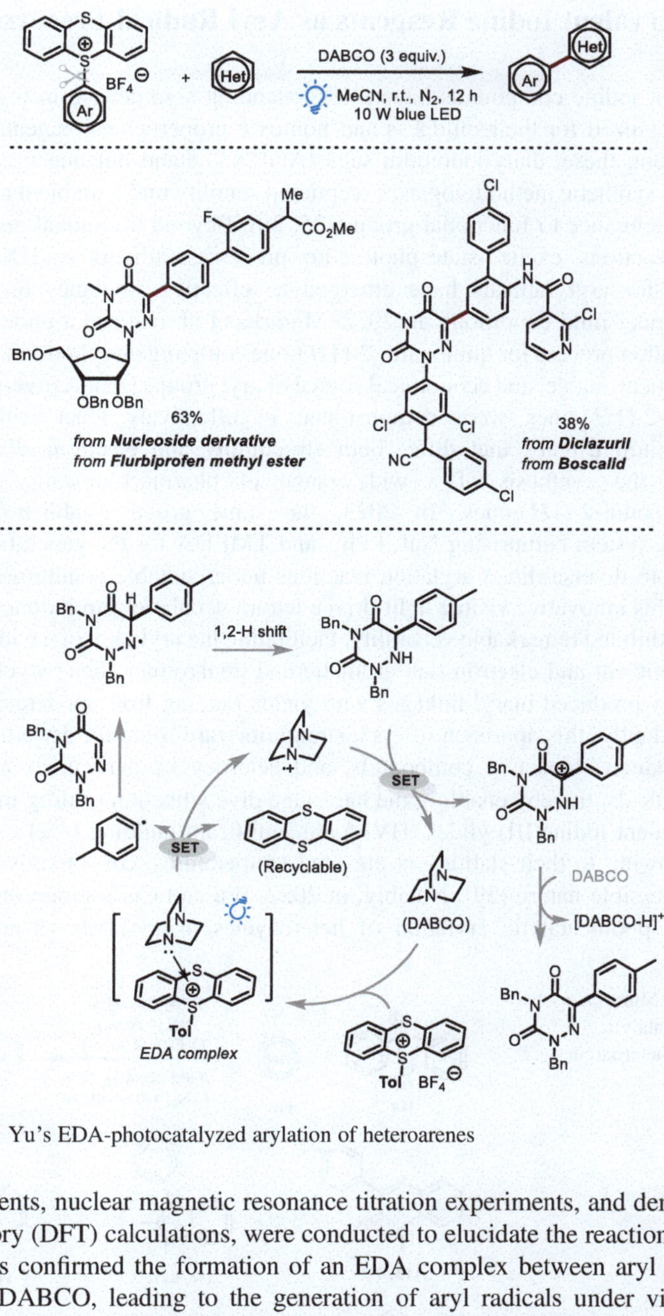

Scheme 10 Yu's EDA-photocatalyzed arylation of heteroarenes

measurements, nuclear magnetic resonance titration experiments, and density functional theory (DFT) calculations, were conducted to elucidate the reaction pathway. The results confirmed the formation of an EDA complex between aryl sulfonium salts and DABCO, leading to the generation of aryl radicals under visible-light irradiation. The aryl radical subsequently reacted with the C=N bond of azauracil to form an *N*-centered radical, followed by a 1,2-H shift to generate a carbon radical. A single-electron transfer between $DABCO^{+}$ and the carbon radical produced a radical cation, which then underwent proton loss to yield the final product.

6 Hypervalent Iodine Reagents as Aryl Radical Precursors

Hypervalent iodine compounds have a long-standing significance in organic synthesis, recognized for their mildness and nontoxic properties as reagents and oxidants. Among these, diaryliodonium salts ($Ar_2I^+X^-$) stand out due to their well-established synthetic methodologies, exceptional stability under ambient conditions, and broad tolerance to functional groups [35, 36]. Beyond traditional ground-state arylation reactions, excited-state photoredox processes utilizing $Ar_2I^+X^-$ salts as precursors for aryl radicals have emerged as effective strategies for selective arylation under mild conditions. In 2022, Murarka et al. realized a photocatalyzed direct arylation process for quinoxalin-2-(1*H*)-ones utilizing diaryliodonium triflates as a convenient, stable, and economical source of aryl groups [37]. A diverse array of quinoxalin-2-(1*H*)-ones were demonstrated to effectively react with various diaryliodonium triflates that differ both structurally and electronically, thereby facilitating the synthesis of a wide range of pharmacologically significant 3-arylquinoxalin-2-(1*H*)-ones. In 2023, the same group established another photoredox system comprising NaI, PPh_3, and TMEDA for the generation of aryl radicals from diverse direct arylation reactions under suitable conditions (Scheme 11) [38]. This innovative visible light-driven tetramolecular electron donor-acceptor platform exhibited remarkable versatility, facilitating the arylation of a wide range of electron-deficient and electron-rich aromatic and nonaromatic heterocycles, which successfully produced biaryl linkages with yields ranging from moderate to excellent. Importantly, this approach offers a straightforward route for arylating various natural products, bioactive compounds, and heterocyclic derivatives relevant to pharmaceuticals, thereby enabling the late-stage diversification of drug molecules.

Hypervalent iodine(III) ylides (HVIs) are significant building blocks in organic synthesis owing to their stability at elevated temperatures, cost-effectiveness, and readily accessible nature [39]. Notably, in 2022, Wu and Cui's group reported the pioneering photocatalytic arylation of heterocycles using HVIs as aryl radical

Scheme 11 Murarka's EDA-photocatalytic arylation of heteroarenes

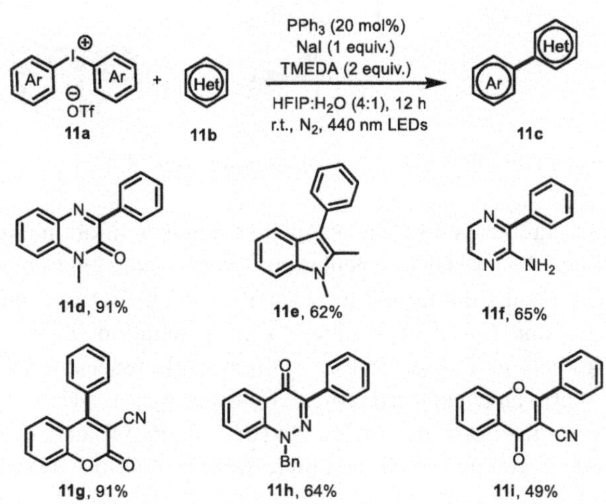

precursors, achieving remarkable substrate diversity and excellent functional group tolerance without the necessity for oxidants or metal additives (Scheme 12) [40]. This method demonstrated applicability to a variety of substrates, including pyrazine, quinoxalin-2(1*H*)-ones, pyridazine, and quinoxaline, and was successfully employed for the late-stage functionalization of complex bioactive molecules. Specifically, the modification of TF-flavone yielded the desired arylated product (**12h**) in 43% yield. Furthermore, substituted coumarins, prevalent in pharmaceuticals such

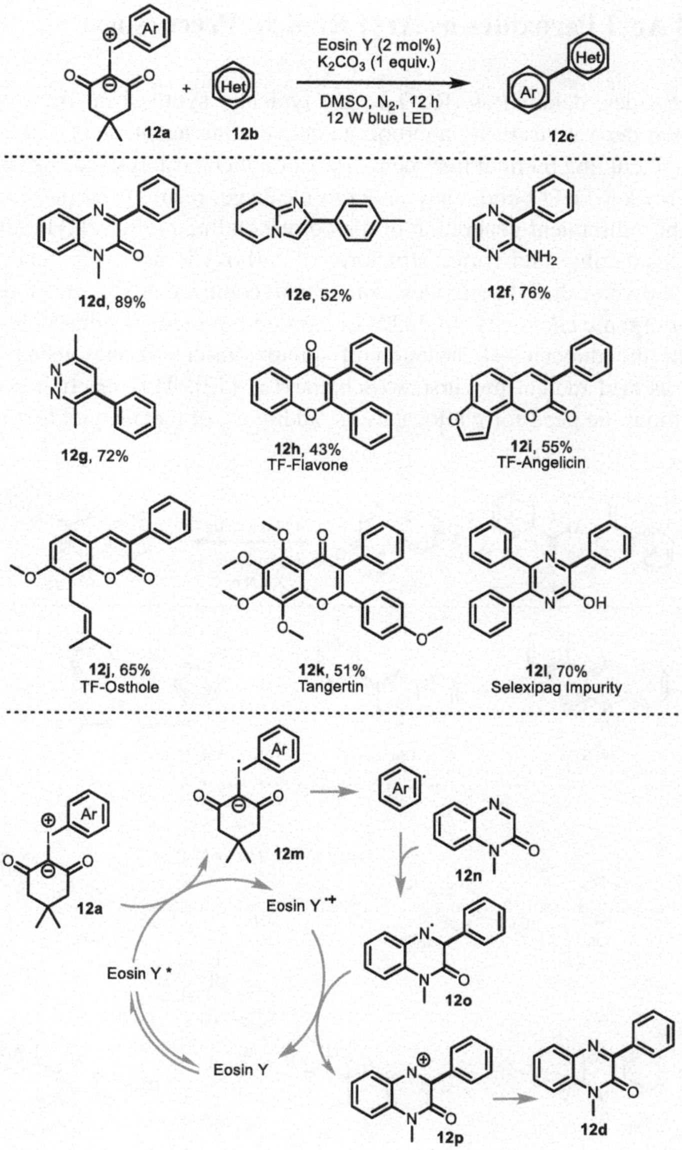

Scheme 12 Wu's eosin Y-photocatalyzed arylation of heteroarenes

as TF-Angelicin (**12i**) and TF-Osthole (**12j**), also exhibited favorable reactivity, producing the desired products in acceptable yields. Importantly, the direct C–H arylation of Tangeretin (**12k**) and Selexipag Impurity (**12l**) resulted in moderate yields of the corresponding products. In this reaction, the single-electron transfer between the excited photocatalyst and HVIs to generate aryl radicals is the focal point of the process.

7 Aryl Acyl Peroxides as Aryl Radical Precursors

Diacyl peroxides, denoted as $(RCO_2)_2$, are typically synthesized from carboxylic acids or their derivatives using appropriate dehydrating agents [41]. Under specific conditions, including thermal treatment, transition metal catalysis, or irradiation, the relatively weaker O–O bonds may undergo cleavage, resulting in the liberation of CO_2 and the subsequent generation of the corresponding aryl or alkyl radicals. The extensive availability and varied structures of carboxylic acids facilitate the facile synthesis of diverse diacyl peroxides, which are recognized as significant reagents in the field of organic chemistry. In 2022, Li's group reported a visible light-mediated method for the direct C–H arylation of quinoxalin-2(1*H*)-ones using aryl acyl peroxides as aryl radical precursors (Scheme 13) [42]. This reaction occurs efficiently without the need for photocatalysts, additives, or metals, yielding a range of

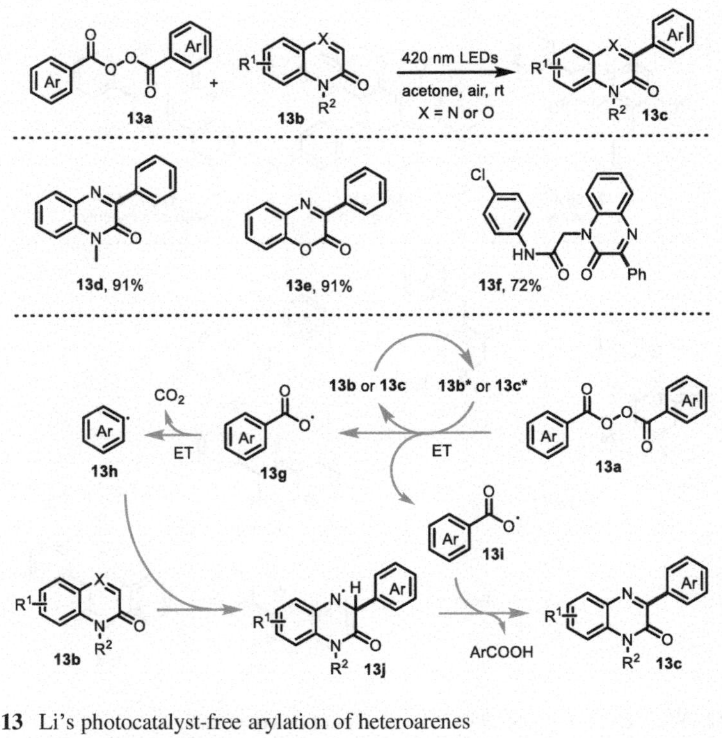

Scheme 13 Li's photocatalyst-free arylation of heteroarenes

3-arylated quinoxalin-2(1*H*)-ones with moderate to good yields. The decarboxylation of the benzoyloxyl radical is particularly easy, requiring only 6.7 kcal mol^{-1}, and the electron transfer process can further improve decarboxylation efficiency.

8 Conclusion

In summary, arylated heterocycles are pivotal structures that significantly contribute to a diverse range of applications within the chemical sciences. Their synthesis remains a critical focus in contemporary organic chemistry. The advent of photocatalysis signifies a transformative shift in the methodologies for arylation, providing an appealing alternative to traditional transition metal-catalyzed cross-coupling techniques. This chapter has highlighted the extensive array of protocols emerging from this modern approach, showcasing their compatibility with various functional groups, adaptability, and user-friendliness.

By systematically categorizing advancements in this field, particularly the innovative visible light-driven arylation methods, we have aimed to illuminate both the breadth of progress and the existing limitations. A thorough understanding of the mechanisms behind aryl radical formation is essential for guiding future research endeavors.

Our goal with this chapter is not only to inform but also to inspire synthetic organic chemists to delve deeper into the photoinduced synthesis of arylated heterocycles. We anticipate that this discussion will spark further interest and innovation, ultimately advancing the field of synthetic chemistry.

Acknowledgements We acknowledge the financial support from the National Natural Science Foundation of China (21971224 and 22171249), the Natural Science Foundation of Henan Province (232300421363), and the Science and Technology Innovation Talents in Universities of Henan Province (23HASTIT003).

References

1. Liu Y, Zuo L, Shang T, Wang Z, Yu B (2022) Recent advances in photocatalytic polyfluoroarylation. Chem Commun 58:12900–12912
2. Wales SM, Saunthwal RK, Clayden J (2022) C(sp3)-arylation by conformationally accelerated intramolecular nucleophilic aromatic substitution (SNAr). Acc Chem Res 55:1731–1747
3. Babu SS, Muthuraja P, Yadav P, Gopinath P (2021) Aryldiazonium salts in photoredox catalysis – recent trends. Adv Synth Catal 363:1782–1809
4. Dutta U, Porey S, Pimparkar S, Mandal A, Grover J, Koodan A, Maiti D (2020) Para-selective arylation of arenes: a direct route to biaryls by norbornene relay palladation. Angew Chem Int Ed 59:20831–20836
5. Pan Y, Zhu Y, Li S, Li G, Ma Z, Qian Y, Huang W (2024) Photoinduced arylation or alkylation of 1,2,4-triazine-3,5(2H,4H)-diones with hydrazines. Chem Commun 60:3665–3668

6. Noor H, Gao P, Guo J, Zhang S, Jia X, Yuan Y (2024) Palladium-catalyzed radical arylation of N-hydroxyphthalimides with triphenyl phosphites as an unusual arylating agent via C(sp2)-O bond cleavage. Org Chem Front 11:1069–1075
7. Kumar R, Sharma A, Sharma A (2024) Mechanochemically induced thianthrenium salts-based arylation of diverse heterocyclic scaffolds. ACS Sustain Chem Eng 12:12808
8. Miyaura N, Suzuki A (1995) Palladium-catalyzed cross-coupling reactions of organoboron compounds. Chem Rev 95:2457–2483
9. Han F-S (2013) Transition-metal-catalyzed Suzuki–Miyaura cross-coupling reactions: a remarkable advance from palladium to nickel catalysts. Chem Soc Rev 42:5270–5298
10. Negishi E (1982) Palladium- or nickel-catalyzed cross coupling. A new selective method for carbon-carbon bond formation. Acc Chem Res 15:340–348
11. Denmark SE, Sweis RF (2002) Design and implementation of new, silicon-based, cross-coupling reactions: importance of silicon–oxygen bonds. Acc Chem Res 35:835–846
12. Milstein D, Stille JK (1978) A general, selective, and facile method for ketone synthesis from acid chlorides and organotin compounds catalyzed by palladium. J Am Chem Soc 100:3636–3638
13. Ghosh I, Marzo L, Das A, Shaikh R, König B (2016) Visible light mediated photoredox catalytic arylation reactions. Acc Chem Res 49:1566–1577
14. Xie L-Y, Peng S, Yang L-H, Peng C, Lin Y-W, Yu X, Cao Z, Peng Y-Y, He W-M (2021) Aryl acyl peroxides for visible-light induced decarboxylative arylation of quinoxalin-2(1H)-ones under additive-, metal catalyst-, and external photosensitizer-free and ambient conditions. Green Chem 23:374–378
15. Song H-Y, Jiang J, Wu C, Hou J-C, Lu Y-H, Wang K-L, Yang T-B, He W-M (2023) Semi-heterogeneous g-C3N4/NaI dual catalytic C–C bond formation under visible light. Green Chem 25:3292–3296
16. Lu Y-H, Wu C, Hou J-C, Wu Z-L, Zhou M-H, Huang X-J, He W-M (2023) Ferrocene-mediated photocatalytic annulation of N-sulfonyl ketimines on a polycrystalline WSe2 semiconductor photocatalyst. ACS Catal 13:13071–13076
17. Li R, Chen X, Wei S, Sun K, Fan L, Liu Y, Qu L, Zhao Y, Yu B (2018) A visible-light-promoted metal-free strategy towards arylphosphonates: organic-dye-catalyzed phosphorylation of arylhydrazines with trialkylphosphites. Adv Synth Catal 360:4807–4813
18. Ghosh T, Santra S, Zyryanov GV, Ranu BC (2023) Recent advances in visible light mediated photoinduced aryl radical generation and its application in synthesis. J Photochem Photobiol 16:100192
19. Kvasovs N, Gevorgyan V (2021) Contemporary methods for generation of aryl radicals. Chem Soc Rev 50:2244–2259
20. Andrieux CP, Pinson J (2003) The standard redox potential of the phenyl radical/anion couple. J Am Chem Soc 125:14801–14806
21. Bhawale RT, Kshirsagar UA (2023) Visible light assisted direct C3–H arylation of pyrido[1,2-a]pyrimidin-4-ones and thiazolo[3,2-a]pyrimidin-5-ones. J Org Chem 88:9537–9542
22. Kalyani D, McMurtrey KB, Neufeldt SR, Sanford MS (2011) Room-temperature C–H arylation: merger of Pd-catalyzed C–H functionalization and visible-light photocatalysis. J Am Chem Soc 133:18566–18569
23. Hari DP, Schroll P, König B (2012) Metal-free, visible-light-mediated direct C–H arylation of heteroarenes with aryl diazonium salts. J Am Chem Soc 134:2958–2961
24. Xue D, Jia Z-H, Zhao C-J, Zhang Y-Y, Wang C, Xiao J (2014) Direct arylation of N-heteroarenes with aryldiazonium salts by photoredox catalysis in water. Chem Eur J 20:2960–2965
25. Liang K, Li N, Zhang Y, Li T, Xia C (2019) Transition-metal-free α-arylation of oxindoles via visible-light-promoted electron transfer. Chem Sci 10:3049–3053
26. Zhu D-L, Jiang S, Young DJ, Wu Q, Li H-Y, Li H-X (2022) Visible-light-driven C(sp2)–H arylation of phenols with arylbromides enabled by electron donor–acceptor excitation. Chem Commun 58:3637–3640

27. Ghosh I, Ghosh T, Bardagi JI, König B (2014) Reduction of aryl halides by consecutive visible light-induced electron transfer processes. Science 346:725–728
28. Candish L, Freitag M, Gensch T, Glorius F (2017) Mild, visible light-mediated decarboxylation of aryl carboxylic acids to access aryl radicals. Chem Sci 8:3618–3622
29. Chateauneuf J, Lusztyk J, Ingold KU (1988) Spectroscopic and kinetic characteristics of aroyloxyl radicals. 2. Benzoyloxyl and ring-substituted aroyloxyl radicals. J Am Chem Soc 110:2886–2893
30. Meng H, Liu M-S, Shu W (2022) Organothianthrenium salts: synthesis and utilization. Chem Sci 13:13690–13707
31. Xu H, Zhang J, Zuo J, Wang F, Lv J, Hun X, Yang D (2022) Recent advances in visible-light-catalyzed C-C bonds and C-heteroatom bonds formation using sulfonium salts. Chin J Org Chem 42:4037–4059
32. Alvarez EM, Karl T, Berger F, Torkowski L, Ritter T (2021) Late-stage heteroarylation of hetero(aryl)sulfonium salts activated by α-amino alkyl radicals. Angew Chem Int Ed 60:13609–13613
33. Aukland MH, Šiaučiulis M, West A, Perry GJP, Procter DJ (2020) Metal-free photoredox-catalysed formal C–H/C–H coupling of arenes enabled by interrupted Pummerer activation. Nat Catal 3:163–169
34. Sun K, Shi A, Liu Y, Chen X, Xiang P, Wang X, Qu L, Yu B (2022) A general electron donor–acceptor complex for photoactivation of arenes via thianthrenation. Chem Sci 13:5659–5666
35. Kikushima K, Elboray EE, Jiménez-Halla JOC, Solorio-Alvarado CR, Dohi T (2022) Diaryliodonium(III) salts in one-pot double functionalization of C–IIII and ortho C–H bonds. Org Biomol Chem 20:3231–3248
36. Yin K, Zhang R (2017) Transition-metal-free direct C–H arylation of quinoxalin-2(1H)-ones with diaryliodonium salts at room temperature. Org Lett 19:1530–1533
37. Samanta RK, Meher P, Murarka S (2022) Visible light photoredox-catalyzed direct C–H arylation of quinoxalin-2(1h)-ones with diaryliodonium salts. J Org Chem 87:10947–10957
38. Meher P, Panda SP, Mahapatra SK, Thombare KR, Roy L, Murarka S (2023) A general electron donor–acceptor photoactivation platform of diaryliodonium reagents: arylation of heterocycles. Org Lett 25:8290–8295
39. Yu D, Yang W, Chen S, Zhou C-Y, Guo Z (2024) Photocatalyst-free visible light-induced C (sp2)–H arylation of quinoxalin-2(1H)-ones and coumarins. Chem Eur J 30:e202401371
40. Ren J, Pi C, Cui X, Wu Y (2022) Divergent C(sp2)–H arylation of heterocycles via organic photoredox catalysis. Green Chem 24:3017–3022
41. Cheng F, Fan L, Lv Q, Chen X, Yu B (2023) Alkyl radicals from diacyl peroxides: metal-/base-/additive-free photocatalytic alkylation of N-heteroaromatics. Green Chem 25:7971–7977
42. Xu J, Zhang H, Zhao J, Ni Z, Zhang P, Shi B-F, Li W (2020) Photocatalyst-, metal- and additive-free, direct C–H arylation of quinoxalin-2(1H)-ones with aryl acyl peroxides induced by visible light. Org Chem Front 7:4031–4042

Top Heterocycl Chem (2025) 62: 219–242
https://doi.org/10.1007/7081_2025_92

Published online: 17 April 2025

Photoinduced Synthesis and Functionalization of Fused Thiophenes

Yuanyuan Sun, Yinsong Zhao, and Xuefeng Jiang

Contents

Abstract Fused thiophenes are highly privileged scaffolds within sulfur heterocycles, widely utilized in pharmacological chemistry and material science. The integration of photoinduced strategies into synthetic chemistry has recently enhanced the efficiency and mildness of synthesizing and modifying fused thiophenes. This book chapter summarizes representative and recent advancements in the photocatalytic synthesis and functionalization of fused thiophenes over the past decade, focusing on photocatalytic cyclization, direct coupling with simple fused thiophenes, and diverse functionalization of thianthrenium salts in these processes.

Keywords Cross-coupling · C–S functionalization · Cyclization · Fused thiophenes · Photocatalysis

---

Y. Sun and Y. Zhao (✉)
Hainan Institute of East China Normal University, State Key Laboratory of Estuarine and Coastal Research, East China Normal University, Shanghai, People's Republic of China
e-mail: yszhao@sklec.ecnu.edu.cn

X. Jiang (✉)
Hainan Institute of East China Normal University, State Key Laboratory of Estuarine and Coastal Research, East China Normal University, Shanghai, People's Republic of China

State Key Laboratory of Petroleum Molecular and Process Engineering, Shanghai Key Laboratory of Green Chemistry and Chemical Processes, School of Chemistry and Molecular Engineering, East China Normal University, Shanghai, People's Republic of China
e-mail: xfjiang@chem.ecnu.edu.cn

1 Introduction

Photocatalytic cyclization of sulfur-containing unsaturated compounds has been achieved for modular constructions of thiophene derivatives [1–8]. Alternatively, approaches involving the photoinduced direct coupling of simple fused thiophenes enable the introduction of synthetically valuable functional groups and π-extended moieties with high regioselectivity [9–11]. Particularly, due to their bench-stable, easy synthesis, broad structural diversity, and remarkable reactivity, thianthrenium salts have been emerging as one class of the most versatile carbon-reagent precursors of alkyl, aryl, vinyl, and alkynyl frameworks [12] and are extensively explored to undergo various photocatalytic coupling reactions to form carbon–carbon [13–15] and carbon–heteroatom bonds [16–20].

This chapter is organized into the following sections: (a) cyclization reactions, (b) coupling reactions with simple fused thiophenes, and (c) coupling reactions with thianthrenium analogues. The authors anticipate that this chapter will inspire continuous innovation of novel photoinduced transformations of fused thiophenes in future.

2 Photoinduced Synthesis of Fused Thiophenes via Cyclization

In 2016, Yuan group [21] developed a potassium hydroxide/dimethyl sulfoxide (KOH/DMSO) superbase-promoted method for the synthesis of 2-substituted benzothiophenes (**1c**) via photoinduced intermolecular annulation of 2-halothioanisoles (**1a**) with terminal alkynes (**1b**) at ambient temperature. The reaction mixture containing KOH (1.2 equiv.), 2-halothioanisole (**1a**), alkyne (**1b**), and dry DMSO was irradiated by xenon lamp (420–780 nm, 1 W cm^{-2}) for 26 h to achieve the desired product (**1c**) with up to 95% yield (Scheme 1). This protocol displayed good functional group tolerance and high regioselectivity, providing a facile and efficient access to the synthesis of benzothiophene derivatives. The mechanistic studies suggested that this photoinduced reaction underwent an aryl radical pathway induced by a visible light-driven SET process.

In 2017, Yan and coworkers [22] reported a visible light-promoted photocatalyst-free reaction of disulfides (**2a**) and alkynes (**2b**) for the synthesis of benzothiophenes (**2c**) (Scheme 2). Under the irradiation of blue LEDs (450 nm, 12 W), a variety of disulfides and alkynes containing electron-donating and electron-withdrawing groups were well compatible, affording the anticipated products with moderate to good yields. The gram-scale reaction was conducted to give the final product in 64% yield, indicating the practical applicability of this methodology.

In the same year, Zhu group [23] realized the synthesis of benzothiophenes via an effective metal-free photoredox-mediated tandem cyclization reaction of free thiophenols (**3a**) with alkynes (**3b**) (Scheme 3). In the presence of an organic

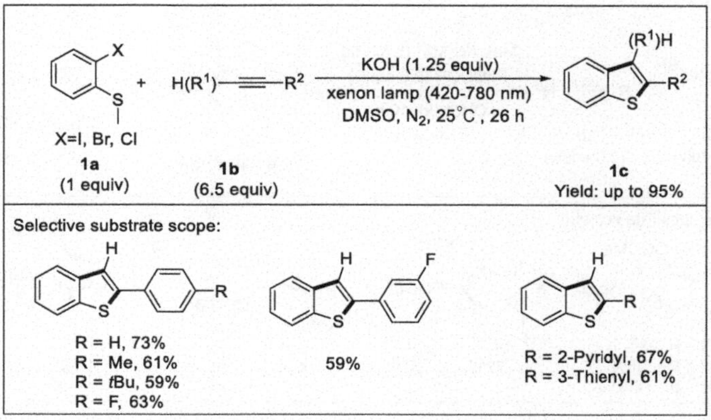

Scheme 1 Photoinduced synthesis of 2-substituted benzothiophenes (Yuan method)

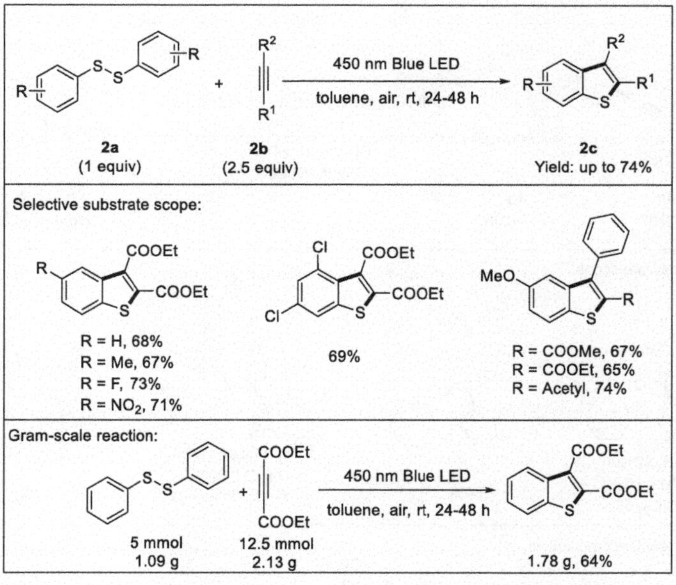

Scheme 2 Visible light-promoted synthesis of benzothiophenes (Yan method)

photocatalyst (Mes-Acr-Me$^+$), substituted thiophenols bearing electron-donating and electron-withdrawing groups could react with alkynes to produce the desired benzothiophenes with up to 98% yield.

In 2018, Song and coworkers [24] devised a visible light-induced cascade annulation reaction of methyl(2-(phenylethynyl) phenyl) sulfane (**4a**) and sulfonyl chlorides (**4b**) for the preparation of 3-sulfonyl benzothiophenes. The reactions employed [IrdF(CF$_3$)(ppy)$_2$(dtbbpy)]PF$_6$ as the photocatalyst and K_2CO_3 as the base, which were conducted under blue LED irradiation (5 W) for 24 h at room temperature, giving **4c** with up to 95% yield (Scheme 4). This protocol features

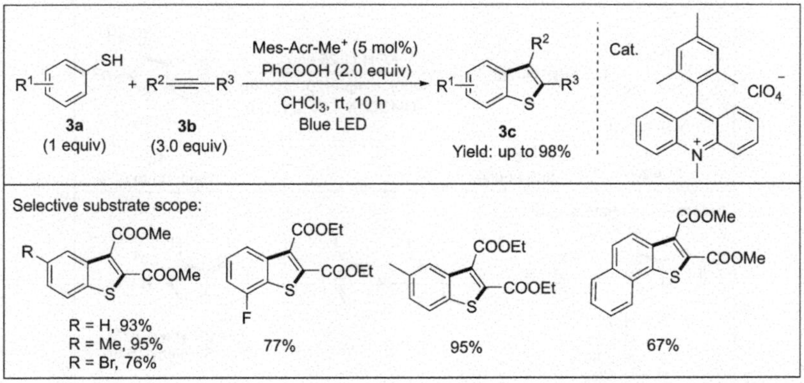

Scheme 3 Photoredox-mediated synthesis of benzothiophenes (Zhu method)

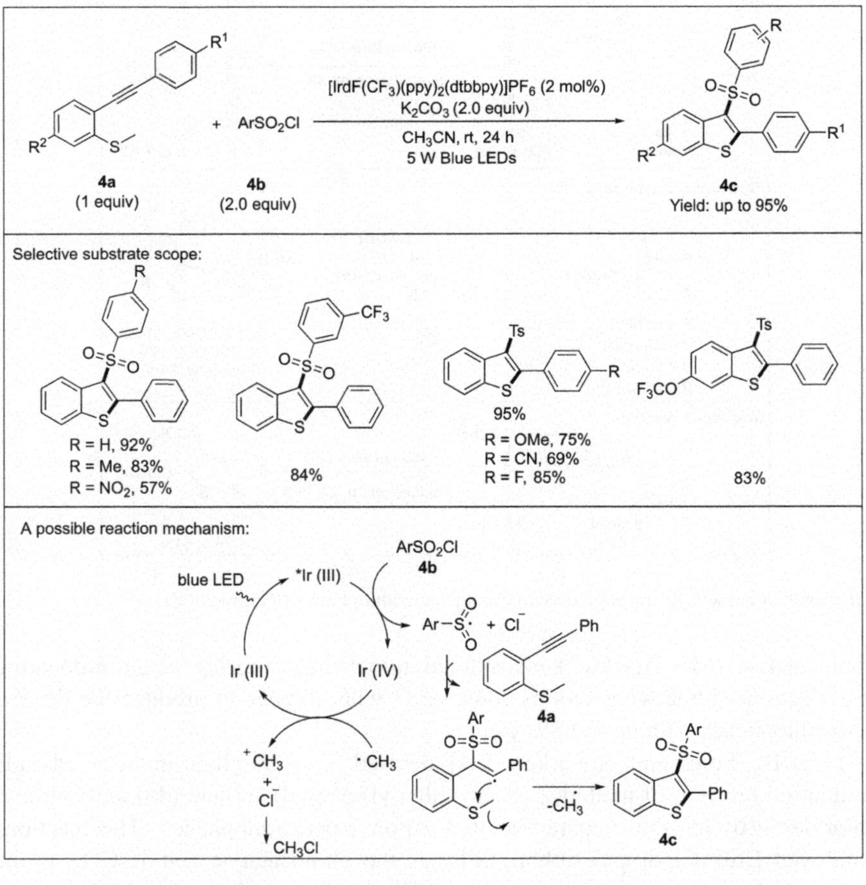

Scheme 4 Photocatalyzed synthesis of 3-sulfonyl benzothiophenes (Song method)

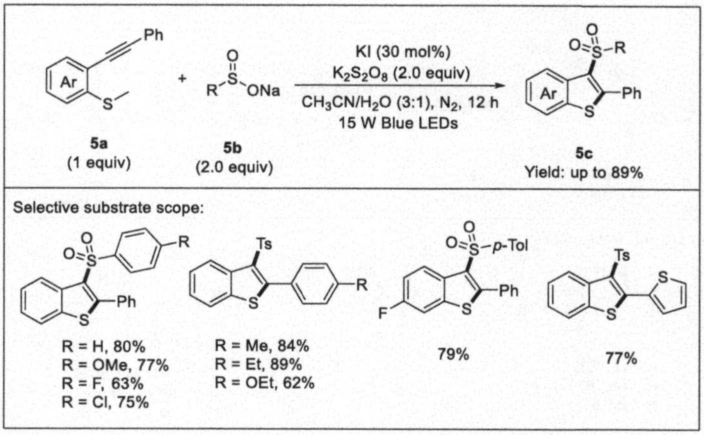

Scheme 5 Visible light-induced synthesis of sulfonated benzothiophenes (Yang method)

simple operation, mild conditions, and good functional group tolerance. Based on the mechanism studies, a catalytic cycle was proposed. The sulfonyl chlorides **4b** underwent a single-electron transfer (SET) process to generate sulfonyl radicals, which reacted with alkynes followed by an intramolecular cyclization to give the final product with the release of methyl radical.

Recently, Yang group [25] developed a straightforward strategy employing KI (30 mol%) as the catalyst, $K_2S_2O_8$ (2 equiv.) as the oxidant, and $CH_3CN{:}H_2O$ (3:1) as the solvent, for the synthesis of 3-sulfonated benzothiophenes (**5c**) from 2-alkynylthioanisoles (**5a**) and sodium sulfinates (**5b**). The desired products (**5c**) achieved up to 89% yield (Scheme 5). A variety of 2-alkynylthioanisoles (**5a**) and sodium sulfinates (**5b**) containing electron-donating groups such as methyl and methoxyl group and electron-withdrawing groups such as fluoro, nitro, and cyano groups were well compatible for this protocol and afforded the anticipated product in 69–92% yields. Moreover, substrates containing a heteroaromatic ring such as pyridinyl and thiophene of 2-alkynylthioanisoles (**5a**) could also undergo the reaction smoothly to afford the corresponding products with reasonable yields. The reaction proceeded to undergo a free radical process, as confirmed by a free radical inhibition experiment.

In 2018, Kshirsagar group [26] established an efficient, mild, and metal-free photoredox-catalyzed approach for the synthesis of 3-thiocyanobenothiophene (**6c**) from the radical annulation of 2-alkynylthioanisoles (**6a**) with ammonium thiocyanate (**6b**) with visible light as a green energy source and oxygen as a green oxidant. It was observed that no product was formed in the absence of photocatalyst or light, and the desired products (**6c**) were achieved with up to 81% yield after 48 h (Scheme 6).

In 2021, Pan and coworkers [27] reported a visible light-induced [3+2] oxidative cyclization of thiavinyl 1,3-dipoles, ketene dithioacetals (**7a**), and alkynes (**7b**) in the presence of an acridine photosensitizer without other additives, affording the desired

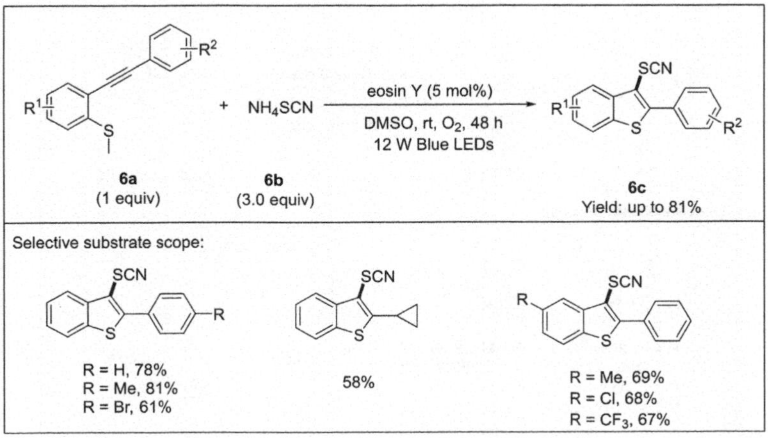

Scheme 6 Photoredox-catalyzed synthesis of 3-thiocyanobenothiophene (Kshirsagar method)

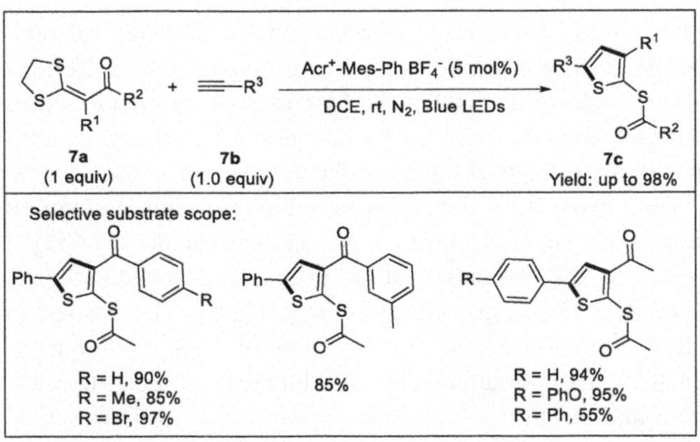

Scheme 7 Visible light-induced synthesis of thiophenes (Pan method)

product (**7c**) with up to 98% yields (Scheme 7). A variety of substrates were well compatible to this protocol, affording the anticipated product with moderate to high yields. Mechanism experiments including light off/on experiment, UV-vis spectroscopic measurements, and Stern-Volmer emission quenching studies revealed that polarized ketene dithioacetals quenched the excited photosensitizer more easily than alkynes.

In 2019, Oyamada group [28] developed a photoinduced intramolecular cyclization reaction of 1,2-bis(2-methylthiophenyl)-ethynes for the synthesis of benzothieno[3,2-*b*]benzothiophene derivatives (BTBTs) (**8b**) (Scheme 8). With the assistance of I_2/PhI(OAc)$_2$, the reaction of 3-iodo-(2-methylthiophenyl)benzo[*b*] thiophenes (**8a**) in DCM was irradiated under a high-pressure Hg lamp (>290 nm, 400 W) for 6 h, giving **8b** with up to 82% yields. Notably, the mixture

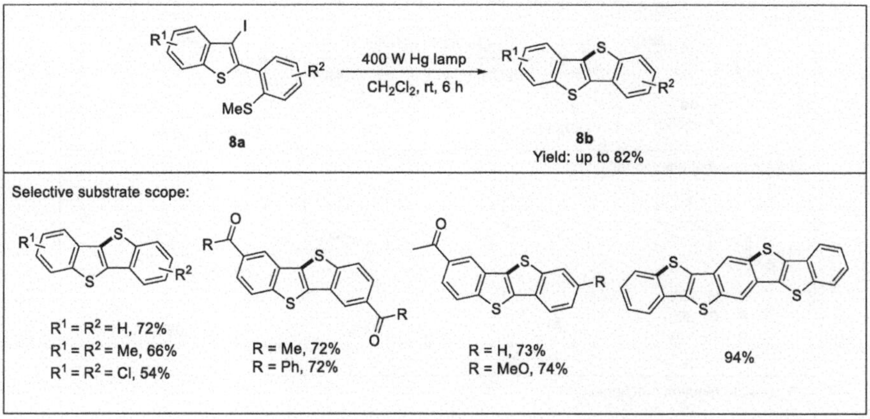

Scheme 8 Visible light-promoted synthesis of BTBTs (Oyamada method)

of regioisomers **8a** and **8a′** provided the same BTBTs **8b**. In addition, π-extended thienoacene such as BBTBDT could also be accessed in 94% yield.

3 Photocatalytic Functionalization of Fused Thiophenes

3.1 Coupling Reactions with Simple Fused Thiophenes

In 2015, Kappe group [29] described a light-induced, catalyst-free one-pot C–H arylation reaction of thiophenes (**9b**) with anilines (**9a**)/*t*BuONO for the preparation of 2-arylthiophenes (**9c**) at room temperature (Scheme 9). The coupling reactions were generally exposed under 60 W CFL for 45 min, furnishing the desired 2-arylthiophenes with up to 80% yield. When the reaction was carried out without a photocatalyst analogous, lower conversion was obtained, demonstrating the essential role of photocatalyst and light. Under optimized conditions, various substituted anilines containing functional groups like esters, nitriles, and nitro groups were investigated and well tolerated. The reactions of anilines with tBuONO could form nitrosamines, in equilibrium with their corresponding diazo anhydrides, which decomposed into aryl radicals slowly under light irradiation (>300 nm). This protocol avoids the handling of hazardous nitrosamine and diazo-type intermediates.

In the proposed catalytic cycle, the reaction of anilines with *t*BuONO afforded nitrosamine **1**, which gave diazo anhydrides **2** via self-condensation. Then a light-induced C–N cleavage of diazo anhydride **2** provided the corresponding aryl radical **3** and an azoxy radical **4** with the release of N_2 gas. The addition of the aryl radical with the thiophenes furnished intermediate **5**, which underwent a single-electron transfer/proton transfer with **4 to** form the desired functional thiophenes and the nitrosamine (diazo hydroxide **1′**).

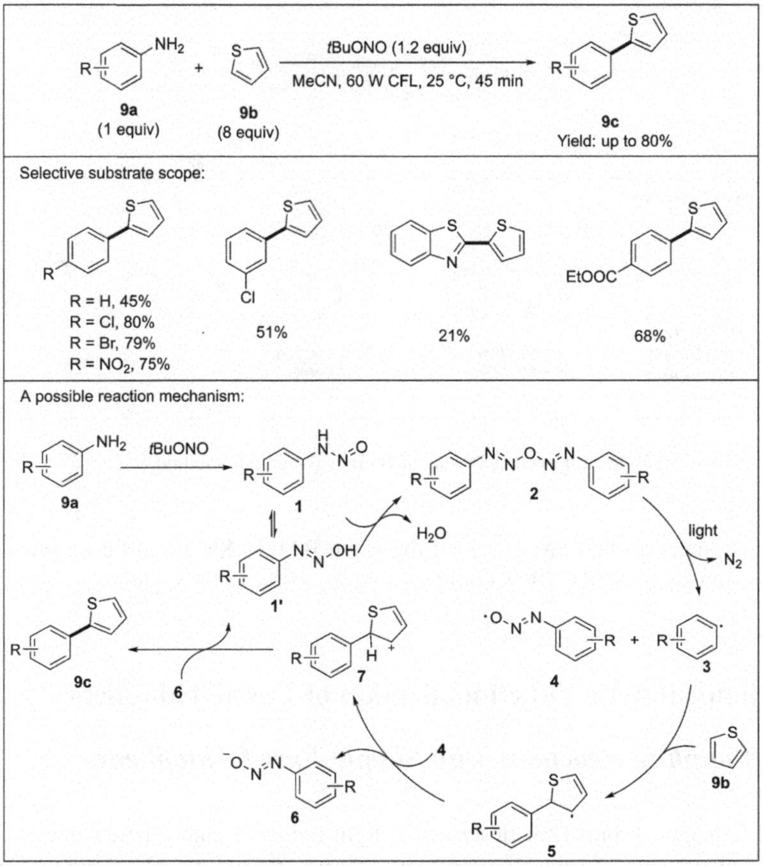

Scheme 9 Light-induced synthesis of bi(hetero)aryls (Kappe method)

Recently, Gupta and coworkers [30] developed a C–H arylation of thiophenes (**10b**) with anilines (**10a**) catalyzed by light and Zin(II)porphyrin. The reaction mixture in mixed DMSO/ACN was stirred at room temperature and irradiated by Smartchem Synth Photoreactor (24 W blue LED, 440 nm) to obtain the desired products (**10c**) with up to 57% yields within 30 min (Scheme 10). Only a trace amount of target product was formed when reactions were performed without light or a catalyst, which indicated the essentiality of visible light and catalyst. The substrate scope study suggested that anilines containing electron-donating and -withdrawing groups underwent this transformation very smoothly, affording the desired product with good yields. Moreover, they conducted several controlled experiments, such as radical trapping experiments and HRMS to gain insight into the mechanism and reported that this protocol proceeded to undergo a free radical process, followed by radical generation.

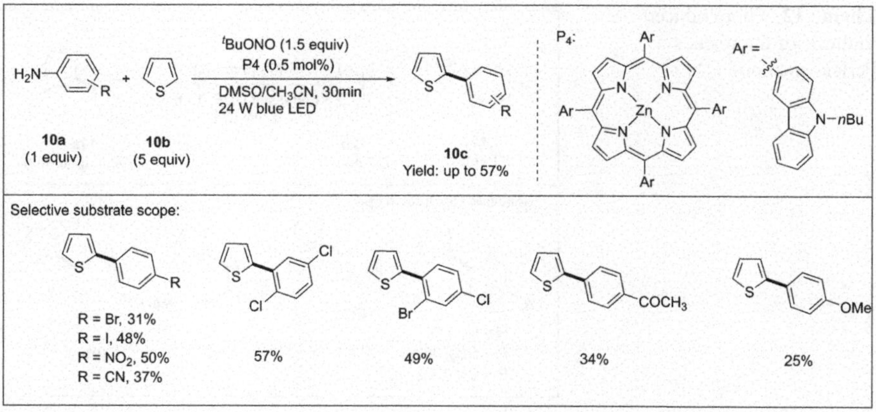

Scheme 10 Visible light-induced synthesis of thiophenes (Gupta method)

Scheme 11 Photoinduced synthesis of (hetero)biaryls (Fagnoni method)

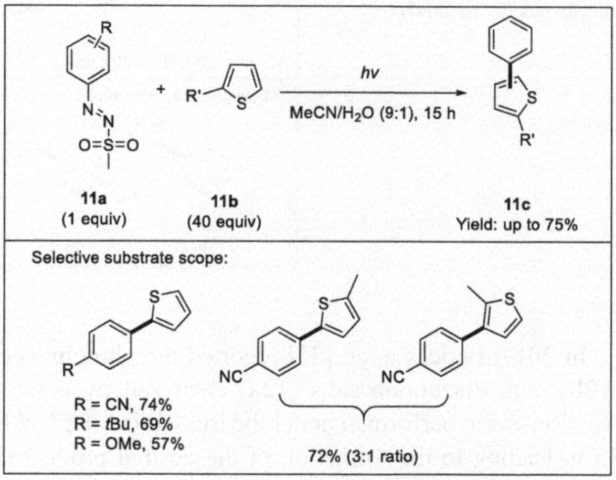

In the same year, Fagnoni group [31] developed a photoinduced, metal-free strategy for the synthesis of (hetero)biaryls (**11c**) from arylazo mesylates (**11a**) with thiophenes (**11b**). No photocatalyst or other additives were used in this reaction. According to this methodology, the authors used MeCN:H_2O (9:1) as solvent, purged for 10 min with nitrogen, capped, and exposed to solar-simulated light (in a Solarbox) or natural sunlight on a window ledge to obtain the desired (hetero)-biaryl (**11c**) product with up to 75% yield within 15 h (Scheme 11). They observed that a variety of electron-rich heteroaromatics such as 2-methylthiophene, *N*-Boc-pyrrole, and unactivated arenes were well compatible for this protocol and afforded the anticipated product with moderate to high yields. Noteworthily, they demonstrated that selective generation of aryl radicals and aryl cations can be attained at will by varying the irradiation conditions (visible light for the former and UVA light for the latter).

Scheme 12 Photoinduced synthesis of thiophenes (Pericàs method)

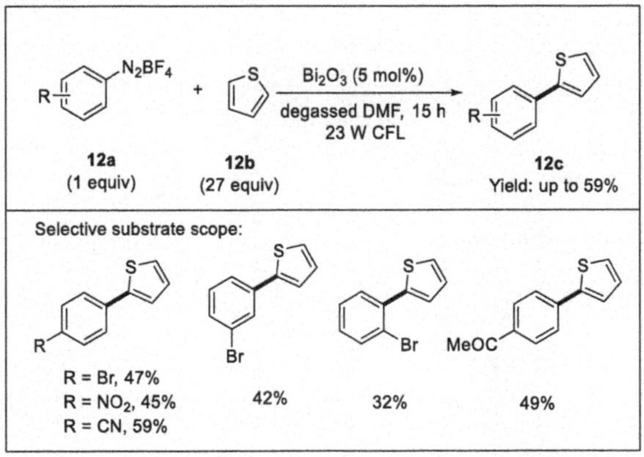

Scheme 13 Visible light-induced synthesis of thiophenes (Zi method)

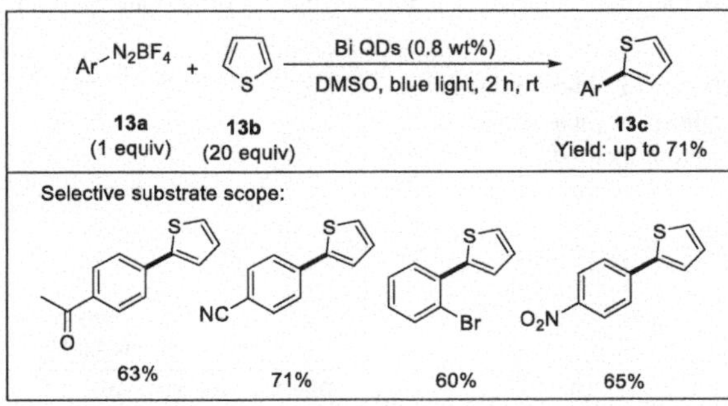

In 2017, Pericàs et al. [32] reported the photoinduced α-arylation of thiophenes (**12b**) with diazonium salts (**12a**) catalyzed by nontoxic bismuth(III) oxide. The reactions were performed under the irradiation of 23 W CFL at room temperature for 15 h, leading to the formation of the desired product (**12c**) with up to 59% yields (Scheme 12). Other photocatalysts such as $BiVO_4$ and Bi_2S_3 showed poorer efficiency. In this protocol, only 5 mol% of catalyst is necessary to prepare a wide range of differently substituted α-arylated heteroarenes in moderate to excellent yields, with the combination of electron-rich heteroarenes with electron-poor diazonium salts.

In 2023, Zi group [33] realized the first example of C–H arylation for the direct synthesis of arylheteroaryl compounds (**13c**) from aryl diazonium salts (**13a**) and thiophenes (**13b**) that employed Bi QDs as a heterogeneous photoredox catalyst. According to this report, the reaction required 0.8 wt% Bi QDs as the photocatalyst in DMSO solvent. The desired product (**13c**) achieved up to 71% yield within 2 h at room temperature (Scheme 13). The generality and versatility were well studied for this transformation. A variety of aryl diazonium salts containing various substituents performed very smoothly in this protocol. Moreover, they investigated the reusability of Bi QD catalyst and found that slight decrease of yield was observed after each

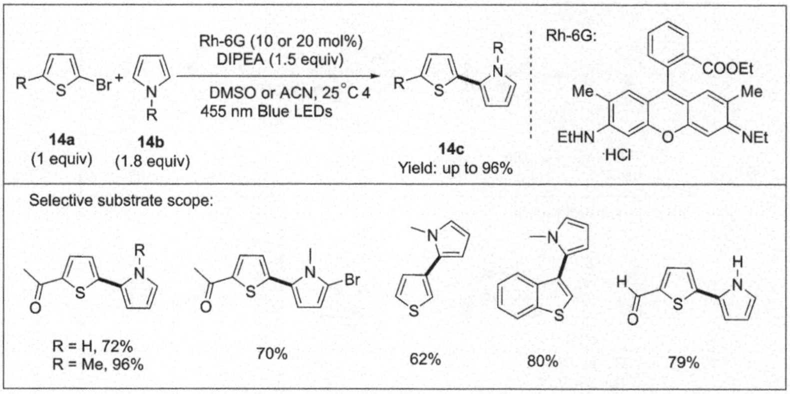

Scheme 14 Photocatalytic synthesis of biheteroaryls (König method)

recycle. The reaction proceeded to undergo a free radical process, as confirmed by radical trapping experiments.

In 2016, König group [34] reported the photocatalytic synthesis of biheteroaryls (**14c**) from the cross-coupling reaction of bromothiophenes (**14a**) with pyrroles (**14b**). These reactions were conducted with 10 or 20 mol% Rh-6G as the photocatalyst, DIPEA (1.5 equiv.) as sacrificial electron donor, and DMSO or ACN as solvent, under the irradiation of 455 nm blue LEDs at room temperature for 24–48 h, achieving the product **14c** with up to 96% yield (Scheme 14). The reaction failed to proceed in the absence of light, catalyst, or DIPEA, confirming that all the components are necessary for the photoredox catalytic reaction. A bunch of bromothiophenes and bromoheteroarenes containing various functional groups were well tolerated, furnishing the desired product with moderate to excellent yields. The methodology was also applied for the formal synthesis of benzimidazole derivative that is an MCH-R_1biheteroaryl antagonist.

In 2017, Itoh et al. [35] developed a novel and concise route to gain α-heteroarylated carbonyls (**15c**) via an iodine-mediated aerobic photooxidative C–H/C–H coupling reaction of thiophenes (**15a**) with 1,3-dicarbonyl compounds (**15b**) (Scheme 15). With the use of 0.2 equivalent of I_2, 1.25 equivalents of K_3PO_4, and mixed MeCN/H_2O solvent, the reaction was carried out under the irradiation of four 22 W compact fluorescent lamps at an ambient temperature for 20 h, achieving the final products with up to 83% yield. A series of aryl-substituted thiophenes were well tolerated in this transformation to furnish the desired product with moderate to excellent yields. 1,3-Dicarbonyls with substituted amide, ketone, and sulfonyl groups at α-position showed competitive high efficiency. In contrast to the traditional methods leading to similar products, the advantages of this methodology are mild reaction conditions, operational simplicity, and an inexpensive and abundant molecular iodine catalyst. The proposed mechanism shows that the mixture of ester and molecular iodine generated a plausible iodinated intermediate **I**. Visible-light irradiation from CFL caused homolytic cleavage of the C–I bond, which generated

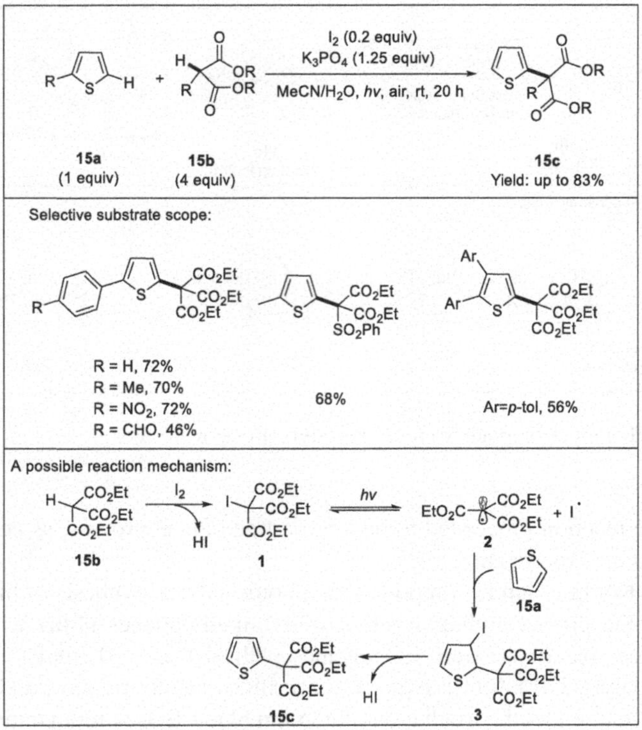

Scheme 15 Photooxidative synthesis of thiophenes (Itoh method)

the C-centered radical species **II**. Addition of **II** to thiophenes and iodination of the generated radical gave intermediate **III**. Finally, aromatization of **3** furnished **15c**.

In 2018, Qin group [36] developed a visible light-induced photocatalytic alkyl addition reaction of (hetero)arene-fused thiophenes (**16b**) using *N*-benzoyl alkyl-sulfinamides (**16a**) as the alkyl radical precursors (Scheme 16). In the presence of Ir-photocatalyst (0.5 mol%) and K_2CO_3 (6.0 equiv.), the reactions were irradiated by blue LEDs (8 W) at 80°C for 24 h, giving the desired products (**16c**) with up to 95% yield. Various substituted *N*-benzoyl alkyl-sulfinamides and (hetero)arene-fused thiophenes successfully underwent this transformation. To explore the scalability of this desulfurative alkyl addition method, the gram-scale reaction of *N*-benzoyl *tert*-butyl-sulfinamide and benzothiophene was performed under the optimized conditions, giving 2-(*tert*-butyl)-2,3-dihydrobenzothiophene in 95% yield.

Deprotonation of the sulfinamide N–H bond in the substrate **16a** with base gave anion **I**. It was subsequently oxidized by the photoexcited Ir^{III} complex to generate the nitrogen-centered radical **II**. Fragmentation of this unique radical species then occurred, yielding *N*-sulfinylbenzamide (**III**) and the corresponding alkyl radical **IV**. The latter could be trapped with (hetero)arene-fused thiophenes. The resultant radical **V** was reduced and protonated to give the final addition product. Meanwhile,

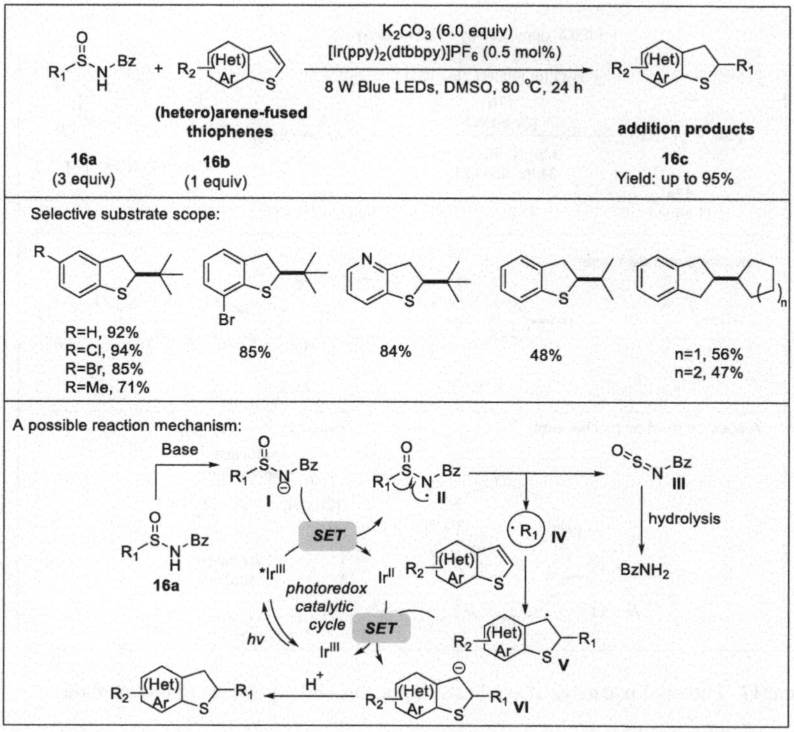

Scheme 16 Visible light-induced synthesis of (hetero)arene-fused thiophenes (Qin method)

oxidation of the Ir^{II} complex via SET regenerated the photocatalyst and closed the catalytic cycle.

3.2 *Functionalization of Thianthrenium Analogues*

Organothianthrenium salts represent a class of fused thioarenes containing a positively charged sulfur atom and a neutral sulfur atom. Over the past years, organothianthrenium salts have been emerging as attractive precursors for a myriad of transformations to forge new C–C and C–X bonds due to their unique structural characteristics and chemical behaviors. In this section, selected representative catalytic processes will be discussed.

In 2019, Ritter and coworkers [37] reported the site-selective synthesis of phenols and aryl ethers (**17c**) through a photoredox and transition metal-catalyzed coupling reactions of aryl thianthrenium salts (**17a**) with water or alcohols/phenols (**17b**) (Scheme 17). With the additions of 1 mol% of $[Ir[dF(CF_3)ppy]_2(dtbpy)PF_6]$ and 1.5 equivalents of CuTC, the reactions were carried out under the irradiation of 34 W blue LED at 30°C for 16 h, achieving **17c** with up to 84% yield. Both electron-rich

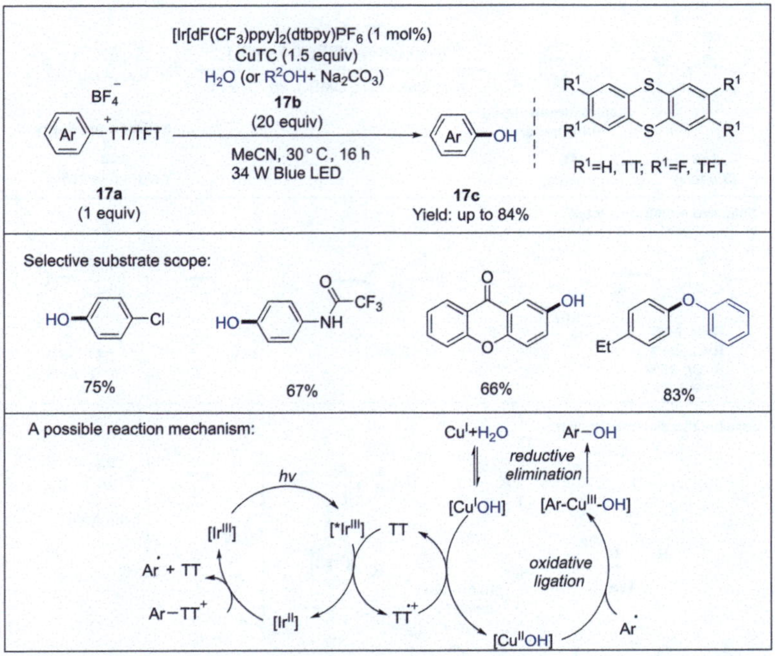

Scheme 17 Photoredox-catalyzed synthesis of phenols and aryl ethers (Ritter method)

and electron-deficient aromatics could be successfully converted into the corresponding phenols and aryl ethers. The protocol differs conceptually from all previous arene oxygenation reactions in that oxygen functionality can be incorporated into complex small molecules at a late stage site-selectively, which has not been shown via aryl halides. The proposed mechanism shows that the excited photocatalyst $*Ir^{III}$ was quenched by thianthrene (TT) to form a thianthrenium radical cation (TT^{+}) and reduced photocatalyst Ir^{II}. Ir^{II} interacted with aryl thianthrenium salt (Ar-TT^{+}) to generate an aryl radical, thianthrene (TT), along with Ir^{III}. On the other hand, $Cu^{I}OH$ could be generated from $Cu^{I}TC$ in the presence of water, which underwent SET with a thianthrenium radical cation ($TT^{\cdot+}$) to form $Cu^{II}OH$. Aryl radicals recombined with $Cu^{II}OH$ species to afford Ar–$Cu^{III}OH$ species, which underwent reductive elimination to furnish the final product by C–O bond formation.

Later, the same group [38] developed a photoinduced coupling reaction of aryl thianthrenium salts (**18a**) with *N*-nucleophiles (**18b**) (Scheme 18). In the presence of 5 mol% of $[Ru(bipy)_3](PF_6)_2$, 1.0 equivalent of $[Cu(MeCN)_4]BF_4$, and 3.0 equivalents of NMe_4OH or NaH, all reactants were taken together in a closed round-bottom flask under the irradiation of 60 W blue LED at 15°C for 8 h, providing the desired product (**18c**) with up to 60% yield. A variety of *N*-containing heterocycles performed smoothly in this coupling reaction. Hydrodefunctionalized starting material was observed as side products in some cases, which can easily be removed by

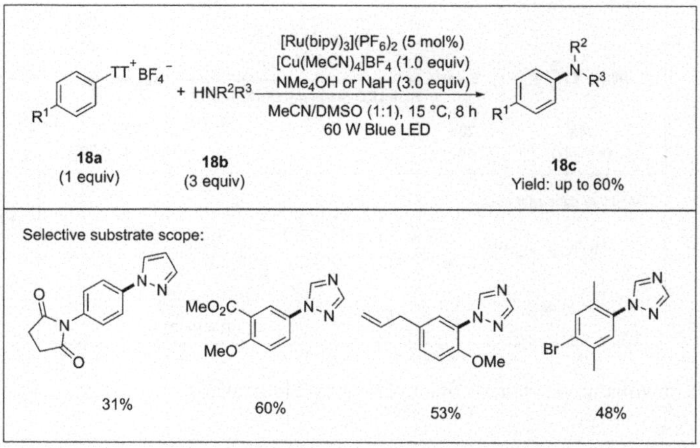

Scheme 18 Photoinduced synthesis of aromatic amine (Ritter method)

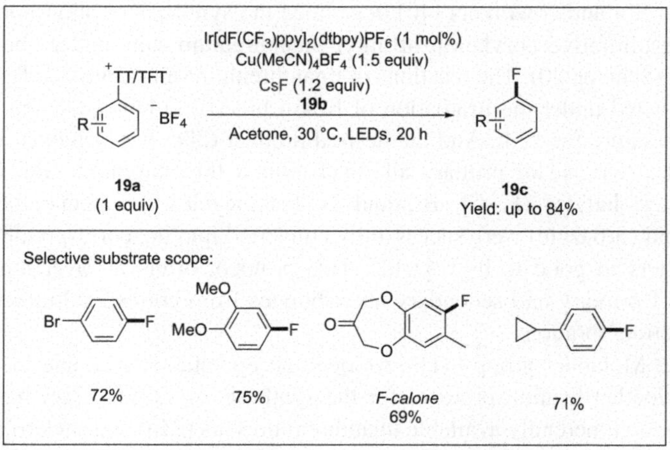

Scheme 19 Photoinduced synthesis of F-containing arenes (Ritter method)

chromatography. Notably, the flurbiprofen methyl ester-derived thianthrenium salt could successfully react with different *N*-nucleophiles for late-stage diversification.

The Ritter group [39] then employed the aryl thianthrenium salts (**19a**) as a mediator to convert aromatics into aryl fluorides (**19c**) through the copper(III) metallophotoredox catalysis (Scheme 19). The reactions of aryl thianthrenium salts (**19a**) and CsF (**19b**) were performed under the irradiation of 68 W blue LEDs at 30° C for 20 h, producing the desired product (**19c**) with up to 84% yield. A variety of electron-deficient, and -rich arenes containing functional groups such as bromo, ester, aldehyde, carbamate, and amide groups were well tolerated, yielding the corresponding aryl fluorides in moderate to good yields. Compared with the

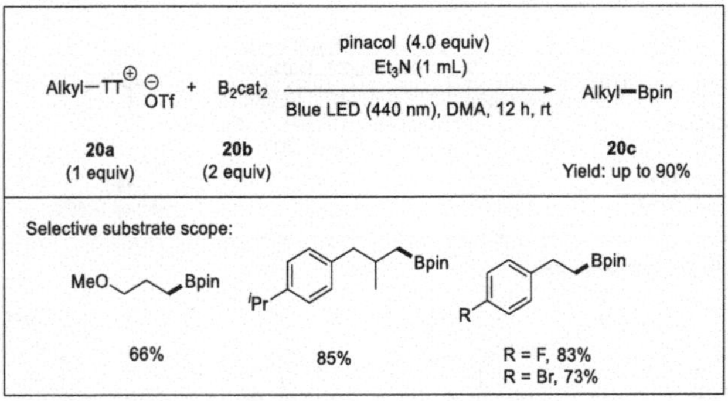

Scheme 20 Photoinduced synthesis of alkyl borates (Shi method)

corresponding halides, aryl thianthrenium salts have higher reduction potentials. The dissociation rate of aromatic radical anions is obviously sped up, and thus the formation process of C–F bond via metallophotoredox catalysis can be realized.

In 2021, Shi and coworkers [40] developed the synthesis of alkyl borates (**20c**) through desulfurative borylation of alkyl thianthrenium salts under photoinduced conditions (Scheme 20). The reactions of thianthrenium salts (**20a**) and B_2cat_2 (**20b**) were conducted under the irradiation of Kessil blue LED bulb (440 nm, 40 W) at room temperature for 12 h. And the desired product (**20c**) was achieved with up to 90% yields. A range of primary alkyl-substituted thianthrenium salts with alkyl chains, ester, halides (F, Cl, Br, and I), heterocycles (thiophene, benzofuran, oxazole, and carbazole) were successfully converted into the corresponding pinacol boronic esters in good to high yields. This protocol offers a novel route for the synthesis of primary and secondary alkyl borates from corresponding alcohols by thianthrenation/boration.

In 2022, Molander group [41] developed an operationally simple visible light-induced thioetherification protocol for the synthesis of sulfide (**21c**) from rapidly accessed or commercially available thianthrenium salts (**21a**) and (hetero)arylthiols) (**21b**) (Scheme 21). Under the irradiation of 427 nm blue Kessil at room temperature for 2–8 h, the desired product (**21c**) could be obtained with up to 89% yields with the addition of 1.5 equivalents of K_2CO_3. The reaction failed to proceed in the absence of base or visible light, confirming the necessity of these conditions to enable the C (sp^2)–S bond formation. A variety of thianthrenium salts containing various substituents were well tolerated in this transformation and furnished the desired product with moderate to excellent yields. In addition, a wide range of functional groups including free boronic acids, carboxylic acids, and halides were compatible in this protocol. The proposed mechanism shows that an EDA complex was formed between aryl thianthrenium salts (**21a**) and in situ-formed aryl thiolate anions, which induced unimolecular nucleophilic substitution ($S_{RN}1$) free radical chain reaction. Irradiation and intramolecular electron transfer of the EDA complex generated disulfides, aryl radicals, and TT. Aryl radicals coupled with another aryl

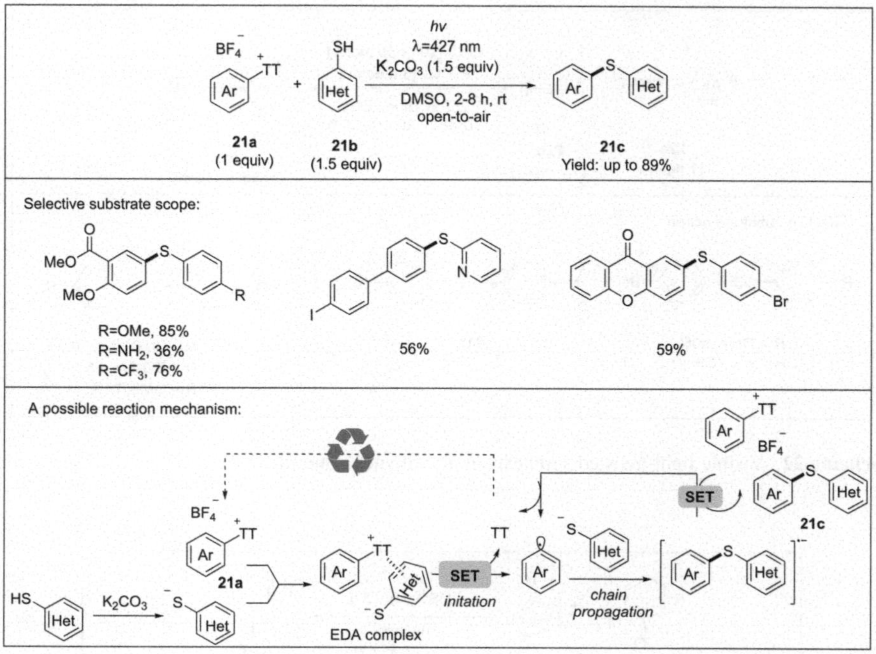

Scheme 21 Visible light-promoted synthesis of sulfide (Molander method)

thiolate anion to afford diaryl sulfide radical anions, which underwent subsequent single-electron reduction of (**21a**) to form the final product (**21c**) and regenerate aryl radicals.

Recently, Guo group [42] reported a visible light-induced copper-catalyzed Sonogashira-type reaction of aryl sulfonium salts (**22a**) and terminal alkynes (**22b**), which offered an efficient approach for the synthesis of arylalkynes (**22c**) (Scheme 22). This method required 10 or 30 mol% CuCl as catalyst, K_2CO_3 (3. 0 equivalents) as base, and DMSO as solvent. The desired product (**22c**) achieved up to 86% yields when the reactions were subjected at room temperature under blue LED irradiation (27 W) for 20 h. The scaled-up synthesis explored the prospective use of this methodology and got up to 78% yield under the same reaction conditions. Moreover, this method can be applied to late-stage alkynylation of drugs, which is difficult to achieve otherwise, indicating the potential of this work in rapid drug modification.

Later, Patureau group [43] developed a simple catalyst-free photoinduced C–H (hetero)arylation cross-coupling of aryl thianthrenium salts (**23a**) with (hetero)-arenes (**23b**) to form (hetero)biaryl products (**23c**) with high efficiency and regioselectivity. The reactions were performed in DMSO under UV-light (254 nm, 144 W) irradiation to give **23c** with up to 64% yield (Scheme 23). Only a trace amount of target product was detected under air, indicating the critical role of anhydrous and oxygen-free atmosphere for this reaction. Aryl thianthrenium salts

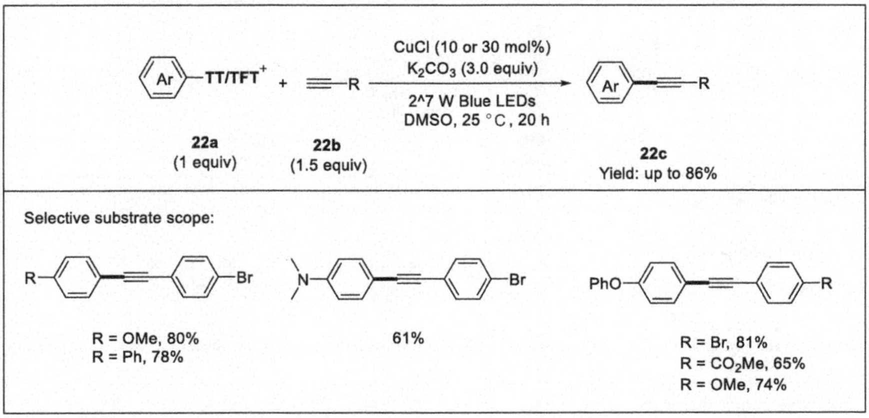

Scheme 22 Visible light-assisted synthesis of arylalkynes (Guo method)

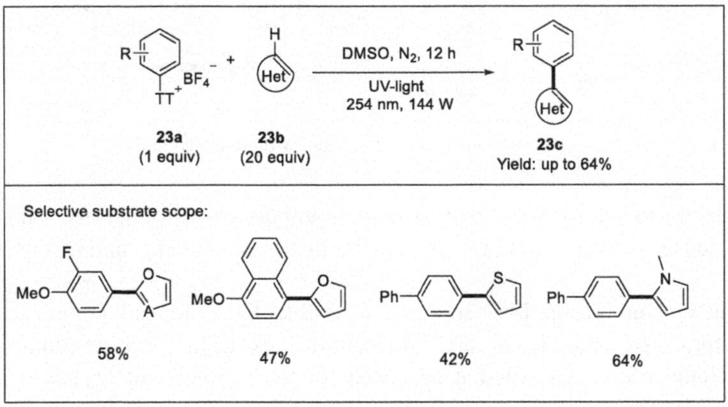

Scheme 23 Photoinduced synthesis of (hetero)biaryls (Patureau method)

containing electron-donating and -withdrawing groups underwent this transformation successfully to afford the desired product in good yields. Furans, thiophenes, and pyrroles could react smoothly to give the target products in good yields. The radical trapping experiments showed that aryl radicals were generated by directly disrupting the C–S bond, which underwent a homolytic aromatic substitution of heteroarenes to form the final products.

In 2022, Yu group [44] discovered a novel mode for the generation of aryl radicals by photoactivation of arenes via thianthrenation. The reactions of aryl thianthrenium salts (**24a**) and *N*-containing heterocycles (**24b**) were conducted with the addition of 3 equivalents of DABCO under the irradiation of blue LED (10 W) for 12 h at room temperature to give the desired product (**24c**) with up to 94%

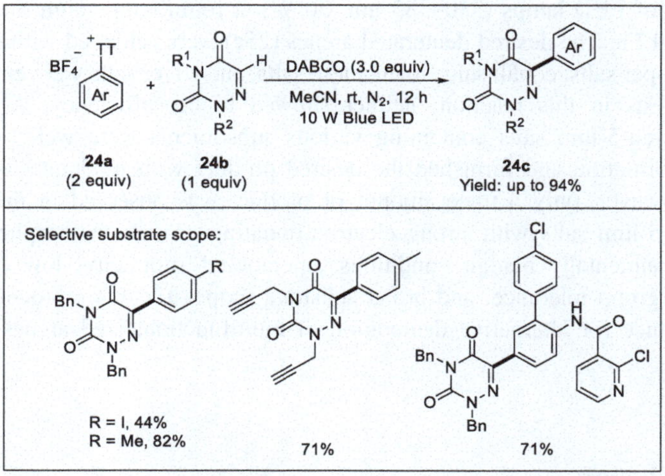

Scheme 24 Photoinduced synthesis of nitrogen-containing compounds (Yu method)

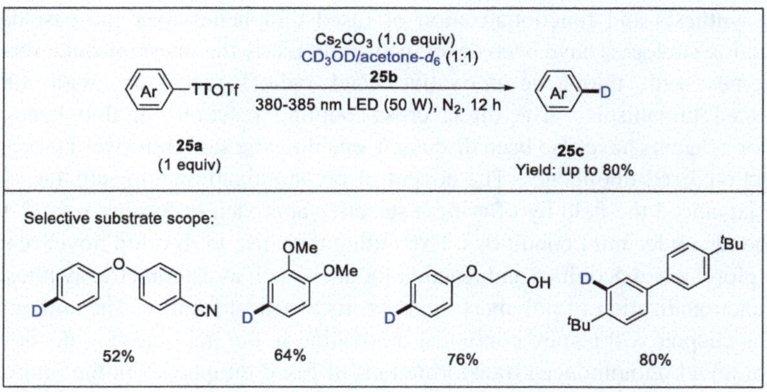

Scheme 25 Photoinduced synthesis of deuterated arenes (Zhang method)

yield (Scheme 24). Different types of substituted aryl thianthrenium salt and (hetero)aryl cycles were well tolerated under optimized reaction conditions. Coumarin, 1-(fluoromethyl) cinnolin-4(1*H*)-one, and 2-phenylimidazo[1,2-*a*] pyridines that participated in this protocol could furnish the target product with moderate to excellent yields. This reaction proceeded through the formation of the EDA complex followed by radical generation, as confirmed by UV-vis absorbance measurements, nuclear magnetic resonance titration experiments, and density functional theory (DFT) calculations.

In 2022, Zhang and coworkers [45] developed a mild and convenient method for deuteration of arylthianthren-5-ium salts (**25a**) with CD_3OD/acetone-d_6 (**25b**) as both solvent and deuterium source(s) (Scheme 25). The reactions were carried out in the presence of 1 equivalent of Cs_2CO_3 in CD_3OD/acetone-d_6 solvent under the irradiation of LED lamps (380–385 nm, 50 W) at room temperature in N_2 atmosphere for 12 h. The desired deuterated arenes (**25c**) were achieved with up to 80% yields. Copper salts, cobalt salts, manganese salts, and silver salts were employed as photocatalysts in this reaction; neither showed higher efficiency. A bunch of arylthianthren-5-ium salts containing various substituents were well tolerated in this transformation and furnished the desired product with moderate to excellent yields. However, only a trace amount of product was observed in the cases of thianthren-5-ium salts with strong electron-donating groups on the phenyl rings. The environmentally benign conditions, operational simplicity, low cost, good functional group tolerance, and broad substrate scopes for this protocol highlight its importance for alternative deuteration of multifunctionalized arenes and drug molecules.

4 Conclusion

In conclusion, this chapter has highlighted the remarkable progress in the photoinduced synthesis and functionalization of fused thiophenes over the past decade. Cyclization strategies have been developed, particularly the intermolecular reactions of alkynes with thiophene derivatives and radical precursors with alkynyl-substituted thioanisoles. The direct cross-coupling reactions of thiophenes with arylation reagents have also been discussed, enabling the straightforward preparation of functionalized thiophenes. The advent of organothianthrenium salts has significantly advanced the field by offering a selective approach to forging new C–C and C–X bonds under mild conditions. Expanding their use to develop novel reactions and explore new reactivities and reaction modes – such as asymmetric synthesis and post-functionalization of polymers – represents a critical frontier. The authors hope that this chapter will inspire continued innovation in the field, driving the development of novel photoinduced transformations of fused thiophenes in the future.

Acknowledgements The authors acknowledge the financial support provided by the NSFC (22125103, 21971065, and 22301077), STCSM (20XD1421500, 20JC1416800, and 22JC140100), and Shanghai Pujiang Program (22PJ1403200).

References

1. Wang Z, Shi J, Li C, Zhao D, Wang Z, Wang H (2010) The synthesis of octyl-substituted hexathia[7]heterohelicene based on dithieno[2,3-*b*:3′,2′-*d*]thiophene via irradiation. Lett Org Chem 7(1):85–89
2. Fukazawa A, Karasawa T, Zhang H, Minemura K, Camacho C, Wang J, Irle S, Yamaguchi S (2013) Photochemical double 5-*exo* cyclization of alkenyl-substituted dithienylacetylenes: efficient synthesis of diarylated dithienofulvalenes. Angew Chem Int Ed 52(40):10519–10523

3. Fukazawa A, Oshima H, Shiota Y, Takahashi S, Yoshizawa K, Yamaguchi S (2013) Thiophene-fused bisdehydro[12]annulene that undergoes transannular alkyne cycloaddition by either light or heat. J Am Chem Soc 135(5):1731–1734
4. Liu X, Yu P, Xu L, Yang J, Shi J, Wang Z, Cheng Y, Wang H (2013) Synthesis for the mesomer and racemate of thiophene-based double helicene under irradiation. J Org Chem 78(12): 6316–6321
5. Gao F, Wang J, Liu L, Ma N, Yang C, Gao Y, Xia W (2017) Synthesis of carbonylated heteroaromatic compounds via visible-light-driven intramolecular decarboxylative cyclization of O-alkynylated carboxylic acids. Chem Commun 53(61):8533–8536
6. Jana S, Verma A, Kadu R, Kumar S (2017) Visible-light-induced oxidant and metal-free dehydrogenative cascade trifluoromethylation and oxidation of 1,6-enynes with water. Chem Sci 8(9):6633–6644
7. Dyachenko NV, Khoroshutin AV, Sotnikova YA, Karnoukhova VA, Tokarev SD, Anisimov AV, Fedorov YV, Fedorova OA (2019) Synthesis of fused heterocyclic systems via the Mallory photoreaction of arylthienylethenes. Photochem Photobiol Sci 18(12):2901–2911
8. Xue J, Bai L, Zhang L, Zhou Y, Lin X, Mou N, Xiao D, Luo Q (2020) One-pot synthesis of 2,4-diacyl thiophenes from α-oxo ketene dithioacetals and propargylic alcohols. J Org Chem 85(15):9761–9775
9. Liu Y, Xue D, Wang J, Zhao C, Zou Q, Wang C, Xiao J (2013) Room-temperature arylation of arenes and heteroarenes with diaryl-iodonium salts by photoredox catalysis. Synlett 24(4): 507–513
10. Shaikh RS, Düsel SJS, König B (2016) Visible-light photo-Arbuzov reaction of aryl bromides and trialkyl phosphites yielding aryl phosphonates. ACS Catal 6(12):8410–8414
11. He T, Wei H, Zhou Y, Jiang L, Baell JB, Yu Y, Huang F (2023) Visible light-induced borylation and arylation of small organic molecules using carbon dots. Org Chem Front 10(12):2918–2926
12. Meng H, Liu M, Shu W (2022) Organothianthrenium salts: synthesis and utilization. Chem Sci 13(46):13690–13707
13. Ye F, Berger F, Jia H, Ford J, Wortman A, Börgel J, Genicot C, Ritter T (2019) Aryl sulfonium salts for site-selective late-stage trifluoromethylation. Angew Chem Int Ed 58(41): 14615–14619
14. Chen C, Wang M, Lu H, Zhao B, Shi Z (2021) Enabling the use of alkyl thianthrenium salts in cross-coupling reactions by copper catalysis. Angew Chem Int Ed 60(40):21756–21760
15. Cai Y, Chatterjee S, Ritter T (2023) Photoinduced copper-catalyzed late-stage azidoarylation of alkenes via arylthianthrenium salts. J Am Chem Soc 145(25):13542–13548
16. Berger F, Plutschack MB, Riegger J, Yu W, Speicher S, Ho M, Frank N, Ritter T (2019) Site-selective and versatile aromatic C–H functionalization by thianthrenation. Nature 567(7747): 223–228
17. Jia H, Häring AP, Berger F, Zhang L, Ritter T (2021) Trifluoromethyl thianthrenium triflate: a readily available trifluoromethylating reagent with formal CF_3^+, $CF_3^\cdot$, and CF_3^- reactivity. J Am Chem Soc 143(20):7623–7628
18. Kreijger SD, Troian-Gautier L (2022) Sulfonylation enabled through the photoactivation of EDA complexes. Chem Catal 2(4):647–666
19. He F, Bao P, Tang Z, Yu F, Deng W, Wu J (2022) Photoredox-catalyzed α-sulfonylation of ketones from sulfur dioxide and thianthrenium salts. Org Lett 24(15):2955–2960
20. Li Q, Huang J, Cao Z, Zhang J, Wu J (2022) Photoredox-catalyzed reaction of thianthrenium salts, sulfur dioxide and hydrazines. Org Chem Front 9(14):3781–3785
21. Gao L, Chang B, Qiu W, Wang L, Fu X, Yuan R (2016) Potassium hydroxide/dimethyl sulfoxide superbase-promoted transition metal-free synthesis of 2-substituted benzothiophenes under visible light. Adv Synth Catal 358(8):1202–1207
22. Ye L, Qian L, Chen Y, Zhang X, Yan M (2017) A practical synthesis of benzothiophenes via visible-light-promoted cyclization of disulfides and alkynes. Org Biomol Chem 15(3):550–554

23. Xia X, Zhang G, Zhu S (2017) Visible-light-induced synthesis of benzothiophenes and benzoselenophenes via the annulation of thiophenols or 1,2-diphenyldiselane with alkynes. Tetrahedron 73(19):2727–2730
24. Yan J, Xu J, Zhou Y, Chen J, Song Q (2018) Photoredox-catalyzed cascade annulation of methyl(2-(phenylethynyl)phenyl)sulfanes and methyl(2-(phenylethynyl)phenyl)selanes with sulfonyl chlorides: synthesis of benzothiophenes and benzoselenophenes. Org Chem Front 5(9):1483–1487
25. Dong H, Chen C, Zhao J, Ji Y, Yang W (2023) Photoinduced photocatalyst-free cascade cyclization of alkynes with sodium sulfinates for the synthesis of benzothiophenes and thioflavones. Molecules 28(11):4436–4448
26. Tambe SD, Jadhav MS, Rohokale RS, Kshirsagar UA (2018) Metal-free synthesis of 3-thiocyanatobenzothiophenes by eosin Y photoredox-catalyzed cascade radical annulation of 2-alkynylthioanisoles. Eur J Org Chem 2018(35):4867–4873
27. Zheng B, Li X, Song Y, Meng S, Li Y, Liu Q, Pan L (2021) Visible-light-induced formation of thiavinyl 1,3-dipoles: a metal-free [3+2] oxidative cyclization with alkynes as easy access to thiophenes. Org Lett 23(9):3453–3459
28. Kitamura T, Morita K, Nakamori H, Oyamada J (2019) Synthesis of [1] benzothieno[3,2-*b*][1] benzothiophene derivatives via successive iodocyclization/photocyclization of alkynes. J Org Chem 84(7):4191–4199
29. Cantillo D, Mateos C, Rincon JA, de Frutos O, Kappe CO (2015) Light-induced C-H arylation of (hetero)arenes by in situ generated diazo anhydrides. Chem Eur J 21(37):12894–12898
30. Janaagal A, Sanyam, Mondal A, Gupta I (2023) Robust zinc(II) porphyrin catalyst for visible light induced C-H arylation of heteroarenes. J Org Chem 88(13):9424–9431
31. Crespi S, Protti S, Fagnoni M (2016) Wavelength selective generation of aryl radicals and aryl cations for metal-free photoarylations. J Org Chem 81(20):9612–9619
32. Buglioni L, Riente P, Palomares E, Pericàs MA (2017) Visible-light-promoted arylation reactions photocatalyzed by bismuth (III) oxide. Eur J Org Chem 2017(46):6986–6990
33. Huang Z, Zhu J, Dong Q, Huang Q, Wang M, Zhu Y, Huang W, Zi Y (2023) Photoredox C-H arylation of heteroarenes by high-efficiency bismuth quantum dots. Tetrahedron Lett 127: 154667–154672
34. Marzo L, Ghosh I, Esteban F, König B (2016) Metal-free photocatalyzed cross coupling of bromoheteroarenes with pyrroles. ACS Catal 6(10):6780–6784
35. Sudo Y, Yamaguchi E, Itoh A (2017) Photo-oxidative cross-dehydrogenative coupling-type reaction of thiophenes with α-position of carbonyls using a catalytic amount of molecular iodine. Org Lett 19(7):1610–1613
36. Di J, He H, Wang F, Xue F, Liu X, Qin Y (2018) Regiospecific alkyl addition of (hetero)arene-fused thiophenes enabled by a visible-light-mediated photocatalytic desulfuration approach. Chem Commun 54(37):4692–4695
37. Sang R, Korkis SE, Su W, Ye F, Engl PS, Berger F, Ritter T (2019) Site-selective C-H oxygenation via aryl sulfonium salts. Angew Chem Int Ed 58(45):16161–16166
38. Engl PS, Häring AP, Berger F, Berger G, Pérez-Bitrián A, Ritter T (2019) C-N cross-couplings for site-selective late-stage diversification via aryl sulfonium salts. J Am Chem Soc 141(34): 13346–13351
39. Li J, Chen J, Sang R, Ham WS, Plutschack MB, Berger F, Chabbra S, Schnegg A, Genicot C, Ritter T (2020) Photoredox catalysis with aryl sulfonium salts enables site-selective late-stage fluorination. Nat Chem 12(1):56–62
40. Chen C, Wang Z, Lu H, Zhao Y, Shi Z (2021) Generation of non-stabilized alkyl radicals from thianthrenium salts for C-B and C-C bond formation. Nat Commun 12(1):4526–4534
41. Cabrera-Afonso J, Granados A, Molander GA (2022) Sustainable thioetherification via electron donor-acceptor photoactivation using thianthrenium salts. Angew Chem Int Ed 61(22): e202202706
42. Liang L, Niu H, Li R, Wang Y, Yan J, Li C, Guo H (2020) Photoinduced copper-catalyzed site-selective C(sp^2)-C(sp) cross-coupling via aryl sulfonium salts. Org Lett 22(17):6842–6846

43. Zhao Y, Yu C, Liang W, Patureau FW (2021) Photochemical (hetero-)arylation of aryl sulfonium salts. Org Lett 23(16):6232–6236
44. Sun K, Shi A, Liu Y, Chen X, Xiang P, Wang X, Qu L, Yu B (2022) A general electron donor-acceptor complex for photoactivation of arenes via thianthrenation. Chem Sci 13(19): 5659–5666
45. Lin Z, Yao Y, Zhang C (2022) Deuteration of arylthianthren-5-ium salts in CD_3OD. Org Lett 24(45):8417–8422

Top Heterocycl Chem (2025) 62: 243–258
https://doi.org/10.1007/7081_2025_84

Published online: 25 April 2025

Visible Light-Driven Construction of Axially Chiral Heterocyclic Compounds

Dong Liang, Zi-Wei He, Jia-Rong Chen, and Wen-Jing Xiao

Contents

Abstract The use of visible light-driven photocatalysis has risen to prominence, offering an efficient and cost-effective synthetic strategy for accessing a broad spectrum of radical species and intermediates. This method also allows for efficient control over the stereoselectivity of reactions, when merging with asymmetric catalysis under mild conditions. On the other hand, the catalytic asymmetric construction of axially chiral heterocyclic compounds has emerged as a burgeoning field of study. This is attributable to their ubiquitous occurrence in natural products, bioactive molecules, advanced materials, and their role as chiral ligands and catalysts. With the advancement of photochemistry, the application of visible light-

D. Liang
School of Chemistry and Materials Engineering, Fuyang Normal University, Fuyang, China

Z.-W. He
Engineering Research Center of Photoenergy Utilization for Pollution Control and Carbon Reduction, Ministry of Education, College of Chemistry, Central China Normal University, Wuhan, China

J.-R. Chen (✉) and W.-J. Xiao (✉)
Engineering Research Center of Photoenergy Utilization for Pollution Control and Carbon Reduction, Ministry of Education, College of Chemistry, Central China Normal University, Wuhan, China

Wuhan Institute of Photochemistry and Technology, Wuhan, China
e-mail: chenjiarong@mail.ccnu.edu.cn; wxiao@ccnu.edu.cn

driven reactions for the synthesis of axially chiral heterocyclic compounds has recently garnered considerable interest. In this chapter, we provide a comprehensive summary and highlight the latest advancements in this rapidly developing area, focusing particularly on the elucidation of reaction mechanisms and the exploration of substrate versatility.

Keywords Atropisomers · Axially chiral · Chiral allenes · Heterocyclic compounds · Visible-light photoredox catalysis

1 Introduction

Chirality is an inherent characteristic of the natural world [1]. While central chirality, encompassing carbon and heteroatom chirality, has traditionally dominated the discourse, alternative chirality types – such as axial, planar, and spiral chirality – have been garnering growing interest in recent years (Fig. 1a) [2, 3]. Among them, axial chirality of molecules refers to the stereoisomerism created by the nonplanar arrangement of four ortho groups around a chiral axis [4, 5]. Axially chiral compounds, particularly those of axially chiral heterocyclic variety, have become increasingly recognized and appreciated in many scientific disciplines. This is largely due to their widespread existence in biologically active molecules, natural products, and materials that exhibit optical purity. They also exhibit a broad spectrum of applications as chiral ligands and organocatalysts in asymmetric catalysis (Fig. 1b) [6–11]. Over the past decade, the rapid evolution of asymmetric catalysis

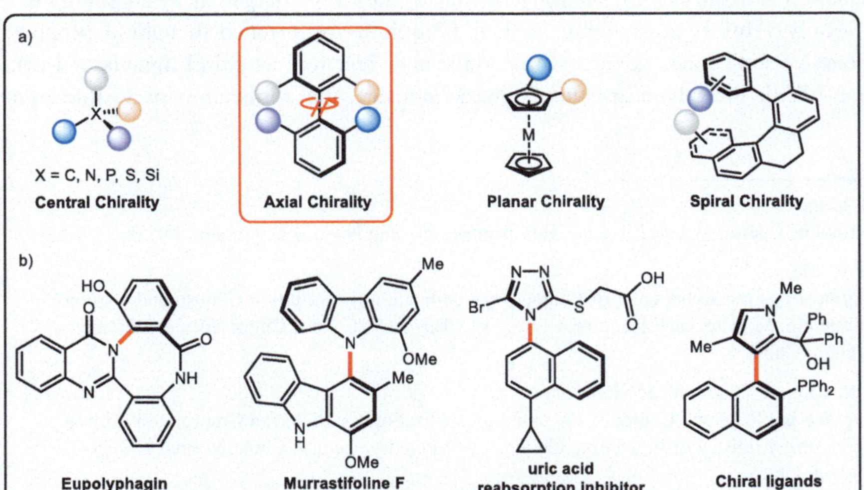

Fig. 1 (**a**) Common chiral elements. (**b**) Representative axially chiral heterocyclic natural products, catalysts, and ligands

has led to the emergence of various methods for the enantioselective synthesis of axially chiral heterocyclic compounds. These include organocatalyzed, transition metal-catalyzed, and Lewis acid-catalyzed strategies [12–22]. Despite these remarkable strides, there is ongoing interest in the pursuit of innovative, economical, efficient, sustainable, and mechanistically unique catalytic systems to expand the repertoire of axially chiral heterocyclic compounds.

Given the unique reactivity modes and structural diversity of radicals, the radical-involved construction strategy of axially chiral heterocyclic compounds would provide an intriguing and robust platform [23]. However, compared to the conventional ionic approaches, the construction of axially chiral heterocyclic compounds via radical pathways remains substantially underdeveloped. This is largely attributed to the challenges of generating highly reactive radical intermediates with precise control and attaining stereoselectivity within the radical processes [24–26]. In recent times, the visible light-driven photoredox-catalyzed reactions have been established as a highly effective tool for the controlled production of various radical species and intermediates under mild conditions [27–31]. Particularly, the field of catalytic asymmetric radical processes has seen significant progress due to the integration of photocatalysis with asymmetric organocatalysis, metal catalysis, and enzyme catalysis [32–35]. These strategic advancements have therefore prompted the emergence of various radical-involved approaches for the synthesis of axially chiral heterocyclic compounds, which are often difficult to achieve using conventional ionic methods. These significant advances have paved the way for new opportunities in harnessing enantioselective radical reactions, thereby broadening the horizons for an innovative approach to the asymmetric synthesis of axially chiral heterocyclic compounds.

Accordingly, this chapter concentrates on reviewing the latest and significant advancements in the realm of visible light-driven synthesis of axially chiral heterocyclic compounds. To facilitate a clearer understanding for readers, the transformations are categorized primarily based on the various types of syntheses for axially chiral heterocyclic compounds. Wherever feasible, particular attention is also given to the discussion of reaction mechanisms and the breadth of substrate applicability.

2 Visible Light-Driven Construction of Heterobiaryl Atropisomers

Atropisomers, which result from the restricted rotation of an axially constrained σ-bond, represent the most prevalent subclass within the realm of axial chirality [36]. Ever since their initial discovery by Christie and Kenner in 1922, atropisomers have emerged as the most prominent chiral skeletons in the field of chiral chemistry [37]. Based on the distinct substituents attached at the termini of the shaft, the majority of known heterocyclic atropisomers can be classified into two main groups: heterobiaryl C–C atropisomers and heterobiaryl C–N atropisomers. Accordingly, in this section, we delve into all the methodologies for the visible light-driven construction of these two categories of heterocyclic atropisomers.

2.1 *Visible Light-Driven Construction of Heterobiaryl C–C Atropisomers*

In 2022, the Xiao and Lu group developed a visible light-driven metallaphotoredox-catalyzed dynamic kinetic asymmetric cross-coupling of racemic heterobiaryl triflates **1** with 1,4-dihydropyridine (DHP) derivatives **2** by the merger of organic photoredox catalysis and asymmetric cobalt catalysis (Scheme 1a) [38]. The crucial factor for the success of this reaction lies in the formation of an axially chiral aryl cobalt species through oxidation addition. This species reduces the rotational energy barrier due to the coordination of pyridine, thereby enabling racemization and the subsequent selective capture of the radicals generated by photocatalysis. Under the optimal reaction conditions, a wide variety of aryl and heteroaryl-substituted heterobiaryl triflates **1** and DHP reagents **2** bearing electron-donating or electron-withdrawing groups were well tolerated to give the corresponding heterobiaryl atropisomers **3** in generally high yields (up to 99% yield) with excellent enantioselectivity (up to 97% ee). In addition, for the asymmetric reduction coupling catalytic system, the reaction also showed good compatibility with benzyl halides **4**, and a series of biaryl axial chiral products **3** containing isoquinoline scaffolds were obtained with high enantioselectivity (up to 95% ee) (Scheme 1b). Furthermore, these axially chiral heterobiaryl products **3** exhibit a significant aggregation-induced emission (AIE) effect.

Employing the same activation strategy, in 2023, Xiao and co-workers developed a fascinating photoinduced cobalt-catalyzed dynamic kinetic reductive conjugate addition of racemic heterobiaryl triflates **1** with conjugated alkenes **5**. This innovative approach facilitated the synthesis of heterobiaryls possessing axial chirality (Scheme 2) [39]. This protocol features mild reaction conditions and good functional group tolerance, giving a series of axially chiral heterobiaryl compounds **6** in high yields with excellent enantioselectivity (up to 91% yield and 97% ee). Significantly, these reaction conditions may substantially reduce the use of valuable metal catalysts and consequently minimize the generation of metal waste. Detailed mechanistic investigations suggested that the reaction initiates with single-electron transfer (SET) oxidation of Hantzsch ester by the excited state photocatalyst 4CzIPN* under visible-light irradiation. This process generates the reduced $4CzIPN^{\cdot -}$ photocatalyst, which subsequently reduces the Co(II)/L complex to the low-valent Co(I)/L species through the SET process. Subsequently, the racemic heterobiaryl triflate **1** was oxidatively added to Co(I)/**L** species to generate Co(III)/L intermediate **1-A**, which could be further reduced by the reduced photocatalyst to Co(II)/**L** species. As mentioned in Scheme 1, the coordination with nitrogen reduced the rotational energy barrier of the axially chiral Co(II)/**L** intermediate, leading to the

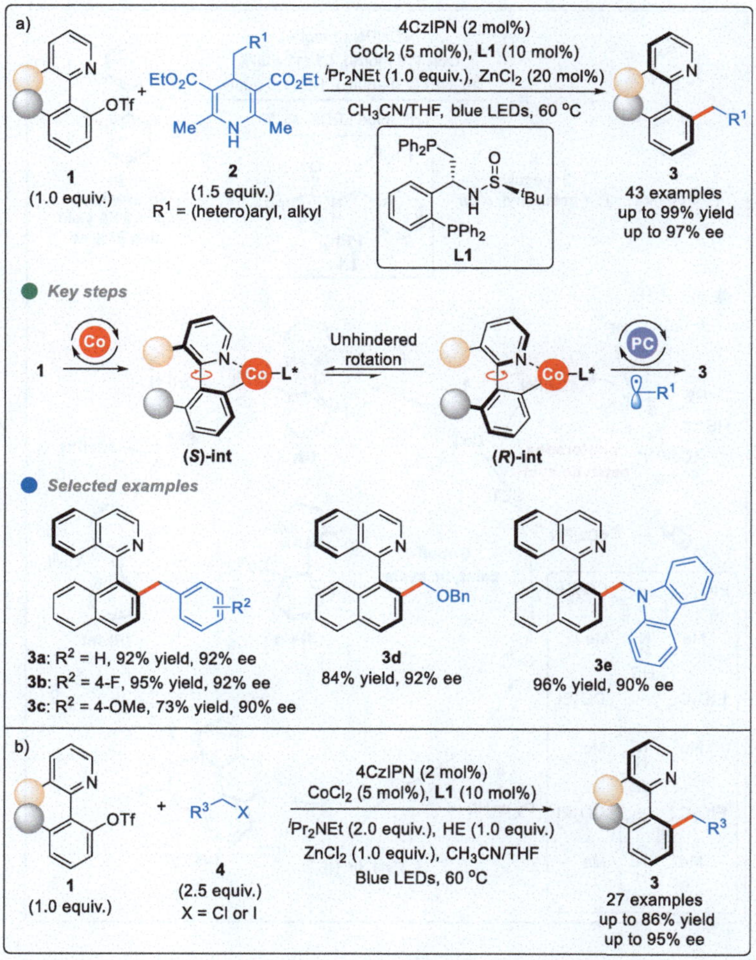

Scheme 1 Cobalt/photoredox-catalyzed asymmetric radical coupling reaction for the construction of heterobiaryl atropisomers (Xiao/Lu method). (**a**) Asymmetric cross-coupling of racemic heterobiaryl triflates with 1,4-dihydropyridine (DHP) derivatives. (**b**) Asymmetric reduction coupling reaction with benzyl halides

transformation of the **(*S*)-int** into the dominant **(*R*)-int**. After migration of alkene **5** into chiral cobalt species, Co(II)/**L** intermediate **1-B** was generated and could be protonated, yielding the axially chiral product **6**. Concurrently, this process facilitated the regeneration of the Co(II)/**L** catalyst. Notably, further kinetic studies have unveiled that the migratory insertion of the alkene **5** into the chiral Co(II)/**L** species constitutes the rate-determining step of the reaction.

Originally described in 1971, the Minisci reaction is the radical substitution process involving electron-deficient nitrogen-containing aromatic heterocycles and nucleophilic carbon radicals [40]. Given its ability to directly incorporate a range of

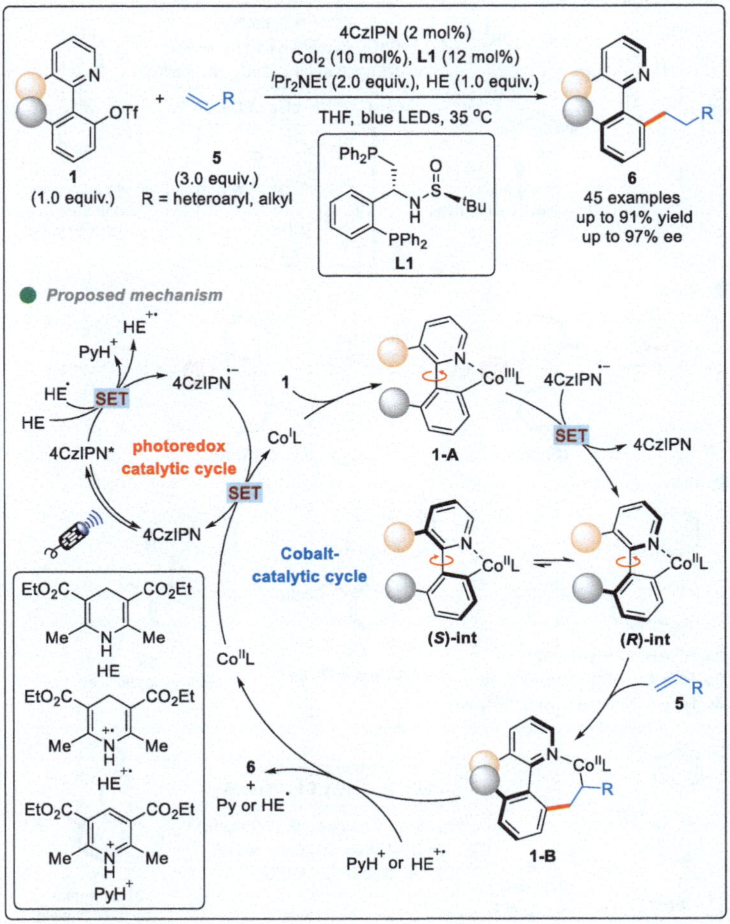

Scheme 2 Cobalt/photoredox-catalyzed dynamic kinetic reductive conjugate addition for construction of heterobiaryl atropisomers (Xiao method)

substituents, including alkyl, aryl, acyl, trifluoromethyl, and hydroxymethyl groups into electron-deficient heterocycles, the Minisci reaction serves as a potent tool for the synthesis of diverse aza-heteroaromatic compounds and the late-stage functionalization of complex aza-heterocyclic systems [41, 42]. In 2022, the Xiao group reported a pioneering visible light-driven Minisci reaction of 5-aryl pyrimidines **7** and α-amino acid-derived redox-active esters **8** under mild conditions (Scheme 3) [43]. This approach offered a new method for concurrent construction of centrally and axially chiral heterobiaryls via a photocatalytic radical pathway. It delivered the desired products **9**, which comprise a centrally chiral α-branched amine and an axially chiral heterobiaryl moiety, with outstanding diastereo- and enantioselectivity (up to >99% ee and >19:1 d.r.). It is particularly noteworthy that the TBS group of the heterobiaryl product could be readily removed, yielding

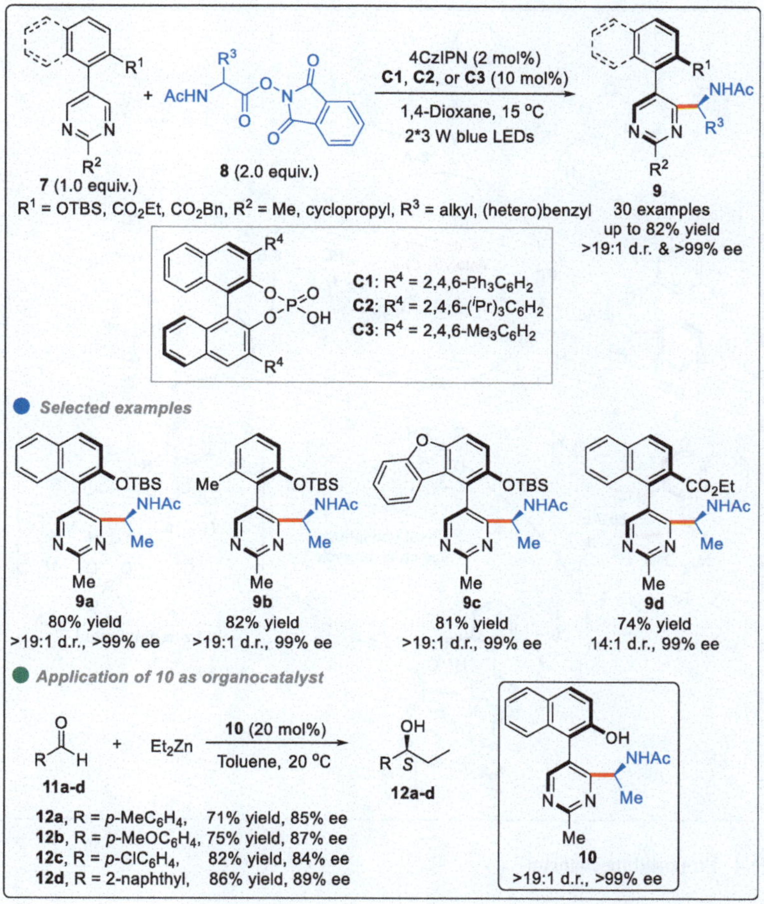

Scheme 3 Visible light-promoted catalytic asymmetric construction of axially and centrally chiral heterobiaryls by Minisci reaction (Xiao method)

hydroxy-free compound **10**. This compound can then serve as a novel chiral protonic ligand, effectively promoting the catalytic asymmetric addition of diethylzinc to aldehydes. This process results in chiral secondary alcohols **12** with good yields and exceptional enantioselectivity.

Drawing from comprehensive mechanistic experiments, a plausible mechanism involving visible light-driven radical process for this reaction was proposed. As shown in Scheme 4, the reaction began with off-cycle single-electron transfer oxidation of the mixture of **7** and chiral phosphoric acid by the excited state photocatalyst, resulting in a reduced-state photocatalyst that initiates the primary cycle. Subsequently, the α-amino acid-derived redox-active ester **8** could be reduced by the reduced-state photocatalyst to form α-aminoalkyl radical **8-B** and the ground-state photocatalyst. Subsequently, the radical intermediate **8-B** engaged in a radical addition to the pyrimidine **7**, facilitated by a chiral phosphoric acid acting as the

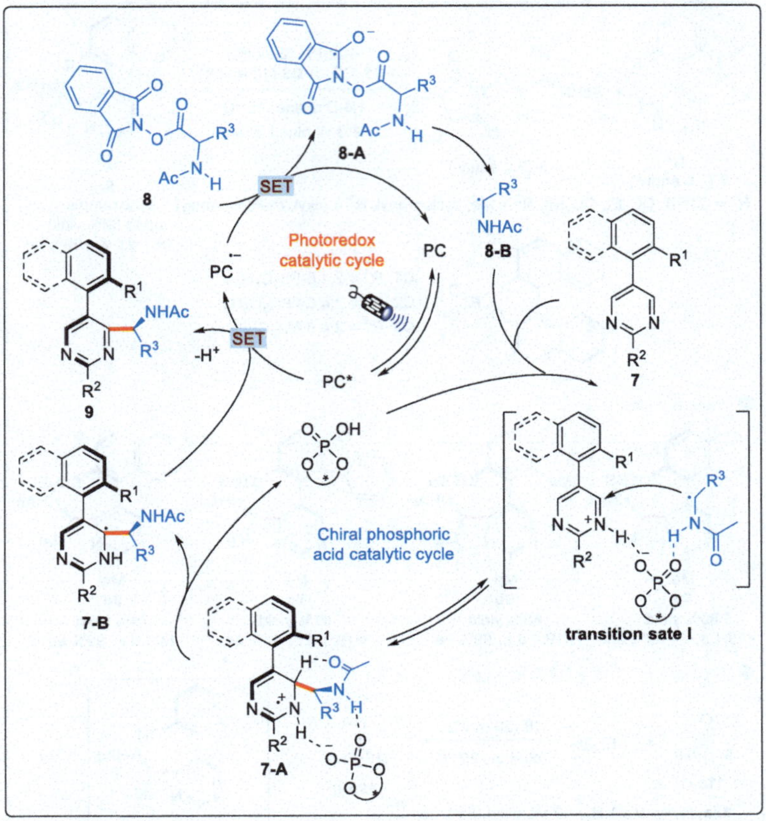

Scheme 4 Proposed mechanism

chiral organocatalyst. This led to the formation of the radical cation **7-A** via transition state **I**. The radical cation **7-A** was then deprotonated by the internal amide carbonyl group, leading to the generation of the radical **7-B**. This radical was further oxidized by the excited-state photocatalyst and underwent rearomatization, culminating in the production of the product **9**. Meanwhile, the reduced-state photocatalyst was regenerated, closing the photoredox catalytic cycle.

2.2 *Visible Light-Driven Construction of Heterobiaryl C–N Atropisomers*

Compared to the C–C axially chiral heterocyclic structures, the development of catalytic asymmetric methods for constructing C–N axially chiral heterocyclic compounds has been significantly lagging until recent advancements [44–47]. Specifically, C–N atropisomeric indole derivatives, which are a distinctive type of

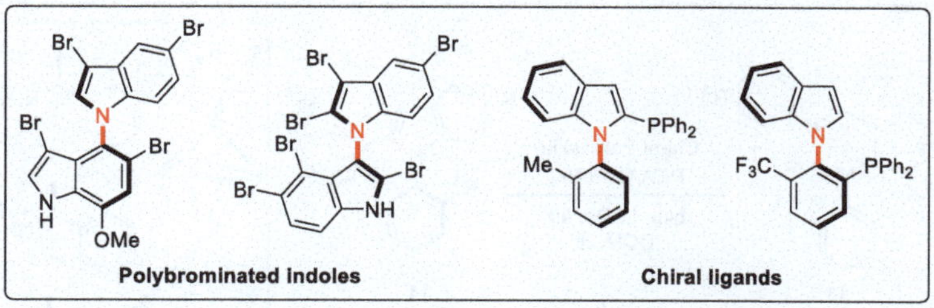

Scheme 5 Representative examples of C–N axially chiral indole derivatives

5-membered heterocyclic ring, are frequently encountered in biologically active natural products and chiral ligands (Scheme 5) [48–51]. However, the high rotational freedom of the C–N axis and the reduced size of the five-membered ring, which leads to a lower rotational barrier, have made the catalytic stereoselective synthesis of such C–N axially chiral indole frameworks a significant challenge. To date, only a limited number of examples have been reported [52, 53].

In 2023, the Yan group demonstrated that a heterogeneous metal-halide perovskite nanocrystal catalyst (NC) could effectively catalyze the asymmetric synthesis of indole-based C–N axially chiral compounds **14**. This was achieved through the intramolecular radical cyclization of aryl-2-vinylanilines **13** under visible-light conditions [54]. In this approach, the authors used chiral 1-phenylethylamine (PEA)-hybridized perovskite PEA/$CsPbBr_3$ NC as the photocatalyst in conjunction with chiral organocatalysts. The reaction was carried out under the irradiation of a 456 nm Kessil blue LED and in an air atmosphere, yielding the desired axially chiral *N*-arylindoles **14** (Scheme 6). A diverse range of aryl-2-vinylanilines **13**, containing various functional groups such as MeO, NO_2, F, CF_3, and CN, were well tolerated under these conditions, affording excellent yields and enantioselectivity, with results reaching up to 72% yield and 99% ee. Mechanistic investigations have disclosed that the 2-vinylaniline substrates stereoselectively bind to the chiral NC surface, achieving enantiodifferentiation of approximately −2.4 kcal/mol between the two enantiomers. It is noteworthy that several conventional photocatalysts, including Ru(phen)$_3$ and Ir(ppy)$_3$, could initiate this visible-induced radical cyclization process, leading to the formation of the desired target compounds. However, even in the presence of free chiral PEA, the reaction fails to achieve enantioselectivity. These findings underscore the specificity and high efficiency of this heterogeneous catalyst in the atroposelective synthesis of pharmaceutical scaffolds, as well as the potential for asymmetric synthesis using chiral solid catalysts directly. Moreover, the chiral PEA-hybridized perovskite PEA/$CsPbBr_3$ NC, serving as a heterogeneous catalyst, could be readily recovered and reused up to four times, yielding favorable results.

Quinolines and indoles are pivotal building blocks in the realm of physiologically active compounds and materials [55, 56]. In 2023, the Akiyama group disclosed a visible light-driven Minisci reaction strategy, achieving the enantioselective synthesis of axially chiral 2-alkyl-3-(*N*-indolyl)quinolines **17** (Scheme 7) [57]. This approach allows for access to target molecules possessing both axial and central

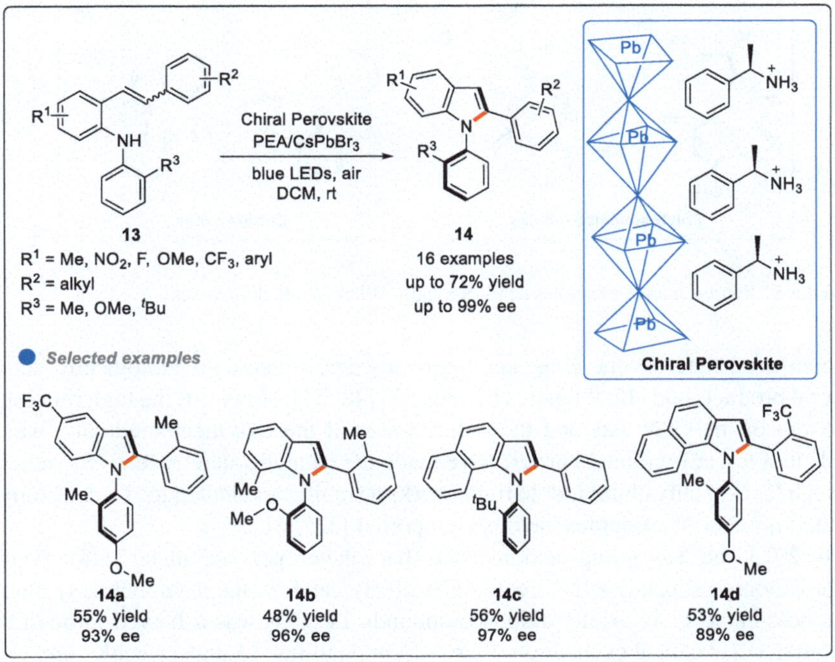

Scheme 6 Visible light-driven chiral perovskite PEA/$CsPbBr_3$-catalyzed asymmetric synthesis of indole-based C–N axially chiral compounds (Yan method)

chirality through the synergistic application of chiral lithium phosphate **C4**-Li and iridium photocatalyst. The reaction proceeded smoothly under mild conditions, affording the desired addition products **17**, which feature both C–N axial chirality and central carbon chirality. These products were obtained in high yields and with high to excellent optical purity (up to 92% yield, >19:1 d.r., and 99% ee) by judiciously selecting an appropriate chiral lithium phosphate catalyst and a photocatalyst.

3 Visible Light-Driven Construction of Chiral Allenes

Allenes are classified as 1,2-propadiene derivatives, distinguished by the distinctive feature of having two consecutive carbon-carbon double bonds. The perpendicular orientation of the neighboring π systems within the allene framework enables the potential for axial chirality upon appropriate substitution (Scheme 8a) [58]. As illustrated in Scheme 8b, the allene-based scaffolds are commonly found in biologically active natural compounds such as laballenic acid and citroside A [59–62]. Furthermore, due to the unique reactivity of the orthogonally aligned π systems, allenes have emerged as highly versatile building blocks, capable of engaging in an array of

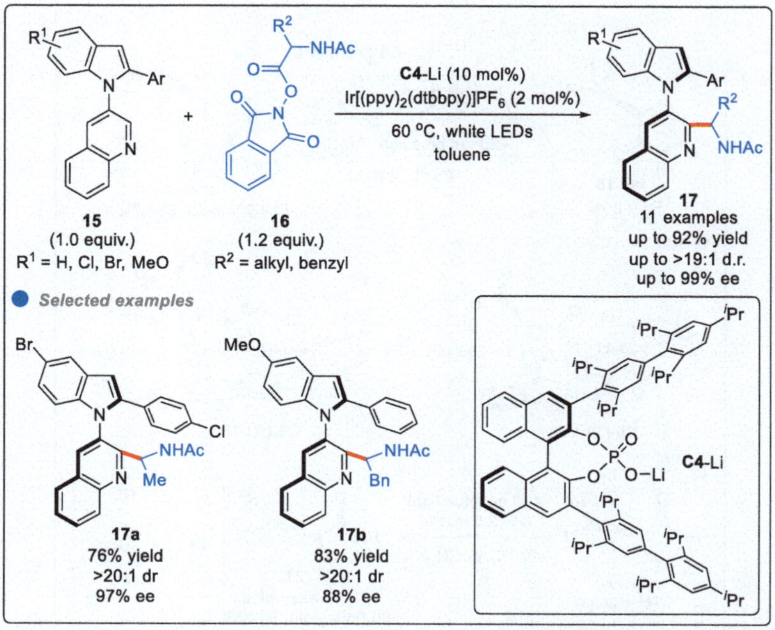

Scheme 7 Visible light-induced synthesis of axially chiral 3-(*N*-indolyl)quinolines (Akiyama method)

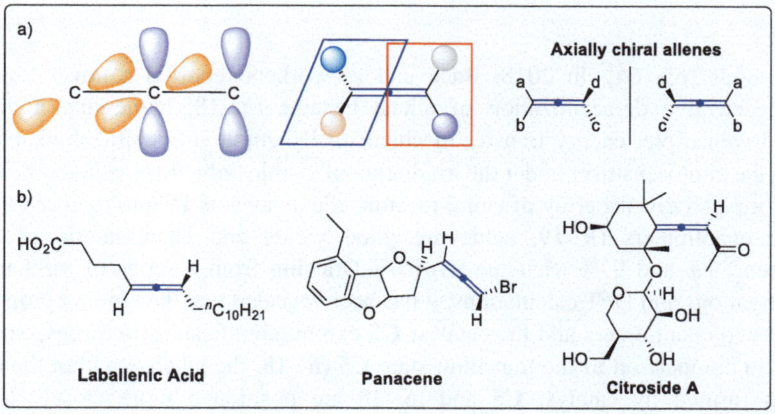

Scheme 8 (**a**) Axially chiral allenes. (**b**) Selected examples of bioactive molecules containing axially chiral allene scaffold

potent chemical transformations [63, 64]. Consequently, the pursuit of synthesizing axially chiral allenes through visible light-induced photocatalysis is gaining significant momentum in contemporary research.

Deracemization is a selective process that transforms a racemic mixture into a single enantiomer, offering a direct and economical approach to producing chiral

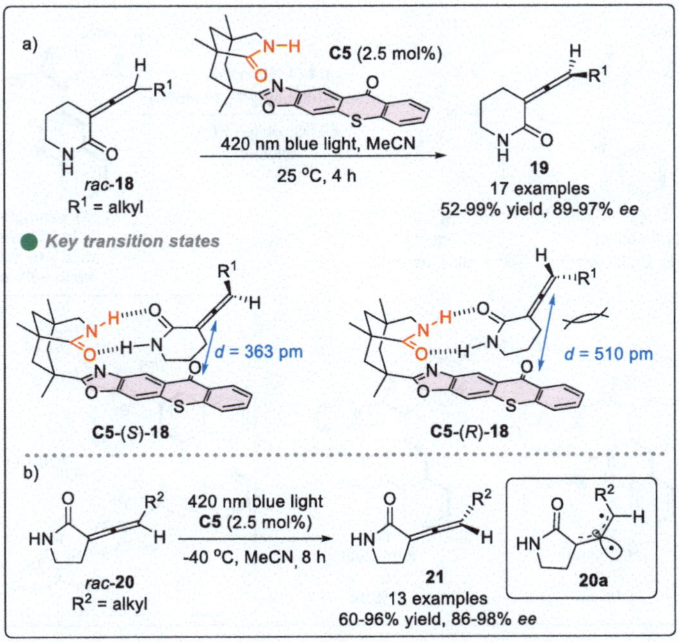

Scheme 9 Visible light-driven deracemization of axially chiral allenes by triplet energy transfer (Bach method). (**a**) Photochemical deracemization reaction and its key transition states. (**b**) Photochemical deracemization of five-membered ring allene lactams and allene amides

compounds [65, 66]. In 2018, Bach and co-workers reported a novel catalytic enantioselective deracemization of allene lactams *rac*-**18**, harnessing a visible light-driven triplet energy transfer mechanism. By employing chiral thioxanthone **C5** as the photosensitizer under the irradiation of visible light, this method efficiently transformed a diverse array of chiral racemic allene lactams **18** into their respective single enantiomers (*R*)-**19**, achieving good yields and enantiomeric excesses between 89% and 97% (Scheme 9a) [67]. Drawing from a series of mechanistic investigations and DFT calculations, it has been revealed that the binding properties of the two enantiomers and the catalyst **C5** exhibit significant differences. Specifically, in comparison to the transition state **C5**-(*R*)-**18**, the alkenes within the complexes formed by catalyst **C5** and (*S*)-**18** are positioned more closely to the thioxanthone carbonyl chromophore, leading to a more rapid triplet energy transfer. Therefore, the transition state **C5**-(*S*)-**18** can proceed more rapidly and transform into another enantiomer via a visible light-driven triplet energy transfer mechanism. Subsequently, the Bach group expanded this methodology to encompass five-membered ring allene lactams and allene amides (Scheme 9b) [68]. Additionally, Yu's group has recently developed a novel photoexcited palladium-catalyzed deracemization reaction for disubstituted allenes [69]. Utilizing an appropriate chiral palladium complex as a catalyst, a variety of axially chiral allenes can be synthesized, typically achieving high yields and excellent enantioselectivities under visible-light irradiation.

4 Conclusion

With the evolvement of visible light-driven photocatalysis, the photochemical asymmetric construction of axially chiral heterocyclic compounds has witnessed remarkable advances over the past few years. In this section, we highlight the recent and representative achievements according to different types of axially chiral heterocycles and catalytic activation modes. Notwithstanding these significant advances, however, this field is still in its infancy. Therefore, novel powerful chiral ligands or catalysts should be designed, and new asymmetric catalytic modes, especially asymmetric dual-catalytic strategies, should be further explored to expand the reaction profiles and application to the asymmetric synthesis of complex axially chiral heterocyclic compounds. We hope that this chapter will be of great interest to the experts in this area as well as stimulates continued innovation in the development of new methods for the construction of axial chiral heterocyclic compounds through visible light-driven reactions.

Acknowledgements We are grateful to the financial support from the Scientific Research Start Fund of Fuyang Normal University (No. 2023KYQD0035) and the College Students' Innovative Entrepreneurial Training Plan Program of Anhui Province (No. S202410371081).

References

1. Liu M, Zhang L, Wang T (2015) Supramolecular chirality in self-assembled systems. Chem Rev 115(15):7304–7397
2. Eliel EL, Wilen SH (1994) Stereochemistry of organic compounds. John Wiley & Sons
3. Wang Z, Meng L, Liu X, Zhang L, Yu Z, Wu G (2022) Recent progress toward developing axial chirality bioactive compounds. Eur J Med Chem 243:114700–114717
4. Adams R, Yuan HC (1933) The stereochemistry of diphenyls and analogous compounds. Chem Rev 12(2):261–338
5. Smyth JE, Butler NM, Keller PA (2015) A twist of nature–the significance of atropisomers in biological systems. Nat Prod Rep 32(11):1562–1583
6. Fu W, Tang W (2016) Chiral monophosphorus ligands for asymmetric catalytic reactions. ACS Catal 6(8):4814–4858
7. Bringmann G, Gulder T, Gulder TAM, Breuning M (2011) Atroposelective total synthesis of axially chiral biaryl natural products. Chem Rev 111(2):563–639
8. Glunz PW (2018) Recent encounters with atropisomerism in drug discovery. Bioorg Med Chem Lett 28(2):53–60
9. LaPlante SR, Fader LD, Fandrick KR, Fandrick DR, Hucke O, Kemper R, Miller SPF, Edwards PJ (2011) Assessing atropisomer axial chirality in drug discovery and development. J Med Chem 54(20):7005–7022

10. LaPlante SR, Edwards PJ, Fader LD, Jakalian A, Hucke O (2011) Revealing atropisomer axial chirality in drug discovery. ChemMedChem 6(3):505–513
11. Clayden J, Moran MJ, Edwards PJ, LaPlante SR (2009) The challenge of atropisomerism in drug discovery. Angew Chem Int Ed 48(35):6398–6401
12. Wang YB, Tan B (2018) Construction of axially chiral compounds via asymmetric organocatalysis. Acc Chem Res 51(2):534–547
13. Wencel-Delord J, Panossian A, Leroux FR, Colobert F (2015) Recent advances and new concepts for the synthesis of axially stereoenriched biaryls. Chem Soc Rev 44(11):3418–3430
14. Liao G, Zhou T, Yao QJ, Shi BF (2019) Recent advances in the synthesis of axially chiral biaryls via transition metal-catalysed asymmetric C-H functionalization. Chem Commun 55(59):8514–8523
15. Carmona JA, Franco CR, Fernández R, Hornillos V, Lassaletta JM (2021) Atroposelective transformation of axially chiral (hetero)biaryls. From desymmetrization to modern resolution strategies. Chem Soc Rev 50(5):2968–2983
16. Chu W-D, Zhang Y, Wang J (2017) Recent advances in catalytic asymmetric synthesis of allenes. Cat Sci Technol 7(20):4570–4579
17. Rodríguez-Salamanca P, Fernández R, Hornillos V, Lassaletta JM (2022) Chem Eur J 28(28): e202104442
18. Min XL, Zhang XL, Shen R, Zhang Q, He Y (2022) Recent advances in the catalytic asymmetric construction of atropisomers by central-to-axial chirality transfer. Org Chem Front 9(8):2280–2292
19. Wang X, Chen X, Lin W, Li P, Li W (2022) Recent advances in organocatalytic enantioselective synthesis of axially chiral allenes. Adv Synth Catal 364(7):1212–1222
20. He X-L, Wang C, Wen Y-W, Wang Z, Qian S (2021) Recent advances in catalytic atroposelective construction of pentatomic heterobiaryl scaffolds. ChemCatChem 13(16): 3547–3564
21. Mei G-J, Koay WL, Guan C-Y, Lu Y (2022) Atropisomers beyond the C-C axial chirality: advances in catalytic asymmetric synthesis. Chem 8(7):1855–1893
22. Zhang H-H, Li T-Z, Liu S-J, Shi F (2024) Catalytic asymmetric synthesis of atropisomers bearing multiple chiral elements: an emerging field. Angew Chem Int Ed 63(3):e202311053
23. Yan M, Lo JC, Edwards JT, Baran PS (2016) Radicals: reactive intermediates with translational potential. J Am Chem Soc 138(39):12692–12714
24. Sibi MP, Manyem S, Zimmerman J (2003) Enantioselective radical processes. Chem Rev 103(8):3263–3295
25. Xiong T, Zhang Q (2021) Recent advances in the direct construction of enantioenriched stereocenters through addition of radicals to internal alkenes. Chem Soc Rev 50(16):8857–8873
26. Mondal S, Dumur F, Gigmes D, Sibi MP, Bertrand MP, Nechab M (2022) Enantioselective radical reactions using chiral catalysts. Chem Rev 122(6):5842–5976
27. Prier CK, Rankic DA, MacMillan DWC (2013) Visible light photoredox catalysis with transition metal complexes: applications in organic synthesis. Chem Rev 113(7):5322–5363
28. Romero NA, Nicewicz DA (2016) Organic photoredox catalysis. Chem Rev 116(17): 10075–10166
29. Marzo L, Pagire SK, Reiser O, König B (2018) Visible-light photocatalysis: does it make a difference in organic synthesis? Angew Chem Int Ed 57(32):10034–10072
30. Zhou Q-Q, Zou Y-Q, Lu L-Q, Xiao W-J (2019) Visible-light-induced organic photochemical reactions via energy transfer pathways. Angew Chem Int Ed 58(6):1586–1604
31. Festa AA, Voskressensky LG, Van der Eycken EV (2019) Visible light-mediated chemistry of indoles and related heterocycles. Chem Soc Rev 48(16):4401–4423
32. Brimioulle R, Lenhart D, Maturi MM, Bach T (2015) Enantioselective catalysis of photochemical reactions. Angew Chem Int Ed 54(13):3872–3890
33. Wang C, Lu Z (2015) Catalytic enantioselective organic transformations via visible light photocatalysis. Org Chem Front 2(2):179–190

34. Jiang C, Chen W, Zheng W-H, Lu H (2019) Advances in asymmetric visible-light photocatalysis, 2015-2019. Org Biomol Chem 17(38):8673–8689
35. Saha D (2020) Catalytic enantioselective radical transformations enabled by visible-light. Chem Asian J 15(14):2129–2152
36. Cheng JK, Xiang SH, Li SY, Liu Y, Tan B (2021) Recent advances in catalytic asymmetric construction of atropisomers. Chem Rev 121(8):4805–4902
37. Christie GH, Kenner J (1922) LXXI – The molecular configurations of polynuclear aromatic compounds. Part I. The resolution of γ-6:6′-dinitro- and 4:6:4′:6′-tetranitrodiphenic acids into optically active components. J Chem Soc Trans 121:614–620
38. Jiang X, Xiong W, Deng S, Lu F-D, Jia Y, Yang Q, Xue L-Y, Qi X, Tunge JA, Lu L-Q, Xiao W-J (2022) Construction of axial chirality via asymmetric radical trapping by cobalt under visible light. Nat Catal 5:788–797
39. Xiong W, Jiang X, Wang W-C, Cheng Y, Lu L-Q, Gao K, Xiao W-J (2023) Dynamic kinetic reductive conjugate addition for construction of axial chirality enabled by synergistic photoredox/cobalt catalysis. J Am Chem Soc 145(14):7983–7991
40. Minisci F, Bernardi R, Bertini F, Galli R, Perchinummo M (1971) Nucleophilic character of alkyl radicals-VI: a new convenient selective alkylation of heteroaromatic bases. Tetrahedron 27(15):3575–3579
41. Duncton MAJ (2011) Minisci reactions: versatile CH-functionalizations for medicinal chemists. Med Chem Commun 2(12):1135–1161
42. Proctor RSJ, Phipps RJ (2019) Recent advances in Minisci-type reactions. Angew Chem Int Ed 58(39):13666–13699
43. Liang D, Chen JR, Tan LP, He ZW, Xiao WJ (2022) Catalytic asymmetric construction of axially and centrally chiral heterobiaryls by Minisci reaction. J Am Chem Soc 144(13): 6040–6049
44. Kitagawa O (2021) Chiral Pd-catalyzed enantioselective syntheses of various N-C axially chiral compounds and their synthetic applications. Acc Chem Res 54(3):719–730
45. Frey J, Malekafzali A, Delso I, Choppin S, Colobert F, Wencel-Delord J (2020) Angew Chem Int Ed 59(23):8844–8848
46. Li S-L, Yang C, Wu Q, Zheng H-L, Li X, Cheng J-P (2018) J Am Chem Soc 140(40): 12836–12843
47. An Q-J, Xia W, Ding W-Y, Liu H-H, Xiang S-H, Wang Y-B, Zhong G, Tan B (2021) Nitrosobenzene-enabled chiral phosphoric acid catalyzed enantioselective construction of atropisomeric N-arylbenzimidazoles. Angew Chem Int Ed 60(47):24888–24893
48. Wang Z-S, Zhu L-J, Li C-T, Liu B-Y, Hong X, Ye L-W (2022) Synthesis of axially chiral N-arylindoles via atroposelective cyclization of ynamides catalyzed by chiral Brønsted acids. Angew Chem Int Ed 61(20):e202201436
49. Xia W, An Q, Xiang S, Li S, Wang Y, Tan B (2020) Chiral phosphoric acid catalyzed atroposelective C-H amination of arenes. Angew Chem Int Ed 59(17):6775–6779
50. Wang F, Jing J, Zhao Y, Zhu X, Zhang X-P, Zhao L, Hu P, Deng W-Q, Li X (2021) Rhodium-catalyzed C-H activation-based construction of axially and centrally chiral indenes through two discrete insertions. Angew Chem Int Ed 60(30):16628–16633
51. Sun L, Chen H, Liu B, Chang J, Kong L, Wang F, Lan Y, Li X (2021) Rhodium-catalyzed atroposelective construction of indoles via C-H bond activation. Angew Chem Int Ed 60(15): 8391–8395
52. Zhou L, Li Y, Li S, Shi Z, Zhang X, Tung C-H, Xu Z (2023) Asymmetric rhodium-catalyzed click cycloaddition to access C–N axially chiral N-triazolyl indoles. Chem Sci 14(19): 5182–5187
53. Li T-Z, Liu S-J, Tan W, Shi F (2020) Catalytic asymmetric construction of axially chiral indole-based frameworks: an emerging area. Chem Eur J 26(68):15779–15792
54. Mishra K, Guyon D, Martin JS, Yan Y (2023) Chiral perovskite nanocrystals for asymmetric reactions: a highly enantioselective strategy for photocatalytic synthesis of N-C axially chiral heterocycles. J Am Chem Soc 145(31):17242–17252

55. Burikhanov R, Sviripa VM, Hebbar N, Zhang W, Layton WJ, Hamza A, Zhan CG, Watt DS, Liu C, Rangnekar VM (2014) Arylquins target vimentin to trigger Par-4 secretion for tumor cell apoptosis. Nat Chem Biol 10:924–926
56. Sviripa VM, Burikhanov R, Obiero JM, Yuan Y, Nickell JR, Dwoskin LP, Zhan CG, Liu C, Tsodikov OV, Rangnekar VM, Watt DS (2016) Par-4 secretion: stoichiometry of 3-arylquinoline binding to vimentin. Org Biomol Chem 14(1):74–84
57. Yamanomoto K, Yamamoto K, Yoshida S, Sato S, Akiyama T (2024) Enantioselective synthesis of 3-(N-indolyl)quinolines containing axial and central chiralities. Chem Commun 60(5):582–585
58. Maitland P, Mills WH (1935) Experimental demonstration of the allene asymmetry. Nature 135:994–995
59. Yu S, Ma S (2012) Allenes in catalytic asymmetric synthesis and natural product syntheses. Angew Chem Int Ed 51(13):3074–3112
60. Rivera-Fuentes P, Diederich F (2012) Allenes in molecular materials. Angew Chem Int Ed 51(12):2818–2828
61. Pu X, Qi X, Ready JM (2009) Allenes in asymmetric catalysis: asymmetric ring opening of meso-epoxides catalyzed by allene-containing phosphine oxides. J Am Chem Soc 131(30): 10364–10365
62. Cai F, Pu X, Qi X, Lynch V, Radha A, Ready JM (2011) Chiral allene-containing phosphines in asymmetric catalysis. J Am Chem Soc 133(45):18066–18069
63. Lu T, Lu Z, Ma Z-X, Zhang Y, Hsung RP (2013) Allenamides: a powerful and versatile building block in organic synthesis. Chem Rev 113(7):4862–4904
64. Alonso JM, Almendros P (2021) Deciphering the chameleonic chemistry of allenols: breaking the taboo of a onetime esoteric functionality. Chem Rev 121(7):4193–4252
65. Rachwalski M, Vermue N, Rutjes FPJT (2013) Recent advances in enzymatic and chemical deracemisation of racemic compounds. Chem Soc Rev 42(24):9268–9282
66. Wang P-Z, Xiao W-J, Chen J-R (2023) Light-empowered contra-thermodynamic stereochemical editing. Nat Rev Chem 7(1):35–50
67. Hölzl-Hobmeier A, Bauer A, Silva AV, Huber SM, Bannwarth C, Bach T (2018) Catalytic deracemization of chiral allenes by sensitized excitation with visible light. Nature 564:240–243
68. Plaza M, Jandl C, Bach T (2020) Photochemical deracemization of allenes and subsequent chirality transfer. Angew Chem Int Ed 59(31):12785–12788
69. Song C, Bai X, Li B, Dang Y, Yu S (2024) Photoexcited palladium-catalyzed deracemization of allenes. J Am Chem Soc 146(30):21137–21146

Top Heterocycl Chem (2025) 62: 259–282
https://doi.org/10.1007/7081_2025_85

Published online: 13 April 2025

Biocatalytic Synthesis and Functionalization of Heterocyclic Compounds Fueled by Light

Yu Fu, Guojiao Wu, and Fangrui Zhong

Contents

Abstract The biocatalytic transformation has garnered increasing attention in the realm of green chemical synthesis due to the wide range of enzymes available, their high catalytic activities and specificities, as well as the mild reaction conditions. The concept of utilizing light energy in chemical synthesis by combining photocatalysis and biocatalysis, known as photobiocatalysis, presents a promising opportunity to further enhance the sustainability of the process. Additionally, the integration of both approaches opens up possibilities to unlock new synthetic reactivities. This chapter aims to provide an overview of the state-of-the-art advances in photobiocatalysis toward synthesis and functionalization of heterocyclic compounds.

Y. Fu
College of Pharmacy, Shenzhen Technology University, Shenzhen, China

G. Wu and F. Zhong (✉)
Hubei Engineering Research Center for Biomaterials and Medical Protective Materials, Hubei Key Laboratory of Bioinorganic Chemistry and Materia Medica, School of Chemistry and Chemical Engineering, Huazhong University of Science and Technology (HUST), Wuhan, China
e-mail: chemzfr@hust.edu.cn

Keywords Artificial photoenzyme · Directed evolution · Enantioselective synthesis · Heterocyclic compounds · Photobiocatalysis

1 Introduction

The concept of sustainable chemistry is gaining traction, with a focus on developing eco-friendly asymmetric catalytic systems that are garnering increasing attention [1]. Enzymes, as biological macromolecule catalysts produced by living cells, exhibit high efficiency and selectivity in catalyzing a wide range of chemical reactions under mild conditions, making them ideal green catalysts [2]. Despite their century-long use in organic synthesis, the inherent specificity of natural enzymes, shaped by natural evolution and selection, often restricts their compatibility to natural substrates and biochemical reactions. Recent progress in synthetic biology, molecular biology, and enzyme engineering has significantly reduced the cost of enzyme production and enabled precise reshaping of protein structures at the molecular level [3, 4]. In this context, artificially engineered enzymes have been significantly developed and widely used in organic synthesis for unnatural reactions. For example, natural hemoglobin has been evolved to catalyze carbene and nitrogen transfer reactions, enabling the asymmetric alkylation [5] and amination [6] of C–H bonds.

Heterocyclic compounds represent one of the most prominent classes of organic compounds. They are ubiquitously found in the skeletal structures of drug molecules, natural products, and pesticides, contributing significantly to the biological profiles of these molecules. The efficient and environmentally friendly synthesis of heterocyclic compounds is of immense interest to the synthetic community, and numerous synthetic strategies relying on metal complexes or organic catalysts have been documented. Photocatalysis has emerged as a powerful tool in organic synthesis. The merging of biocatalysis with photocatalysis, known as photobiocatalysis, represents an effective combination of the advantages of the precise selectivity of enzymatic reactions and the rich photochemical reactivity, offering an attractive tool for chiral selectivity control in radical reactions. In this chapter, we will review the applications of photobiocatalysis in the synthesis and functionalization of heterocyclic compounds. This text is organized based on the patterns of coupling biocatalysis and photochemistry: (a) enzyme-photocatalyst-coupled photobiocatalysis, (b) photobiocatalysis driven by excitation of electron donor-acceptor (EDA) complex, (c) photobiocatalysis mediated by redox-active natural enzymes, and (d) photobiocatalysis mediated by artificial photoenzymes.

2 Enzyme-Photocatalyst-Coupled Photobiocatalysis for the Synthesis of Heterocyclic Compounds

The integration of exogenous photocatalysts with natural enzyme catalysis offers a straightforward approach to develop photobiocatalytic systems. These systems are characterized by the presence of two synergistic or sequential catalytic cycles. Typically, the highly reactive species generated from the photocatalytic cycle act as the driving force, facilitating the enzymatic catalytic cycle through processes such as electron transfer or transfer of radical intermediate species. Additionally, the exogenous photocatalyst cycle can also yield stable intermediate products that can serve as specific substrates for enzymatic catalysis in subsequent transformations. This eliminates the need for additional separation and purification steps, thereby enhancing the overall efficiency of the reaction. This section will delve into the synthesis of heterocyclic compounds mediated by enzyme-photocatalyst-coupled photobiocatalysis in different scenarios: synergistic catalysis connected by transferring electrons or radical intermediates and tandem catalysis connected by transferring intermediate product.

2.1 *Enzyme-Photocatalyst Synergistic Catalysis*

Native flavin-dependent "*ene*"-reductases are highly selective enzymes for the asymmetric reduction of activated alkenes [7]. This function is applicable for enone, enoate, and nitroalkene substrates through the native hydride transfer mechanism from two-electron reduced hydroquinone FMN_{hq} to the carbon–carbon double bond. However, electron-deficient heteroaromatic olefins are not suitable substrates for these enzymes likely due to their modest electrophilicity. In 2020, Hyster group developed a synergistic catalytic system utilizing $Ru(bpy)_3Cl_2$ and flavin-dependent alkene reductase NostocER for the asymmetric hydrogenation of vinyl pyridine-based heterocyclic compound [8]. Upon light excitation, the excited $Ru(bpy)_3^{2+}$ complex gains electrons, resulting in the formation of $Ru(bpy)_3^{+}$. This complex subsequently undergoes single-electron transfer to the substrate molecule, generating the corresponding radical anion. This radical intermediate undergoes enantioselective hydrogen atom transfer (HAT) to produce chiral reduced products, resulting in a series of chiral pyridine analogues (Fig. 1). This strategy involves introducing an exogenous photoredox catalyst to capture photon energy, thereby overcoming the thermodynamic barrier of electron transfer in the redox process. This broadens the catalytic promiscuity of natural *ene*-reductases and demonstrates the significant potential of natural oxidoreductase-catalyzed biocatalytic reactions for the synthesis of pyridine-substituted chiral tertiary alkanes driven by light.

In the above system, the excited-state photocatalyst initially acquires an electron from reduced form of flavin to generate a reduced species. In addition to this pathway, the excited-state photocatalyst can also directly be reduced by substrate

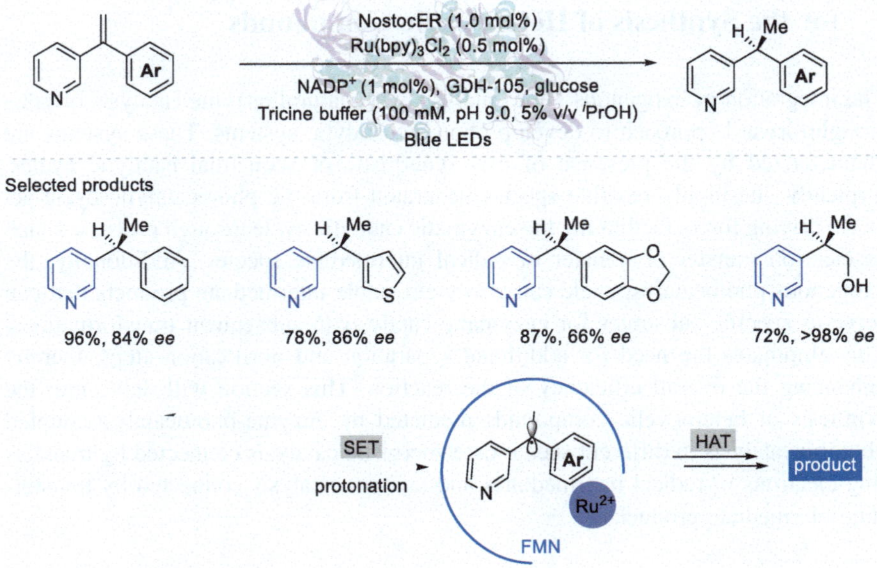

Fig. 1 Photoenzymatic asymmetric hydrogenation of heteroaromatic olefins

via single-electron transfer, generating an active radical intermediate that initiates subsequent asymmetric addition reactions. Recently, combing the exogenous photoredox catalyst $Ru(bpy)_3Cl_2$ with old yellow enzymes, Hyster group developed a redox-neutral decarboxylative coupling reaction between amino acids and vinyl pyridines. Based on mechanistic studies, it is proposed that $Ru(bpy)_3^{2+}$ binds to the protein pocket and helps to localize radical formation to the enzyme's active site. Notably, stereo-complementary enzyme GluER-T36A-Y343D was found to provide access to the opposite enantiomer of the products with high selectivity. This approach resulted in highly enantioselective decarboxylative alkylation reactions, with yields up to 92% and 98% *ee*, enabling the synthesis of a series of pyridine-containing heterocyclic compounds (Fig. 2) [9].

Nitrogen radicals, renowned for their high reactivity and synthetic versatility, have been employed extensively in the synthesis of nitrogen-containing heterocyclic compounds [10, 11]. In 2022, Hyster group reported an enantioselective hydroamination for heterocyclic synthesis based on nitrogen-centered radicals by utilizing the synergistic catalysis of exogenous $Ru(bpy)_3Cl_2$ and the flavin-dependent alkene reductase YqjM (*ene*-reductase from *Bacillus subtilis*) (Fig. 3) [12]. This novel biocatalytic reaction was initiated by the SET of the triplet excited state of $Ru(bpy)_3^{2+}$ by the cofactor FMN_{sq}, generating $Ru(bpy)_3^{+}$. Subsequently, the oxime ester substrate molecule receives electrons from the latter, resulting in the elimination of benzoate anions and the formation of a nitrogen radical. Then the nitrogen radical undergoes 6-*endo*-trig cyclization with the intramolecular alkene,

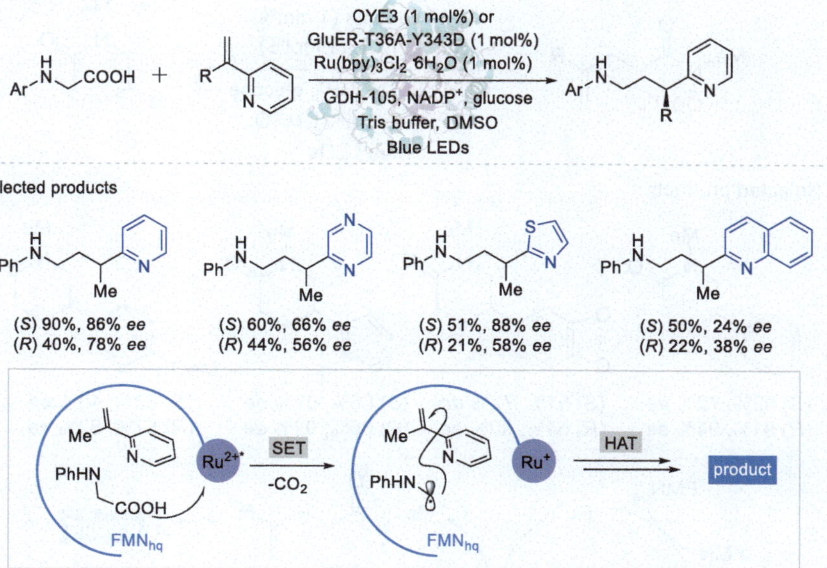

Fig. 2 Asymmetric redox-neutral decarboxylative coupling reaction of heteroaromatic olefins

forming a benzyl radical intermediate. Finally, the enzyme abstracts a hydrogen atom from the flavin cofactor structure, which ultimately leads to the intramolecular asymmetric hydroamination reaction. Through directed evolution of the enzyme cavity, various mutants were obtained that enable intramolecular cyclization reactions beyond 6-endo-trig, including 5-*exo*-trig, 7-*endo*-trig, and 8-*endo*-trig. Through direct evolution, different YqjM mutants led to the formation of both enantiomers of products, despite lower selectivities recorded for the S-enantiomer. Moreover, the utilization of the alkene reductase GluER (*ene*-reductase from *Gluconobacter oxydans*) also enabled the intermolecular asymmetric hydroamination reactions.

Thiamine diphosphate (ThDP) is an essential native cofactor for a number of enzymes widely found in living organisms, such as transketolase and benzaldehyde lyase. It enables the formation of nucleophilic carbon-negative-enamine intermediates through umpolung of carbonyl substrates (Breslow intermediates) from and nucleophilic addition to different electrophiles with carbon–carbon bond coupling [13]. In 2024, a collaborative study by Huang, Liang, and Tian reported a novel system of synergistic photobiocatalysis utilizing the organic dye eosin Y and thiamine diphosphate (ThDP)-dependent enzymes. Protein engineering was employed to modify the ThDP-dependent benzaldehyde lyase into a radical acyltransferase (RAT), which was then applied to asymmetric radical acylation reactions between aromatic aldehydes and phthalimide-active esters. This led to the successful production of a series of functionalized enantioenriched acylheteroarene such as pyrazole, pyridine, benzothiophene, and indole. On the

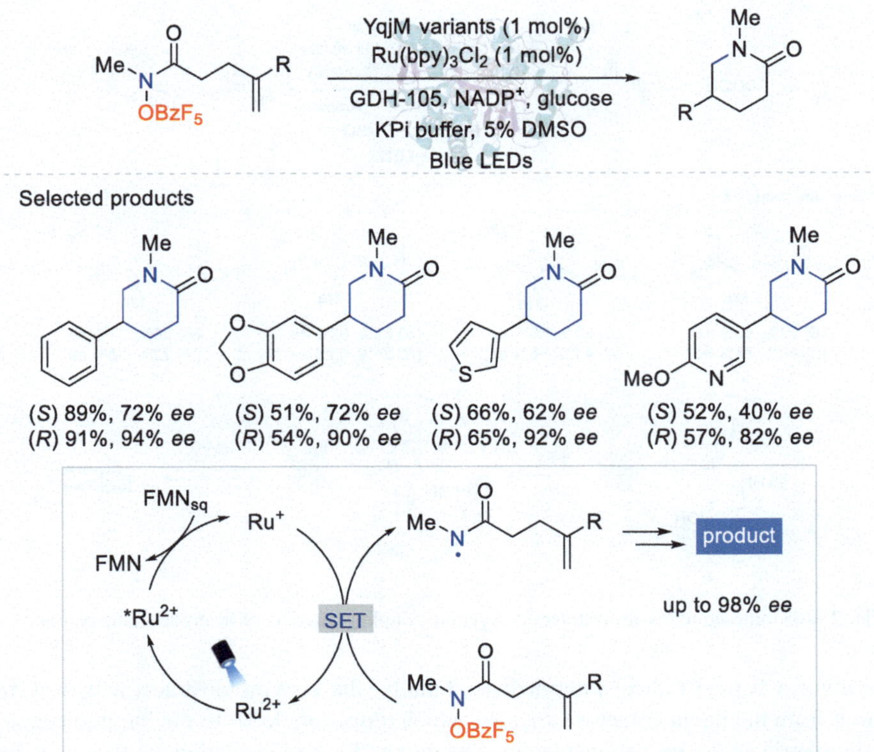

Fig. 3 Photobiocatalytic asymmetric intramolecular hydroamination reaction

basis of mechanistic investigations such as control reactions, low-temperature electron paramagnetic resonance, and computational studies, the authors proposed that the excited state of eosin Y could mediate single-electron oxidation of the Breslow intermediate to form an active radical anion. This radical anion coupled with the alkyl radical generated by single-electron reduction of the phthalimide-active ester, thus enabling a highly enantioselective radical–radical coupling reaction (Fig. 4) [14].

In addition to the aforementioned photobiocatalytic C–C and C–N coupling reactions, radical C–O bond formation was also established. In 2024, Huang and colleagues disclosed a dual bio/photocatalytic system comprising an *ene*-reductase and an organic dye, enabling stereoselective lactonizations of various alkenyl carboxylic acids [15]. By employing the focused rational iterative site-directed mutagenesis (FRISM) strategy for directed evolution, the optimal mutant ER-M5 of olefin reductase GluER was obtained after point mutations and three rounds of iterative combinatorial mutations. In the presence of the photosensitizer rhodamine 6G (Rh6G) and ER-M5, a diverse array of enantioenriched five- and six-membered lactones bearing vicinal stereocenters and even quaternary stereocenters were produced with up to 99% *ee*, up to 12.9:1 diastereomeric ratio (Fig. 5). Mechanistically,

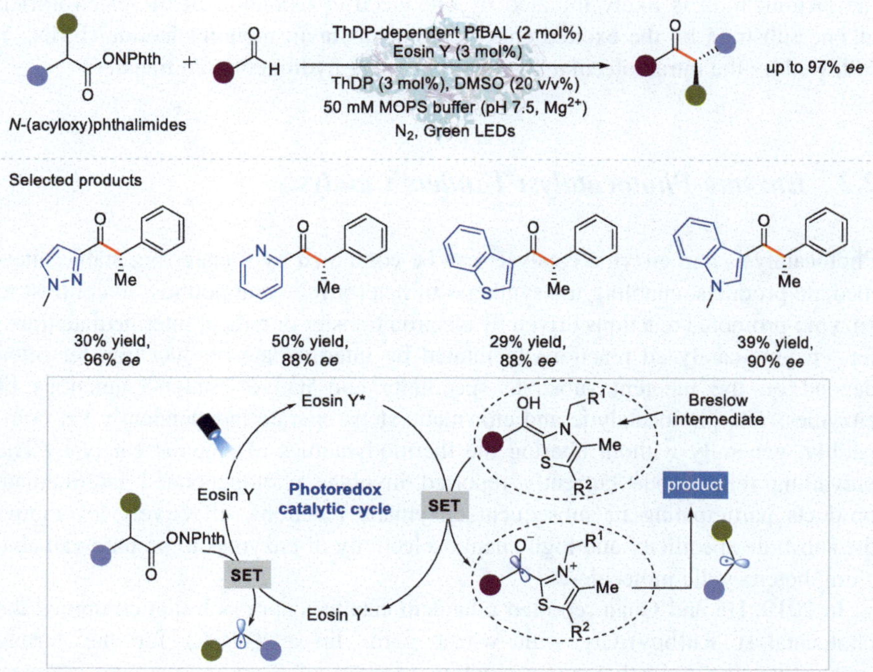

Fig. 4 Synergistic catalysis with photocatalyst and benzaldehyde lyase for radical–radical coupling reaction

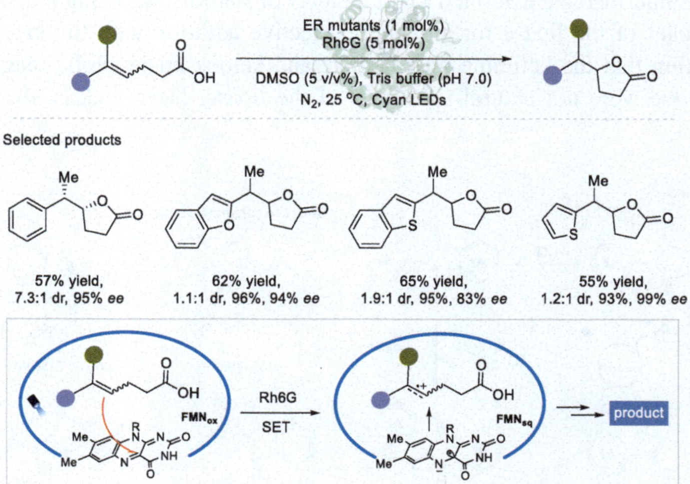

Fig. 5 Synergistic photo/biocatalysis for asymmetric C–O bond formations

the lactonization is likely initiated by one-electron oxidation of the electron-rich alkene substrate by the excited enzyme cofactor flavin mononucleotide (FMN_{ox}), followed by the intramolecular C–O coupling and hydrogen atom transfer.

2.2 *Enzyme-Photocatalyst Tandem Catalysis*

Photocatalysis and enzyme catalysis can be combined by transferring stable intermediate products, enabling the synthesis of heterocyclic compounds. In contrast to enzyme-promoted reactions driven by electron transfer or radical intermediate transfer, enzyme-catalyzed reactions facilitated by intermediate product transfer often depend on the inherent substrate specificity and native catalytic functions of enzymes. The photocatalytic and enzymatic steps operate independently yet compatibly, generally without altering the thermodynamics of enzyme catalysis. The prevailing relay mode currently reported involves photogenerated intermediate products participating in subsequent enzymatic reactions, effectively leveraging the substrate specificity and high enantioselectivity of enzymes to prepare valuable chiral heterocyclic molecules.

In 2019, He and Guan reported a tandem catalysis approach that combined the photocatalyst $Ru(bpy)_3Cl_2$ with wheat germ lipase (WGL) for the formal enantioselective oxidative cross-coupling reaction of 2-arylindoles and ketones, resulting in the synthesis of chiral 2,2-disubstituted indol-3-ones (Fig. 6) [16]. Initially, the Ru^{2+} photocatalyst facilitated the peroxidation of 2-phenylindole in an oxygen atmosphere, yielding a 3-pseudoindoline hydroperoxide intermediate that underwent further dehydration to generate a highly electrophilic ketimine intermediate. This intermediate exhibited a certain level of stability, allowing it to access the active pocket of the lipase for an enantioselective addition with the ketones. It is worth noting that the ketimine intermediate and ketone nucleophilic reagent mentioned above were not natural substrates of the lipase. Nevertheless, the catalytic

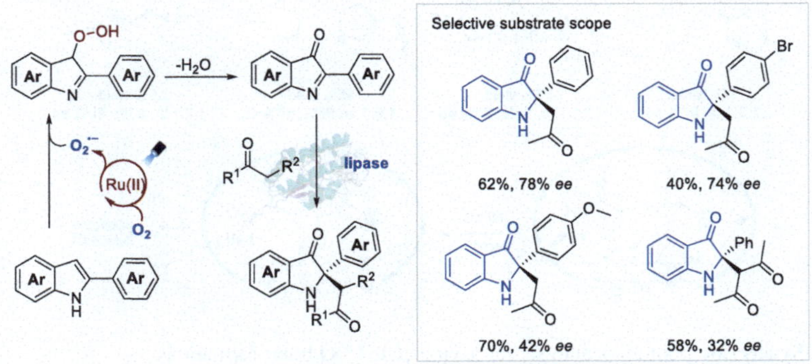

Fig. 6 Photocatalyst-lipase relay for enantioselective oxidative addition to indoles

triad of the lipase could activate the enolization of the ketone through hydrogen bonding, independent of any natural coenzymes.

3 Photobiocatalysis Driven by Excitation of Electron Donor-Acceptor Complex

The formation of electron donor-acceptor (EDA) complexes is the result of dipole-dipole interactions between electron-deficient acceptors and electron-rich donors. The majority of electron donors comprise compounds with lone pairs of electrons or π-electron-rich compounds. In contrast, electron acceptors are commonly found in electron-deficient heterocycles, halogenated hydrocarbons, and others. When these two types of molecules meet, they exhibit significant absorption peaks at longer wavelengths. Upon illumination at the corresponding wavelength, electron transfer takes place, leading to the formation of two reactive species: radical cations and radical anions. This discovery in 2013 by Melchiorre [17] represents a crucial photochemical radical reaction strategy [18]. The process of EDA photocatalysis enables the generation of photogenerated radicals under mild conditions through the unique electron transfer mechanism of EDA pairs. In recent years, there has been continuous development in photobiocatalytic asymmetric reactions driven by the excitation of EDA complexes [19]. This section provides a summary of the application of EDA photobiocatalysis strategies for the synthesis and modification of heterocyclic compounds.

In 2016, Hyster first reported the asymmetric photobiocatalytic radical dehalogenative hydrogenation of α-halolactones catalyzed by the nicotinamide-dependent short-chain dehydrogenase LKADH under visible-light conditions (Fig. 7) [20]. Mechanistic studies have demonstrated that α-halolactones can form EDA complexes with the nicotinamide cofactor NADPH. In the presence of visible light, the carbon–halogen bond is reduced by NADPH, resulting in the cleavage and generation of α-alkyl radicals and the departure of halogen anions. This radical intermediate terminates the reaction by abstracting a hydrogen atom from the cofactor $NADPH^{\bullet+}$. As both the formation of EDA and the hydrogen transfer take place within the chiral cavity of the enzyme, the stereoselectivity of the reaction is determined by the hydrogen transfer from $NADPH^{\bullet+}$, leading to excellent enantioselectivity and good substrate scope. This system paves a way to access a number of functionalized lactones from racemic bromide precursors in yields up to 85% and with excellent enantioselectivities (up to 96% *ee*). Interestingly, no kinetic resolution of α-halolactones was observed. By employing another dehydrogenase, RasADH, the hydrogenation products with opposite absolute configuration were also successfully achieved. This study demonstrates that nicotinamide-dependent dehydrogenases have a reactivity that is distinct from their natural catalytic functions.

In 2019, Hyster et al. further developed visible light-mediated intramolecular asymmetric radical cyclization of α-chloroamide by using flavoenzyme as the

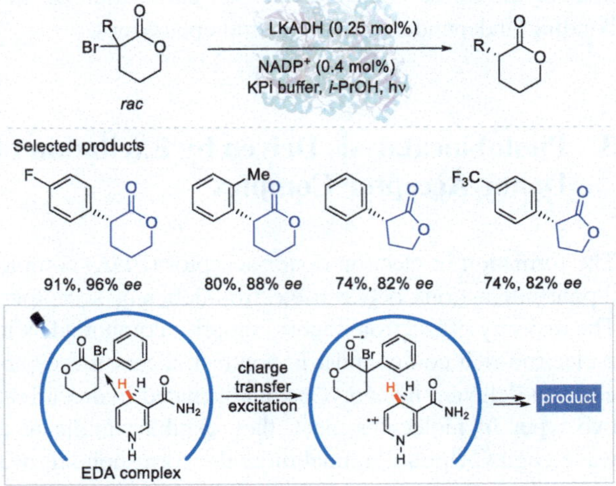

Fig. 7 Enantioselective radical dehalogenation of halogenated lactones

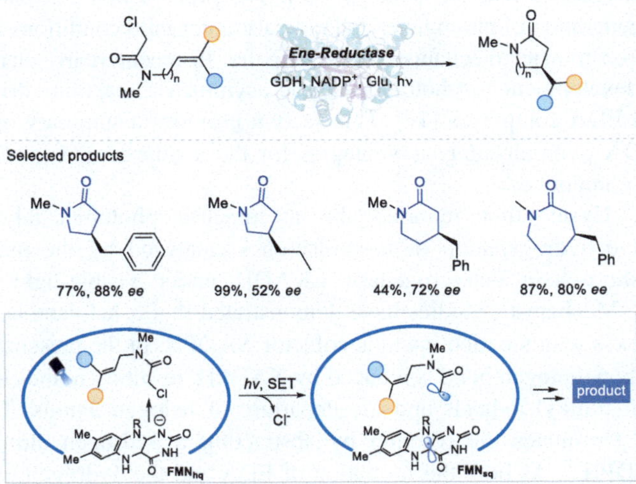

Fig. 8 Intramolecular asymmetric hydroalkylation

catalyst. This resulted in the selective synthesis of a series of five-, six-, seven-, and eight-membered lactam heterocyclic compounds with high yields (up to 99%) and 98% *ee* (Fig. 8) [21]. Like the case of radical dehalogenative hydrogenation of α-halolactones described above, here the flavin-dependent "ene"-reductases (EREDs) form an EDA complex with the α-chloroamide, which undergoes charge transfer excitation under light irradiation to generate a carbon radical. Subsequent intramolecular radical addition of the olefin moiety forms a lactam radical intermediate, which then abstracts a hydrogen atom from the reduced form of flavin to

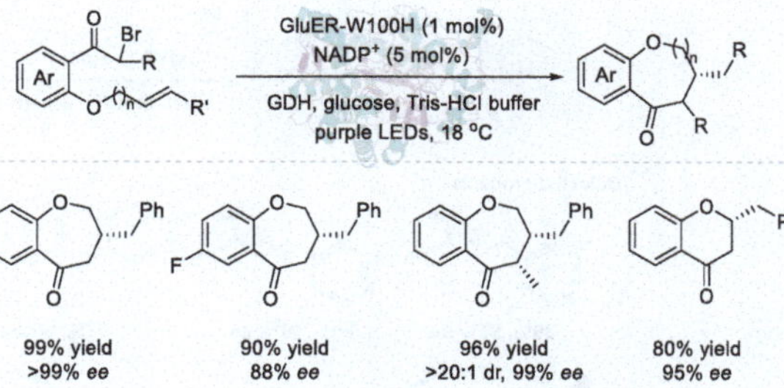

Fig. 9 Asymmetric synthesis of chiral benzooxygenated heterocyclic compounds

complete the asymmetric hydroalkylation of the intramolecular olefin, yielding a variety of β-chiral lactam product.

Inspired by the seminal works of Hyster, Yuan, Zhou, and Rao successfully achieved *ene*-reductase-catalyzed enantioselective intramolecular radical hydroalkylation reactions for the synthesis of chiral oxygen-containing benzo-fused heterocycles. Upon rational site-directed mutagenesis, the enzyme variant GluER-W100H was identified that could efficiently promote the reductive radical addition to produce various benzoxepinones, chromanone, and indanone with different benzo-fused rings benzooxygenated heterocyclic compounds in yield up to 99% and with up to 99% *ee* (Fig. 9) [22]. The synthetic utility of this photoenzymatic reaction was demonstrated by a gram-scale synthesis, which afforded a benzoxepinone product in 658 mg quantity without compromising yield and enantioselectivity.

In 2024, Zhao et al. reported enantioselective radical hydroalkylation of bromomethyl azaarenes to α-methyl substituted styrenes catalyzed by flavin mononucleotide (FMN)-dependent *ene*-reductase (OYE1). This differs from previously known flavoenzyme-catalyzed radical reactions that mostly form prochiral radical centers in carbonyl group-containing substrates. The use of bromomethyl azaarenes establishes comparable hydrogen bond interactions with the *ene*-reductase and effective substrate recognition and activation through similar mechanism of charge transfer excitation. The natural cofactor FMN_{hq} of OYE1 forms an EDA complex with the electron-deficient bromomethyl azaarene substrate. Upon excitation under visible light, the complex undergoes single electron transfer, generating a relatively stable nitrogen-substituted benzyl radical. This radical then reacts with α-methylstyrene to form a radical intermediate, which undergoes hydrogen atom transfer to yield the desired product. The key to achieving selectivity lies in the chiral hydrogen atom transfer step, where hydrogen bonding and spatial effects induced by residues within the enzyme's active pocket play a crucial role in controlling the outcome. Notably, the chiral pocket of OYE1 effectively modulated the distal prochiral radical centers located more than six chemical bonds, or over 6 Å, from the nitrogen atom, thereby enabling the production of a broad array of azaarenes possessing a remote γ-stereocenter [23] in up to 95% and with 99% *ee* (Fig. 10).

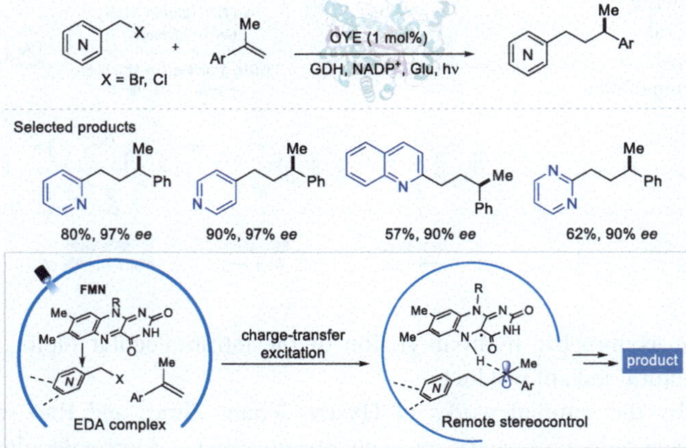

Fig. 10 Enantioselective radical hydroalkylation of bromomethyl azaarenes

4 Photobiocatalysis Mediated by Redox-Active Natural Enzymes

Due to its polycyclic conjugated molecular structure, flavin itself possesses excellent photophysical properties, exhibiting significant absorption in the visible light region spanning 400–520 nm, with an extinction coefficient reaching up to 11,300 L mol^{-1} cm^{-1} at 450 nm [24]. The excited state of flavin exhibits robust redox properties, with the reduction potential of flavin adenine dinucleotide ($FADH_2$*) being −2.26 V vs. standard Calomel electrode (SCE) [25]. And natural selection has, in many instances, favored flavin or nicotinamide as photoactive coenzymes. The oldest known photoenzyme is DNA photolyase [26], a flavin-dependent enzyme that repairs pyrimidine cyclobutane dimers in DNA. In this reaction, the reduced flavin adenine dinucleotide (FAD) cofactor (FAD_{hq}^{-}) acts as a photocatalyst, initiating the reverse [2+2] reaction of the pyrimidine dimer through single-electron transfer reduction, followed by the oxidation of the resulting radical anion by FAD_{sq} [27, 28]. This section will introduce the potential of flavin- and nicotinamide-dependent direct photoexcitation-induced single-electron oxidation or reduction asymmetric radical reactions for the modification and synthesis of heterocyclic compounds.

In 2020, Hyster et al. disclosed a redox-neutral radical cyclization reaction of α-haloamides to produce enantioenriched oxindoles utilizing a flavin-dependent *ene*-reductase (12-oxophytodienoate reductase, OPR1) as a biocatalyst. This protocol led to the synthesis of a series of 3,3-disubstituted oxindoles in yields of up to 97% and with up to 96% *ee* (Fig. 11) [29]. Differing from the prevailing photobiocatalytic approach using flavoenzymes through charge-transfer excitation, here the reaction was initiated with one-electron reductive dehalogenation of substrate by ground-state FMN_{sq}, which is generated from reductive quenching of

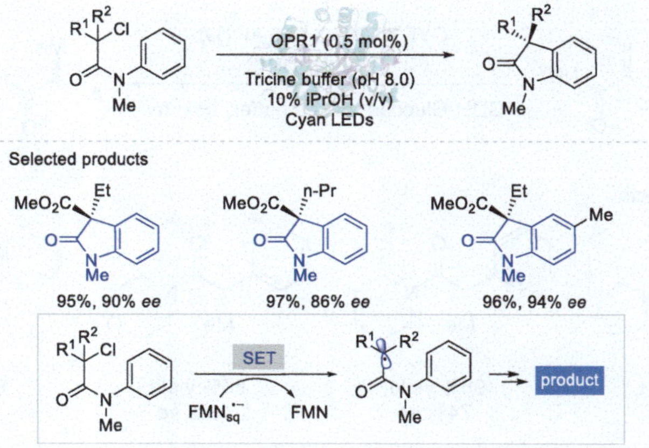

Fig. 11 *ene*-Reductase-catalyzed asymmetric redox-neutral radical cyclization

excited FMN_{hq}* with tricine buffer. This process resulted in the formation of α-acyl radical and FMN. Subsequent intramolecular dearomative radical addition, the stereo-determining step conferred with high enantioselectivity by the protein chiral pocket, generates a chiral reducing vinylogous amido radical. The following oxidation FMN results in the product and regenerates FMN_{sq}. This strategy unveils a novel mechanism for natural photoenzymes to catalyze unnatural reactions and presents a new approach for the asymmetric radical coupling of $C(sp^3)$–$C(sp^2)$ bonds. Notably, FMN herein serves a dual role as an oxidant and reductant in the catalytic cycle, which distinguishes this mechanism from the typical hydrogen atom transfer catalytic mechanisms observed in previous systems.

In another relevant study at the same time, the same group reported a stereoselective radical reduction of acrylamides catalyzed by *ene*-reductases, which provides an attractive approach to access N-heterocycle-containing chiral amides (Fig. 12) [30]. Mechanistically, although acrylamides are structurally similar to the natural substrates of *ene*-reductases, they cannot be reduced by the natural mechanism. The authors hypothesized that such substrates could be recognized and accommodated to the enzyme-active site. Among the enzymes screened, OYE1 demonstrated the highest enantioselectivity, and the addition of exogenous FMN further enhanced the yield of OYE1. Through site-directed saturation mutagenesis, the variant OYE1-F269G was identified, which afforded products with up to 97% yield and 87% *ee*. The reaction commences with the single-electron reduction of the acrylamide olefin by the excited-state FMN_{hq}, leading to the formation of a benzyl carbonyl α-radical anion and neutral FMN_{sq}. Protonation of the radical anion results in a stable α-acyl radical, which undergoes hydrogen atom transfer (HAT) from FMN_{sq} to generate the reduced product. This strategy expands the range of nonnatural olefinic substrates that can be reduced by flavin-dependent *ene*-reductases, thereby promoting the advancement of enzymatic catalytic promiscuity.

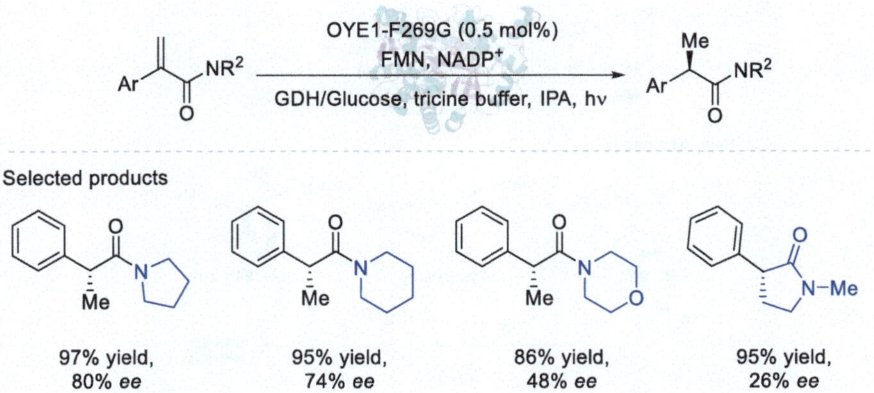

Fig. 12 Direct excitation of *ene*-reductases for asymmetric hydrogenation

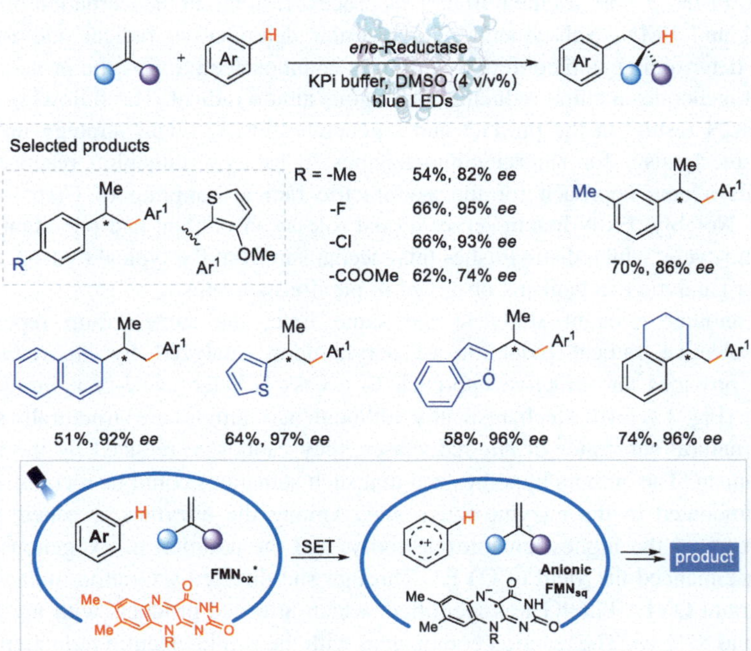

Fig. 13 *ene*-Reductase-catalyzed asymmetric radical hydroarylation of alkenes

In 2023, Huang, Wang, and Tian expanded the repertoire of flavoenzyme photoredox catalysis to asymmetric radical hydroarylation reaction, which could be applied to the modification of various heterocyclic compounds in moderate to good yields and excellent enantioselectivities (up to 98% *ee*) (Fig. 13)

[31]. Mechanistic studies suggested that upon photoirradiation, the flavin cofactor of the *ene*-reductase is excited FMN_{ox}*, which promotes one-electron oxidation of electron-rich arenes to generate aryl radical cations and FMN_{sq}. These radical intermediates then undergo radical addition to olefins within the cavity to form prochiral radical intermediates, which is trapped by FMN_{sq} through electron and proton (or hydrogen atom) transfer. The heterocyclic aromatic rings such as thiophene, indole, pyridine, and pyrroles could be involved either in the electron-rich arene substrates or as substituents of nucleophilic alkenes. Notably, by screening and evolution of different *ene*-reductases, both enantiomers of these heterocycle-containing chiral products could be accessed.

Fatty acid photodecarboxylase (FAP), discovered in 2017 by Beisson in the *Chlorella variabilis*, is a naturally occurring flavoenzyme responsible for the decarboxylation of long-chain fatty acids into alkanes or alkenes [32]. Many synthetic endeavors have been made to expand the substrate scope of native protodecarboxylation activity of FAP, however, with new-to-nature activities under-exploited. Through extensive genome mining, rational engineering, and directed evolution, Yang and Liu successfully repurposed FAP into radical photocyclases for redox-neutral decarboxylative radical cyclization of α,β-unsaturated ester–carboxylic acid substrates [33]. The engineered photoenzymes exhibited prominent reactivity for the *exo*-trig radical cyclization, leading to various α-alkylated esters containing a five- and six-membered saturated tetrahydrofuran and tetrahydropyran in good yields and with excellent chemo-, enantio-, and diastereoselectivities. Notably, two contiguous stereogenic centers were present at the α and β positions in the tetrahydrofuran ester. Empowered by directed evolution, radical photocyclases were engineered to provide mutants (i.e., $CvRAP_2$, $CvRAP_3$, $CvRAP_4$, $CvRAP_5$) that allow access to all of its four possible stereoisomers in up to 89% yield, 90:10 dr, and >98% *ee* (Fig. 14). This represents the first report of a

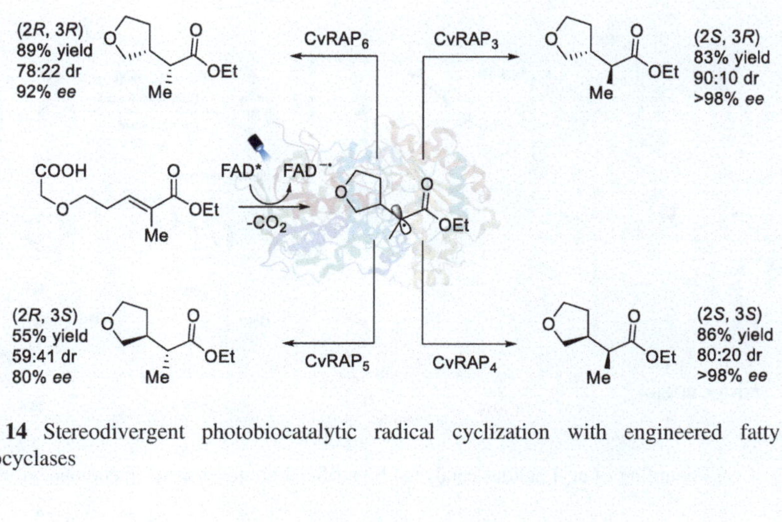

Fig. 14 Stereodivergent photobiocatalytic radical cyclization with engineered fatty acid photocyclases

fully stereodivergent radical annulation in both classical organic chemistry and enzymatically catalyzed asymmetric radical reactions.

5 Photobiocatalysis Mediated by Artificial Photoenzymes

The aforementioned photobiocatalysis predominantly operates through photoinduced electron transfer mechanisms. In parallel to it, energy transfer (EnT) is a distinct catalytic manifold in organic photochemistry that is experiencing rapid development [34, 35]. However, due to the lack of natural triplet photosensitizers, EnT catalysis has not been reported in natural enzymes. Advanced protein engineering technologies such as genetic code expansion and site-selective bioconjugation methods provide opportunities to precisely incorporate nonnatural triplet photosensitizers into protein scaffolds, enabling the rational design and construction of artificial photoenzymes for energy transfer catalysis.

Diaryl ketones represent a prominent class of organic triplet photosensitizer widely used in synthetic photochemistry due to their excellent photophysical properties. In 2021, Fu et al. disclosed artificial photosensitive nickel enzyme for abiological cross-coupling reaction of aryl halides in water [36]. This metallophotoenzyme was designed based on photosensitive protein (PSP) encoding benzoylphenylalanine (BpA) and chemical modification of cysteine residue with a bipyridine-Ni^{2+} complex. Mechanistic studies using transient absorption spectroscopy suggest that these two catalytic entities synergize through energy transfer from excited benzophenone to the nickel ion, thereby facilitating reductive elimination step of the transition metal catalysis. This photobiocatalytic cross-coupling protocol was found to also be amenable to the C–N bond formation, allowing synthesis of hydroxylated indole, quinoline, *N*-arylated pyrroline, and imidazole under mild conditions from the respective aryl halide substrates (Fig. 15) [37]. It is noteworthy

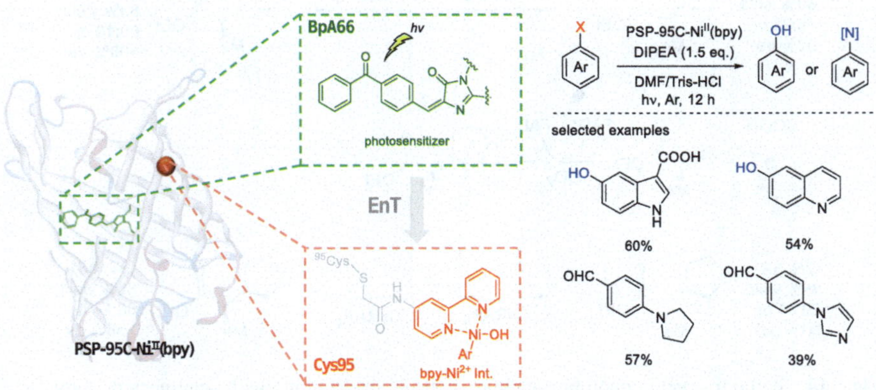

Fig. 15 Cross-coupling of aryl halides catalyzed by artificial photosensitizer metalloenzyme

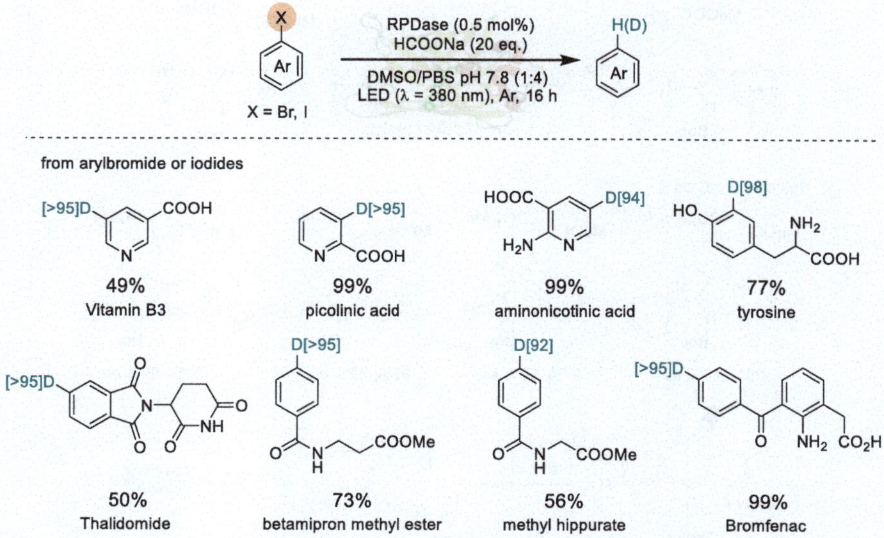

Fig. 16 RPDase-catalyzed dehalogenative deuteration of bioactive molecules

that the genetic code expansion method renders the precise regulation of the spatial distance between the benzophenone and the Ni^{II}(bpy) cofactors, thereby increasing the efficiency of the energy transfer and the yield of the cross-coupling reaction.

The above study showcases that integration of EnT catalysis mode established in synthetic chemistry into proteins could essentially expand the scope of natural biocatalysis. In a subsequent study, Fu et al. demonstrate that a similar artificial BpA-encoded photoenzyme based on a yellow fluorescent protein without metal cofactor, named as reductive photodehalogenase (RPDase), could proficiently catalyze the new-to-nature hydrodehalogenation and deuterodehalogenation of aryl halides (Fig. 16) [38]. Mechanistically, differing from native reductive dehalogenases that typically leverage on metal cofactor activation of specific halide substrates, RPDase operates via a photochemical pathway by harnessing light energy and externally in formate as a sacrificial reductant. The benzophenone chromophore at the excited state is converted into highly reducing radical anion by formate, thus facilitating reductive cleavage of carbon–halogen bond in the aromatic substrates. While exhibiting good substrate generality, this distinct mechanism also allows the use of deuteroformate to mediate the deuterium labeling of bioactive compounds such as nicotinic acid (vitamin B3), aminonicotinic acid, picolinic acid, betamipron methyl ester, methyl hippurate, and marketed drug thalidomide using their halide precursors.

The advent of artificial photoenzyme provides the opportunity to tackle long-standing challenge of enantioselectivity in EnT catalysis [34, 35]. This was realized simultaneously by Wu et al. and the Group laboratory, both utilizing genetically modified artificial triplet photoenzyme for enantioselective [2+2] cycloaddition

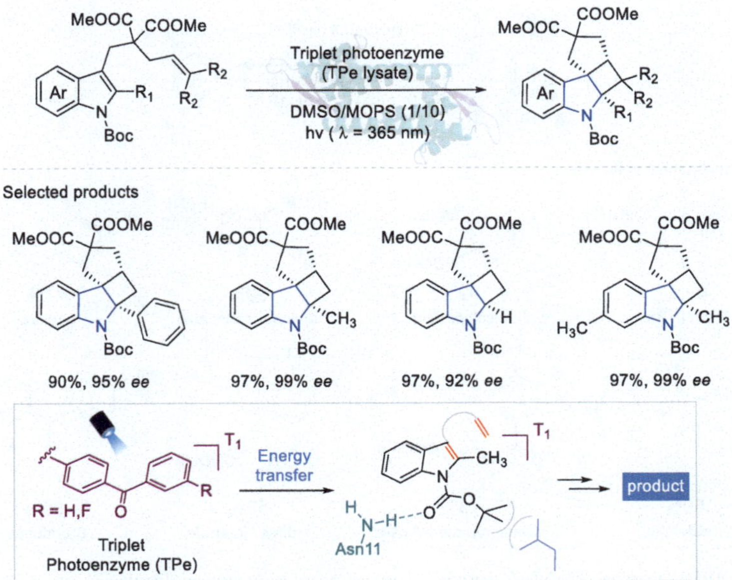

Fig. 17 Triplet photoenzyme-catalyzed enantioselective [2+2] photocycloaddition of indoles

reactions [39, 40]. In the study by Wu et al., the multidrug resistance regulator LmrR was selected as the protein scaffold to encode the BpA photosensitizer. After four rounds of iterative mutagenesis to optimize key residues in the reactive enzyme cavity, the optimal mutant denoted as TPe4.0 was obtained. This mutant is capable of efficiently catalyzing the intramolecular [2+2] photocycloaddition of indole derivatives, resulting in a series of chiral cyclobutane-fused indoline heterocycles with up to 97% yield and 99% *ee* (Fig. 17). The protein crystal structure of photoenzyme complexed with the substrate indicated that the enantioselectivity was originated from synergistic effects of multiple weak bonds formed between the photosensitizer, amino acid residues in the cavity, and substrate.

In the parallel work reported by Green and coworkers, BpA was inserted into the previously reported computer-designed protein scaffold, Diels-Alderase DA_20_00. Initially, the authors constructed the photoenzyme catalyst (EnT1.0) using gene codon expansion technology, and they obtained a highly enantioselective photoenzyme catalyst (EnT1.3) through directed evolution for the intramolecular [2+2] cycloaddition of quinolones as a model reaction. EnT1.3 achieved asymmetric [2+2] photocycloaddition reactions with a turnover number (TON) of up to 300 and 99% *ee* (Fig. 18). Additionally, it has also been extended to intermolecular enantioselective [2+2] cycloadditions of quinolones. This protocol provides an access to optically active functionalized cyclobutane-fused quinolones. Notably, despite the different protein scaffolds used to construct TPe4.0 and EnT1.3 and the significant differences in target substrates, both photoenzyme-catalyzed

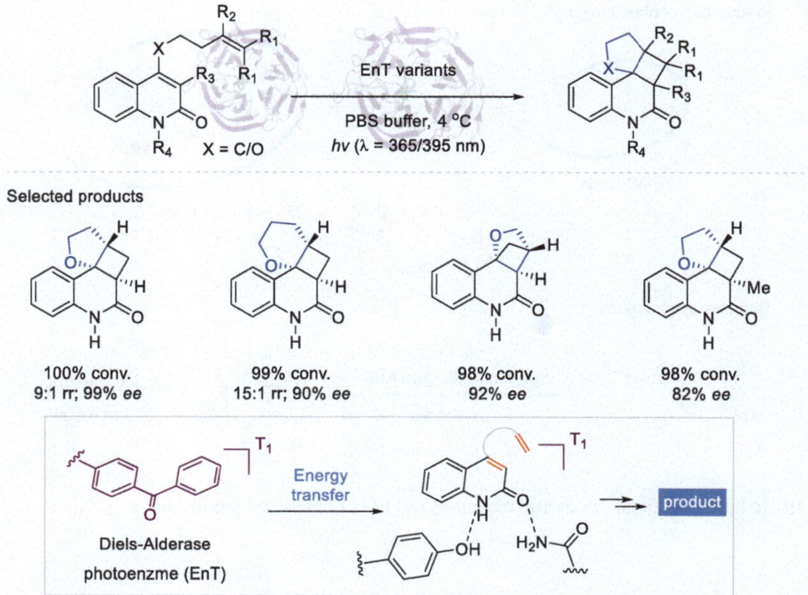

Fig. 18 Triplet photoenzyme-catalyzed enantioselective [2+2] photocycloaddition of quinolones

photocycloaddition reactions exhibited significantly higher air tolerance compared to benzophenone-based chemical catalysts. This suggests that the microenvironment of the protein cavity may play a unique and subtle role in the isolation of oxygen.

In addition to using genetic code expansion method, site-selective chemical modification of protein provides an alternative for the integration of artificial photocatalysts into the protein backbone. The Lewis group used the chemical modification and genetic codon expansion techniques to introduce an Ir(III) bipyridine complex unnatural metal cofactor into the protein scaffold of prolyl oligopeptidase (POP) to construct an artificial metalloenzyme (ArM) [41]. This catalyst was also evaluated in the asymmetric [2+2] cycloaddition of quinolone derivatives through energy transfer (Fig. 19) [42], affording the desired cyclobutane-fused heterocyclic product in 14.5% *ee*.

An essential challenge for artificial enzyme development through chemical modification is to streamline the directed evolution. In 2024, Wu et al. established a chemogenetic modification strategy that facilitated the rapid diversification of artificial triplet photoenzymes. This strategy synergizes cysteine-iodoacetamide bioconjugation chemistry with the iterative mutagenesis in cell lysates. Various artificial photosensitizers could be facilely selectively inserted into natural protein cavity to construct a chemically modified artificial photoenzyme library, designated CMP [43]. Within this library, CMP1.0, featuring a thioxanthone photosensitizer center, was identified as the optimal ancestor photoenzyme for enantioselective [2+2] photocycloaddition of 2-carboxamide indole. Following four rounds of

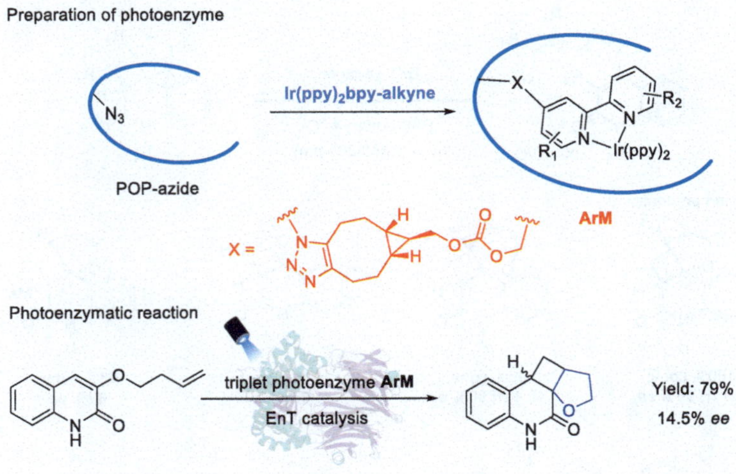

Fig. 19 Ir-based artificial photoenzyme-catalyzed [2+2] photocycloaddition

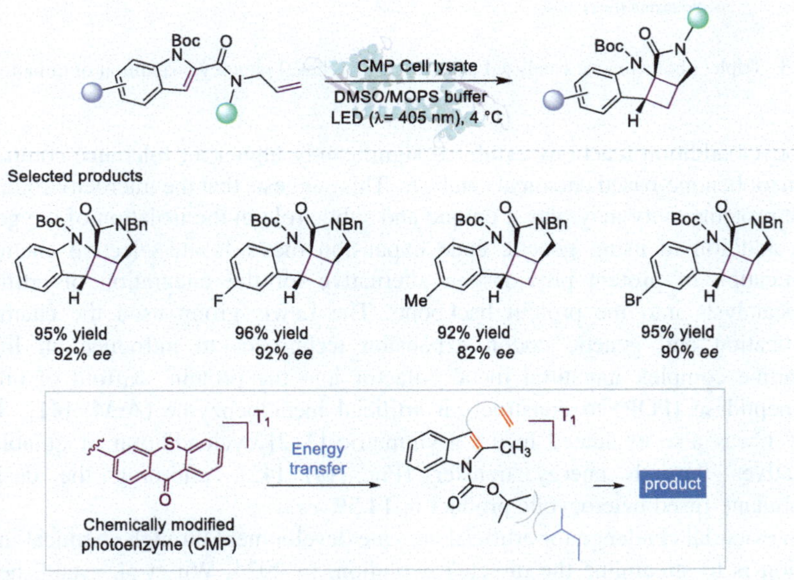

Fig. 20 Enantioselective [2+2] cycloaddition of indole catalyzed by chemogenetically evolved photoenzymes

iterative mutation, CMP4.0 with exceptional catalytic performance was obtained, yielding cyclobutane-indoline in 95% yield and 92% *ee*. Moreover, this photoenzyme was demonstrated to be viable in whole-cell photobiocatalysis, enabling the enantioselective synthesis of cyclobutane-fused indoline tetracycles (Fig. 20).

6 Conclusion

In summary, this chapter highlights the advancements in photobiocatalysis for the functionalization and synthesis of heterocyclic compounds. This innovative approach merges the reactivity of photocatalysis with the selectivity of enzymatic catalysis, presenting a novel strategy for the chiral synthesis of valuable heterocyclic organic compounds. Current strategies for photobiocatalytic synthesis of heterocyclic compounds include coupling chemical photocatalysis with natural enzymatic catalysis, excitation catalysis via natural co-enzyme EDA complexes, direct photoredox of coenzymes, and energy transfer photoenzyme catalysis. Despite being a developing and interdisciplinary field, photobiocatalysis encounters challenges in its application to the synthesis of heterocyclic compounds. The strong reliance on flavoenzymes somewhat limits the reactivity of photoenzymes due to the intrinsic photophysical properties of this cofactor. While the use of exogenous photocatalysts in a synergistic manner or as artificial photocofactors significantly broadens the scope of photobiocatalysis, the range of synthetic photocatalysts currently available remains relatively limited. There is a pressing need to efficiently construct nonnatural photocatalytic sites with diverse reactivities while rapidly achieving high-throughput screening of optimal mutants. On the other hand, the rapidly advancing field of computational protein de novo design is poised to provide groundbreaking tools for the creation of entirely new artificial enzymes. Furthermore, it is anticipated that photoenzymes can be adapted for whole-cell biocatalysis, facilitating the establishment of light-driven microbial cell factories for the sustainable, scalable biosynthesis of valuable heterocyclic compounds.

Acknowledgements F. Zhong and G. Wu acknowledge National Key R&D Program of China (2018YFA0903500, 2019YFA0904101), Key R&D Program of Hubei Province (2021BAA168), and the interdisciplinary research program of Huazhong University of Science and Technology (2023JCYJ001) for financial support. Y. Fu acknowledges National Natural Science Foundation of China (22407091) for financial support.

References

1. Li CJ, Trost BM (2008) Green chemistry for chemical synthesis. Proc Natl Acad Sci 105(36): 13197–13202
2. Rosenthaler L (1908) Durch enzyme bewirkte asymmetrische synthesen. Biochem Z 14(1): 238–253
3. Arnold FH (1996) Directed evolution: creating biocatalysts for the future. Chem Eng Sci 51(23):5091–5102
4. Packer MS, Liu DR (2015) Methods for the directed evolution of proteins. Nat Rev Genet 16(7): 379–394
5. Zhang RK, Chen K, Huang X, Wohlschlager L, Renata H, Arnold FH (2019) Enzymatic assembly of carbon-carbon bonds via iron-catalysed sp^3 C-H functionalization. Nature 565(7737):67–72

6. Prier CK, Zhang RK, Buller AR, Brinkmann-Chen S, Arnold FH (2017) Enantioselective, intermolecular benzylic C-H amination catalysed by an engineered iron-haem enzyme. Nat Chem 9(7):629–634
7. Toogood HS, Scrutton NS (2018) Discovery, characterization, engineering, and applica tions of *ene*-reductases for industrial biocatalysis. ACS Catal 8(4):3532–3549
8. Nakano Y, Black MJ, Meichan AJ, Sandoval BA, Chung MM, Biegasiewicz KF, Zhu T, Hyster TK (2020) Photoenzymatic hydrogenation of heteroaromatic olefins using *ene*-reductases with photoredox catalysts. Angew Chem Int Ed 59(26):10484–10488
9. Sun SZ, Nicholls BT, Bain D, Qiao T, Page CG, Musser AJ, Hyster TK (2024) Enantioselective decarboxylative alkylation using synergistic photoenzymatic catalysis. Nat Catal 7:35–42
10. Kwon K, Simons RT, Nandakumar M, Roizen JL (2021) Strategies to generate nitrogen-centered radicals that may rely on photoredox catalysis: development in reaction methodology and applications in organic synthesis. Chem Rev 122(2):2353–2428
11. Pratley C, Fenner S, Murphy JA (2022) Nitrogen-centered radicals in functionalization of sp^2 systems: generation, reactivity, and applications in synthesis. Chem Rev 122(9):8181–8260
12. Ye Y, Cao J, Oblinsky DG, Verma D, Prier CK, Scholes GD, Hyster TK (2023) Using enzymes to tame nitrogen-centred radicals for enantioselective hydroamination. Nat Chem 15(2): 206–212
13. Kluger R, Tittmann K (2008) Thiamin diphosphate catalysis: enzymic and nonenzymic covalent intermediates. Chem Rev 108(6):1797–1833
14. Xu Y, Chen H, Yu L, Peng X, Zhang J, Xing Z, Bao Y, Liu A, Zhao Y, Tian C, Liang Y, Huang X (2024) A light-driven enzymatic enantioselective radical acylation. Nature 625:74–78
15. Yu J, Zhang Q, Zhao B, Wang T, Zheng Y, Wang B, Zhang Y, Huang X (2024) Repurposing visible-light-excited *ene*-reductases for diastereo-and enantioselective lactones synthesis. Angew Chem Int Ed 63:e202402673
16. Ding X, Dong CL, Guan Z, He YH (2019) Concurrent asymmetric reactions combining photocatalysis and enzyme catalysis: direct enantioselective synthesis of 2, 2-disubstituted indol-3-ones from 2-arylindoles. Angew Chem Int Ed 58(1):118–124
17. Arceo E, Jurberg ID, Álvarez-Fernández A, Melchiorre P (2013) Photochemical activity of a key donor-acceptor complex can drive stereoselective catalytic α-alkylation of aldehydes. Nat Chem 5(9):750–756
18. Wortman AK, Stephenson CR (2023) EDA photochemistry: mechanistic investigations and future opportunities. Chem 9(9):2390–2415
19. Harrison W, Huang X, Zhao H (2022) Photobiocatalysis for abiological transformations. Acc Chem Res 55(8):1087–1096
20. Emmanuel MA, Greenberg NR, Oblinsky DG, Hyster TK (2016) Accessing non-natural reactivity by irradiating nicotinamide-dependent enzymes with light. Nature 540(7633): 414–417
21. Biegasiewicz KF, Cooper SJ, Gao X, Oblinsky DG, Kim JH, Garfinkle SE, Joyce LA, Scholes GD, Hyster TK (2019) Photoexcitation of flavoenzymes enables a stereoselective radical cyclization. Science 364(6446):1166–1169
22. Zhu C, Yuan Z, Deng Z, Yin D, Zhang Y, Zhou J, Rao Y (2023) Photoenzymatic enantioselective synthesis of oxygen-containing benzo-fused heterocycles. Angew Chem Int Ed 62:e202311762
23. Li M, Harrison W, Zhang Z, Yuan Y, Zhao H (2024) Remote stereocontrol with azaarenes via enzymatic hydrogen atom transfer. Nat Chem 16:277–284
24. Siegel LM (1978) Quantitative determination of noncovalently bound flavins: types and methods of analysis. In: Methods in enzymology, vol 53. Academic Press, pp 419–429
25. Grosheva D, Hyster TK (2021) Light-driven flavin-based biocatalysis. In: Flavin-based catalysis: principles and applications. Wiley, pp 291–313
26. Sancar A (2003) Structure and function of DNA photolyase and cryptochrome blue-light photoreceptors. Chem Rev 103(6):2203–2238

27. Zhang M, Wang L, Zhong D (2017) Photolyase: dynamics and electron-transfer mechanisms of DNA repair. Arch Biochem Biophys 632:158–174
28. Conrad KS, Manahan CC, Crane BR (2014) Photochemistry of flavoprotein light sensors. Nat Chem Biol 10(10):801–809
29. Black MJ, Biegasiewicz KF, Meichan AJ, Oblinsky DG, Kudisch B, Scholes GD, Hyster TK (2020) Asymmetric redox-neutral radical cyclization catalysed by flavin-dependent *ene*-reductases. Nat Chem 12(1):71–75
30. Sandoval BA, Clayman PD, Oblinsky DG, Oh S, Nakano Y, Bird M, Scholes GD, Hyster TK (2020) Photoenzymatic reductions enabled by direct excitation of flavin-dependent *ene*-reductases. J Am Chem Soc 143(4):1735–1739
31. Zhao B, Feng J, Yu L, Xing Z, Chen B, Liu A, Liu F, Shi F, Zhao Y, Tian C, Wang B, Huang X (2023) Direct visible-light-excited flavoproteins for redox-neutral asymmetric radical hydroarylation. Nat Catal 6(11):996–1004
32. Sorigue D, Hadjidemetriou K, Blangy S, Gotthard G, Bonvalet A, Coquelle N, Samire P, Aleksandrov A, Antonucci L, Benachir A, Boutet S, Byrdin M, Cammarata M, Carbajo S, Cuiné S, Doak RB, Foucar L, Gorel A, Grünbein M, Hartmann E, Hienerwadel R, Hilpert M, Kloos M, Lane TJ, Légeret B, Legrand P, Li-Beisson Y, Moulin SLY, Nurizzo D, Peltier G, Schirò G, Shoeman RL, Sliwa M, Solinas X, Zhuang B, Barends TRM, Colletier J-P, Joffre M, Royant A, Berthomieu C, Weik M, Domratcheva T, Brettel K, Vos MH, Schlichting I, Arnoux P, Müller P, Beisson F (2021) Mechanism and dynamics of fatty acid photodecarboxylase. Science 372(6538):eabd5687
33. Ju S, Li D, Mai BK, Liu X, Vallota-Eastman A, Wu J, Valentine DL, Liu P, Yang Y (2024) Stereodivergent photobiocatalytic radical cyclization through the repurposing and directed evolution of fatty acid photodecarboxylases. Nat Chem 2024(16):1339–1347
34. Dutta S, Erchinger JE, Strieth-Kalthoff F, Kleinmans R, Glorius F (2024) Energy transfer photocatalysis: exciting modes of reactivity. Chem Soc Rev 53(3):1068–1089
35. Poplata S, Tröster A, Zou YQ, Bach T (2016) Recent advances in the synthesis of cyclobutanes by olefin [2+2] photocycloaddition reactions. Chem Rev 116(17):9748–9815
36. Liu X, Kang F, Hu C, Wang L, Xu Z, Zheng D, Gong W, Lu Y, Ma Y, Wang J (2018) A genetically encoded photosensitizer protein facilitates the rational design of a miniature photocatalytic CO_2-reducing enzyme. Nat Chem 10(12):1201–1206
37. Fu Y, Huang J, Wu Y, Liu X, Zhong F, Wang J (2021) Biocatalytic cross-coupling of aryl halides with a genetically engineered photosensitizer artificial dehalogenase. J Am Chem Soc 143(2):617–622
38. Fu Y, Liu X, Xia Y, Guo X, Guo J, Zhang J, Zhao W, Wu Y, Wang J, Zhong F (2023) Whole-cell-catalyzed hydrogenation/deuteration of aryl halides with a genetically repurposed photodehalogenase. Chem 9:1897–1909
39. Sun N, Huang J, Qian J, Zhou TP, Guo J, Tang L, Zhang W, Deng Y, Zhao W, Wu G, Liao R, Chen X, Wu Y (2022) Enantioselective [2+2]-cycloadditions with triplet photoenzymes. Nature 611(7937):715–720
40. Trimble JS, Crawshaw R, Hardy FJ, Levy CW, Brown MJ, Fuerst DE, Heyes DJ, Obexer R, Green AP (2022) A designed photoenzyme for enantioselective [2+2] cycloadditions. Nature 611(7937):709–714
41. Gu Y, Ellis-Guardiola K, Srivastava P, Lewis JC (2015) Preparation, characterization, and oxygenase activity of a photocatalytic artificial enzyme. ChemBioChem 16(13):1880–1883
42. Liu B, Zubi YS, Lewis JC (2023) Iridium (III) polypyridine artificial metalloenzymes with tunable photophysical properties: a new platform for visible light photocatalysis in aqueous solution. Dalton Trans 52(16):5034–5038
43. Guo J, Qian J, Cai D, Huang J, Yang X, Sun N, Zhang J, Pang T, Zhao W, Wu G, Chen X, Zhong F, Wu Y (2024) Chemogenetic evolution of diversified photoenzymes for enantioselective [2+ 2] cycloadditions in whole cells. J Am Chem Soc 146(28):19030–19041

The manufacturer's authorised representative in the EU is Springer
Nature Customer Service Centre GmbH, Europaplatz 3, 69115 Heidelberg,
Germany. If you have any concerns regarding our products, please
contact ProductSafety@springernature.com

Printed and bound by CPI Group (UK) Ltd, Croydon, CR0 4YY
30/09/2026
02225567-0013